FLORA OF RANGA REDDI DISTRICT

Andhra Pradesh, India

FLORA OF RANGA REDDI DISTRICT
Andhra Pradesh, India

T. Pullaiah
and
M. Silar Mohammed

Department of Botany
Sri Krishnadevaraya University
Anantapur 515003

Regency Publications
New Delhi

ISBN 81-87498-01-3

Published by Regency Publications, 20/36-G, Old Market, West Patel Nagar, New Delhi 110 008 and printed at Radiant Printers, New Delhi 110 008, Phones: 571 2539, 574 0038; Telefax: 578 3571.

Website: *www.regencypublications.com*
Email: *regency@satyam.net.in* and *info@regencypublications.com*

PREFACE

The present study was made possible with the financial assistance from University Grants Commission under the scheme to Prof. T. Pullaiah. The authors are thankful to UGC for sanctioning the scheme.

The authors express their heartfelt thanks to the personnel of Andhra Pradesh forest Department for their help in field trips.

We are also grateful to Dr. M.S. Gayathri, Dr. K. Hanumanthappa, Dr. N. Venkatappa, Dr. P. Rajasekhara Reddy, Dr. K. Sri Rama Murthy and Dr. S. Sandhya Rani for their help in field and Laboratory.

The authors

CONTENTS

Chapter 1

INTRODUCTION

Tropical forests, occupying seven per cent of the land surface, represent a major portion of the biosphere's biomass. These harbour about sixty per cent of the world flora and are depleting at an alarming rate due to unprecedented increase in urbanisation, industrialization, mining, encroachment on vast areas of forest lands for extension of arable expanses.

From time immemorial plants are the source of food, clothing, shelter and medicines for human beings. It is necessary that we should have full knowledge regarding the occurrence, frequency, distribution and phenology of various plants for their proper utilization.

India is one of the twelve mega-biodiversity countries in the world representing 6.5 per cent of world's known wild life and 12 per cent of plant life. Due to over exploitation of the forest resources, atleast ten per cent of India's recorded wild flora is on the verge of extinction and many more wild species are on the threatened list, awaiting a similar fate. Hence there is an urgent need for survey and documentation of all the species occurring in India for conservation and utilisation of plant resources.

In a developing country like India, soil and land resources survey play an important role in economic development. Ramachandran (1978) says "the importance of survey, exploration and exploitation of the floral wealth of the country needs to be reiterated because of its obvious significance in the country's overall socio-economic and cultural development."

The importance of survey of plant resources had also been emphasized by Jain (1978) who says "After independence our planners realised that in an agricultural country like India where the flora is so varied and rich, a proper consensus of the flora of the country and its evaluation for economic exploitation is very important."

The importance of district floras has been emphasized by a number of experts in various fields. Jain (1978) has pointed out that the district Floras will be useful as local identification manuals and in several programmes of rural development and cottage industries.

Such a precise data for Ranga Reddi district has so far not been worked out and with the present policy of conservation, protection and utilization of forest plant resources and subsequently development and uplift of tribal and rural areas, such a detailed data at district level is all the more an urgent need.

Detailed floristic knowledge on district wise basis is in great demand for the various projects in the country, organized by the departments of agriculture, forestry, pharmaceutical industry, plant based industries, Ayurveda/Unani/Homeopathic research organizations etc. and for educational and conservational programmes and University research projects on wild plant species. Hence the present work has been undertaken to survey and document the plant resources of Ranga Reddi district for conservation and judicious utilization.

Ranga Reddi district is one of the ten districts of Telangana region of Andhra Pradesh State. This region was part of Nizam's Dominions, i.e., Hyderabad State till 1948 when it became part of Indian Union. In 1956 the States of India were reorganized and 9 districts of Nizam's dominion became part of Andhra Pradesh. Hyderabad district encompassed this region from then on. In 1978 the city of Hyderabad was separated into a separate district while the rest formed the present Ranga Reddi district. The district encloses the twin cities of Hyderabad-Secunderabad on all sides.

Ranga Reddi district is named after K.V. Ranga Reddi, who was Deputy Chief Minister of Andhra Pradesh. K.V. Ranga Reddi hailed from Peddamangalaram of Chevella taluk in this district.

Chapter 2

TOPOGRAPHY AND GENERAL FEATURES

General Note

Ranga Reddi district lies between 16° 30′ and 18° 20′ of North latitudes and 77° 30′ and 79° 30′ of East longitudes. The total geographical area of the district is 7493 Sq. Km. taking 20th position in the State of Andhra Pradesh. The district has 1055 villages and 5 towns with a population of 25.52 lakhs (15th in position in Andhra Pradesh).

The district is bounded on the north by Medak district on south by Mahaboobnagar district, on east by Nalgonda district of Andhra Pradesh and on west by Bidar district of Karnataka.

The district is divided into 37 revenue mandals and 3 revenue divisions namely Vikarabad, Chevella and Hyderabad East. The main towns of Ranga Reddi district are Hayat Nagar, Tandur, Ibrahimpatnam, Medchal and Vikarabad.

Hills

The district is mostly hilly and sparsely wooded. The general slope of the land is from west to east and south-east. The Rajkonda hill range run in south-eastern direction up to Devarakonda Mandal in Nalgonda district. The Anantagiri range traverses the district from the south in Mahaboobnagar district to the north up to Dharur. A large portion of this range is composed of high grade laterite. Isolated granite hills are also found here and there. Average elevation of the hills is 509.91 m above MSL.

Soils

The major soils of the district may be categorised into red earths comprising loamy sands (dubha), sandy loams (chalaka) and sandy clay loams. The red earth cover 70 per cent of the area while the red and black soil comprising

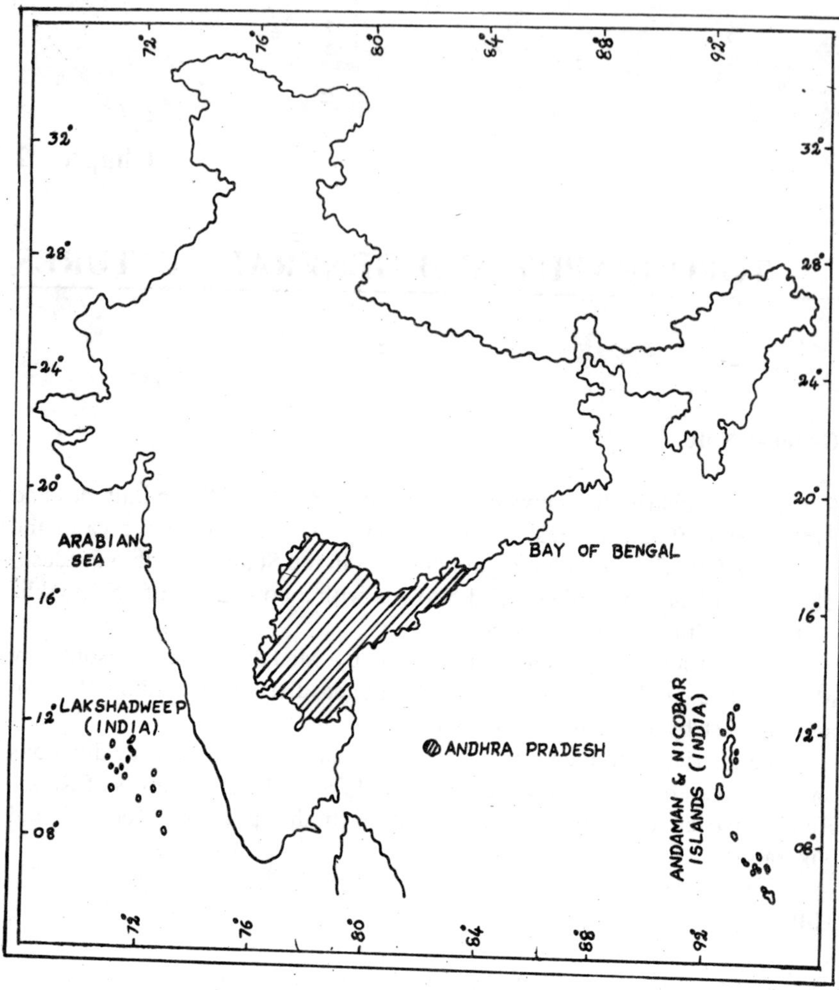

Fig. 1. Map of India showing position of Andhra Pradesh.

clay loams, clays and silty clays cover 20 per cent. The remaining 10 per cent of the area is covered by mixed soils. The red loamy sand which occur on strongly sloping terrain is also found to a large extent.

Minerals

Quartz is the principal mineral of this district and its estimated reserve is about 5 million tonnes. Felspar, clay, amethyat and laterite are also present. Lime stone is found in plenty in Tandur. The estimated reserve of the cement grade lime stone in the block is 173 million tonnes. Granite (building and road material) is also abundant.

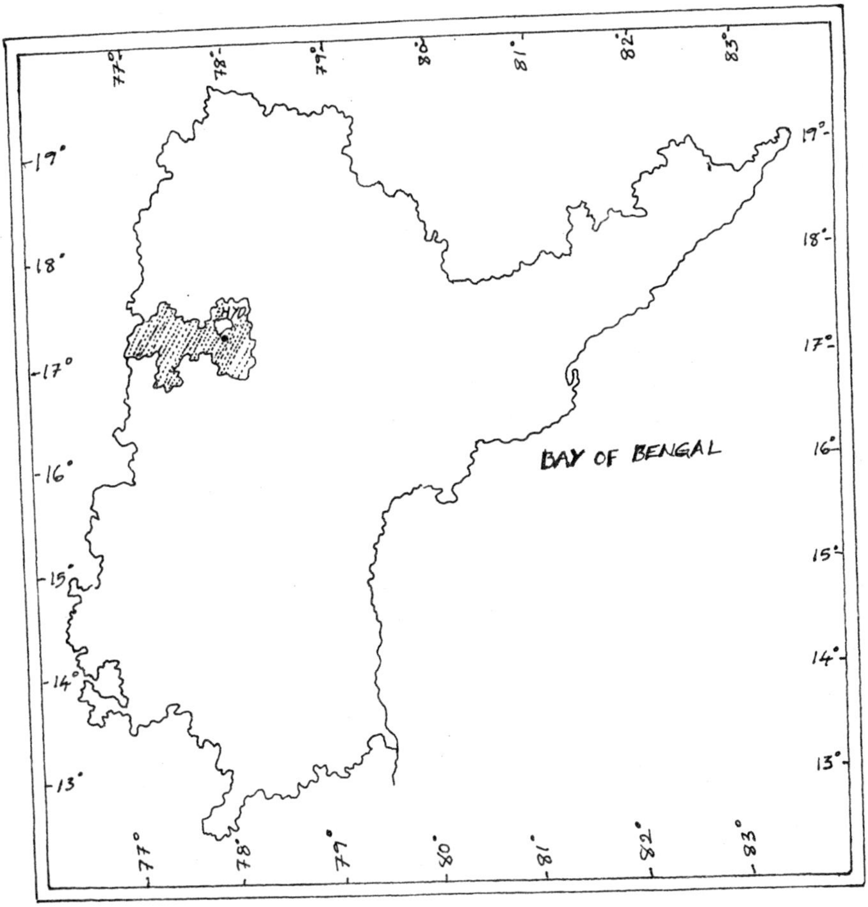

Fig. 2. Map of Andhra Pradesh showing position of Ranga Reddi district.

Climate

The climate of the district is characterized by a hot summer and generally dry weather except during south-west monsoon season. The year may be divided into four seasons, i.e., winter (December–February), summer (March–May), south-west monsoon season (June–September) and post-monsoon season (October–November).

Rainfall

The average annual rainfall in the district is 802.1 mm, bulk of which comes from south-west monsoon from June to September. The month-wise rainfall

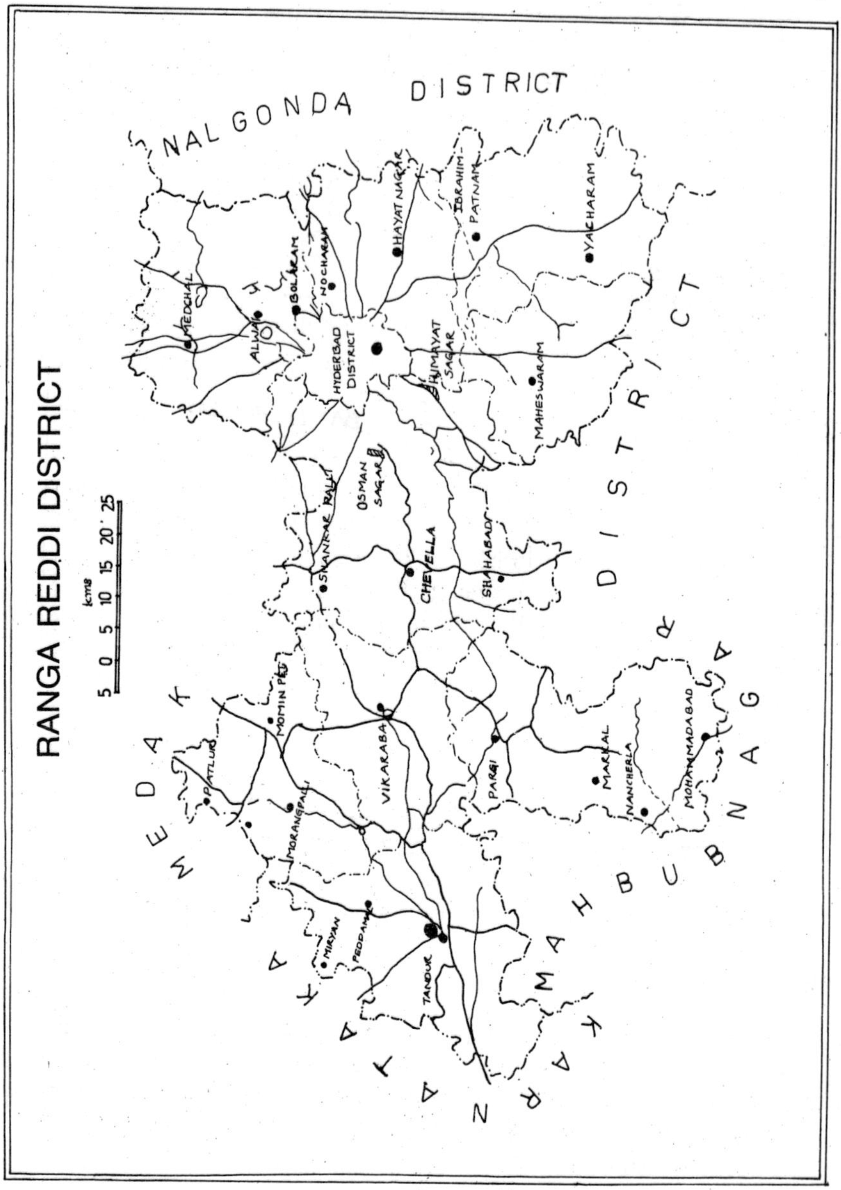

Fig. 3. Map of Ranga Reddi district.

is given in Table 1, while average annual rainfall during the years 1979-80 to 1993-94 is given in Table 2.

Temperature

May is the hottest month with mean daily temperature at 39.6°C. With the onset of monsoon in June there is appreciable drop in temperature. By the beginning of November decrease in day and night temperature is rapid. December is the coldest month with mean daily maximum temperature of 28.6°C and mean daily minimum temperature at 13.6°C. In this season, the night temperature sometimes drops down to 7°C.

Table 1
Rainfall in 1990

Month	Total rainfall in a month
January	5.4 mm
February	—
March	—
April	—
May	—
June	149.1 mm
July	118.0 mm
August	474.2 mm
September	196.2 mm
October	115.8 mm
November	6.2 mm
December	3.3 mm

Table 2
Average Annual Rainfall in Ranga Reddi District

(yearwise) in millimeters	
1979–80	716
1980–81	786
1981–82	1161
1982–83	770
1984–85	804
1985–86	575
1986–87	553
1987–88	975
1988–89	1013
1989–90	833
1990–91	977
1991–92	804
1992–93	820
1993–94	717

Rivers

The chief river of the district is the Musi which arises in the Anantagiri hills near Sivareddipet village in Vikarabad Mandal. It flows almost due east passing through the middle of Hyderabad city into Nalgonda district where it joins the Krishna river near Vadapalle. Another river called Kanga rises in Vikarabad taluk and it drains a number of villages in the erstwhile Tandur and Vikarabad taluks.

Chapter 3

FORESTS AND VEGETATION

Forests

Ranga Reddi district cannot boast of any important timber yielding forests beause of low rainfall and comparatively inferior soil condition aggravated by the ever increasing biotic interference. The soils have very little plant nutrients and so they support only inferior type of forests. The forest area in the district is 72,795.32 hectares against the geographical area of 7,56,289.33 hectares. The forest area covers about 9.7 per cent of the total geographical area of the district as against the minimum 33.3 per cent desirable according to National Forest Policy.

Vegetation

The forests of Ranga Reddi can be classified under Tropical Dry Deciduous forests based on the classification of Champion and Seth (1968). The forests can be categorized into 4 types namely (A) Teak forests, (B) Dry mixed deciduous forests, (C) Scrub jungles and (D) Dry Savannahs.

A. Teak Forests

In these forests *Tectona grandis* is predominant constituting 40 per cent of tree population. The forest blocks showing these forests include Mohammadabad, Kothapalli, Kanmankalva, Kusumasamudram and Anantasagar beats of Mohammadabad range, Anantagiri, Rudraram and Dharur blocks in Vikarabad range. However, due to illegal and indiscriminate felling of Teak trees these forests are being reduced to mixed and open type at an alarming rate.

B. Dry mixed deciduous Forests

These forests are characterized by less proportion of teak with the predominance of other trees like *Anogeissus latifolia, Madhuca indica, Terminalia coriacea, T. bellirica, Dalbergia paniculata, Pterocarpus marsupium* etc.

These types of forests are present in Rangammagudem, Kondapur and Gadirayal beats of Mohammadabad range, Parigi, Tandur and Vikarabad of Vikarabad range.

The species encountered within both dry teak and dry mixed deciduous forests are as follows.

Top Story

This stratum comprises of species like *Anogeissus latifolia, Boswellia serrata, Bombax ceiba, Buchnania lanzan, Careya arborea, Chloroxylon swietenia, Dalbergia paniculata, D. latifolia, Diospyros melanoxylon, Garuga pinnata, Gmelina arborea, Haldinia cordifolia, Lannea coromandelica, Miliusa tomentosa, Pterocarpus marsupium, Semecarpus anacardium, Sterculia urens, Strychnos potatorum, Tectona grandis, Terminalia arjuna, T. alata* and *T. coriacea.*

Middle Story

This stratum comprises of species like *Bauhinia racemosa, Dolichandrone falcata, Butea monosperma, Cassia fistula, Diospyros chloroxylon, D. montana, Holarrhena pubescens, Nyctanthes arbor-tristis, Streblus asper, Acacia leucophloea* and *Cordia obliqua.*

Shrubby Layer

Common shrubs seen in these forests are *Alangium salvifolium, Bridelia retusa, Canthium parviflorum, Catunaregum spinosa, Ixora arborea, Pavetta indica, Premna latifolia, Ziziphus xylopyrus, Mimosa prainiana* etc.

Herbs

Anisochilus carnosus, Abelmoschus ficulneus, Biophytum sensitivum, Blepharis maderaspatana, Borreria articularis, Canscora decurrens, Cissus pallida, Crotalaria spp., *Desmodium* spp. and *Polygala* spp. are common members of ground flora. Plants occurring near water courses include *Hygrophila auriculata, Caesulia axillaris, Commelina* spp., *Cyanotis* spp., *Rotala* spp. etc.

Climbers

Climbers and stragglers constitute a rich flora in the district. These include *Abrus precatorius, Aristolochia indica, Ampelocissus latifolia, Cissus vitiginea, Aspidopterys cordata, Hemidesmus indicus, Cayratia trifolia, Celastrus paniculatus, Cissampelos pariera, Cocculus hirsutus, Dioscorea* spp., *Gymnema sylvestre, Pergularia daemia, Wattakaka volubilis, Olax scandens, Derris scandens, Paracalyx scariosus, Ventilago denticulata, Hewittia scandens, Rivea hypocrateriformis* and *Bauhinia vahlii.*

Grasses and Sedges

Alloteropsis cimicina, Eragrostis atrovirens, Mnesithea granularis, Heteropogon contortus, Lophopogon tridens, Cymbopogon coloratus, C. martinii, Iseilema laxum, Cyperus spp., *Fimbristylis* spp. etc. are common grasses and sedges.

Parasites and Epiphytes

Cassytha filiformis, Dendrophthoe falcata, Scurrula parasitica, Viscum nepalense and *Taxillus bracteatus* are stem parasites while *Striga* spp. are root parasites collected from forests. *Vanda tessellata* is the only epiphyte in these forests.

C. Scrub Jungle

Scrub jungle is situated at the foot of the hills where the surface soil is very shallow and layers of sandy or gravely loam which is eroded in exposed regions. The vegetation of the scrubs is characterised by shrubs and subshrubs like *Capparis zeylanica, Acacia chundra, Mimosa intsia, M. prainiana, Ziziphus oenoplia, Z. xylopyrus, Opuntia stricta, Carissa spinarum, Canthium parviflorum, Lantana camara, Maytenus emarginata, Dichrostachys cinerea* and *Dodonaea viscosa.* These forests are distributed throughout the district, more so in areas bordering Karnataka.

D. Dry Savannahs

These forests are characterised by the presence of species like *Cassia auriculata, Dodonaea viscosa* and *Lantana camara. Cymbopogon coloratus, C. martini, Heteropogon contortus, Aristida* spp. are common in these forests.

Non Forest Vegetation

The non forest vegetation includes terrestial vegetation and aquatic vegetation.

A. Terrestial Vegetation

It includes all waste lands, hedges, roadsides, villages and cultivated lands.

(i) Waste lands, Roadsides and Villages

Ipomoea carnea is predominant in waste lands especially along roadsides and hedges of fields. Other plants which are generally distributed in waste lands are — *Croton bonplandianum, Trid. procumbens, Acalypha indica, Achyranthes aspera, Boerhavia diffusa, Alternanthera pungens, Tribulus terrestris, Cassia occidentalis, Cassia auriculata, Ziziphus mauritiana, Calotropis procera, Amaranthus viridis, Martynia annua, Alangium salvifolium, Solanum surrattense,* and *Prosopis chilensis.*

Along roadsides and villages following trees are generally present. They are *Acacia nilotica, Albizia lebbeck, Annona squamosa, Borassus flabellifer, Cassia siamea, Dalbergia sissoo, Delonix elata, D. regia, Millingtonia hortensis, Pongamia pinnata, Peltophorum pterocarpum, Samanea saman, Phoenix sylvestris, Tamarindus indica, Azadirachta indica.*

Hedges

Abutilon indicum, Caesalpinia pulcherrima, C. bonduc, C. decapetala, Canthium parviflorum, Catunaregam spinosa, Euphorbia tirucalli, Lawsonia inermis, Parkinsonia aculeata and *Prosopis chilensis* are common hedge plants.

Cultivated Field Weeds

Cyperus rotundus, Eclipta prostrata, Digera muricata, Echinochloa colona, Cynodon dactylon, C. barberi, Eleusine indica, Leucas aspera, Cleome spp., *Catharanthus pusillus, Malvastrum coromandelianum, Trianthema portulacastrum* and *Rothia indica* are common weeds encountered in these cultivated fields.

B. Aquatic Vegetation

The rivers, tanks, ponds, ditches, streams, canals, stagnent water and moist places are the site of aquatic plants which may be classified as follows.

1. Submerged hydrophytes

Hydrilla verticillata, Limnophila indica, Ottelia alismoides, Aponogeton natans, Ceratophyllum demersum and *Monochoria vaginalis* come under this category.

2. Floating hydrophytes

Nymphaea nouchali, Nelumbo nucifera, Pistia stratiotes, Utricularia stellaris and *Ipomoea aquatica* come under this category.

3. Emergent hydrophytes

Aeschynomene indica, Ammannia baccifera, Bacopa monnieri, Cleome chelidonii, Echinochloa crusgalli, Hygrophila auriculata, Limnophytum obtusifolium, Marsilea quadrifolia, Polygonum glabrum, Rotala spp., *Typha angustata, Sopubia delphinifolia, Schoenoplectus articulatus, S. juncoides* etc. come under this category.

Wet land hydrophytes

Centella asiatica, Caesulia axillaris, Bacopa monnieri, Commelina spp., *Cyperus* spp., *Lobelia alsinoides, Lindernia* spp., *Ludwigia perennis, Murdannia* spp. *Panicum trypheron, Hoppea dichotoma, Eclipta prostrata, Ageratum conyzoides, Verbascum chinense* come under this category.

Besides these *Terminalia arjuna* is common along river banks. *Eleocharis* spp., *Eriocaulon* spp., *Ischaemum pilosum, Oryza rufipogon, Sacciolepis interrupta, Schoenoplectus* spp. are common in marshy localities.

Chapter 4

PAST AND PRESENT WORK

Past Work

Ranga Reddi district received scant attention from plant explorers. Systematic survey and documentation was not carried out in any part of the district except that of Seshagiri Rao (1986). Whatever little was done was in respect of erstwhile Nizam's Dominions, also known as Hyderabad State.

The study of Flora of Hyderabad State dates back to the earlier part of the 19th century when Walker (1849) and Bradley (1849) published their pioneer work, which included agricultural, medicinal and other economically important species of Daulatabad and Warangal districts of the Hyderabad State. Campbell in 1898 in his "Glimpses of the Nizam's Dominion" included a list of forest plants of Hyderabad State". Subsequently some information on forest plants of State is available in the "Imperial Gazetter of India", provincial series, Hyderabad State (1909). Around this time Briscoe, a forest officer listed 128 forest plants. Patridge (1911) published a book entitled "Forest Flora of Hyderabad State", wherein he described 450 species belonging to 69 families and provided keys to taxa, information on local names and economic importance of plants. This book was later revised by Khan (1953), who added some more information on the vegetation and described 567 wild and cultivated species. Sayeeduddin (1935, 1938, 1941a,b, 1954) published a series of papers on the flora of Hyderabad State and reported a total of 370 species. Suxena (1947) listed 115 grasses from Hyderabad State.

Botanical explorations were revived with the reorganisation of Botanical Survey of India. These include Santapau (1954), Sebastine *et al.* (1960), Thothathri (1964), Sebastine and Henry (1966), Subba Rao and Kumari (1967), Ramayya and Rajagopal (1973), Kapoor and Kapoor (1973), Ramachandrachary (1980), Pullaiah *et al.* (1992, 1998) and Pullaiah and Ravi Prasad Rao (1995). Rajagopal (1973) studied Flora of Hyderabad city which is now surrounded on all sides by Ranga Reddi district. It included a study of the foliar epidermal characters for 400 species providing additional identifying characters. Seshagiri Rao (1986) explored the plant resources of the campus of University of Hyderabad and listed 419 species. This campus is now a part of Ranga Reddi district.

Recently Pullaiah (1997), Pullaiah and Ali Moulali (1997), Pullaiah and Chennaiah (1997) and Pullaiah and Surya Prakash Babu (1998) brought out "Flora of Andhra Pradesh" in 4 volumes which included collections from Ranga Reddi district.

Present Work

Intensive explorations were conducted in different parts of the district during different seasons of the year for a period of six years (1990, 1991, 1993–1998). Exploration trips were regularly made in every season and in all the representative localities of the district from floristic point of view, covered all topographical zones and noted the seasonal changes of the vegetation.

Plants were collected in either flowering or fruiting stages and if available both the stages with good vegetative growth were collected. Standard methods described by Santapau (1955) and Jain and Rao (1977) have been followed. Each collection of individual species was labelled with field numbers in quadruplicate after noting the details of plant description such as habit, habitat, flower colour, smell, relative abundance of the plant and its distribution on a standard field note book. Details about the altitude, soil and rocks are also entered in the field note book. Care was also taken in collecting the bulbs, corms, rhizomes, tubers etc. if the plant possess them. In case of plants in vegetative stages in a particular field trip localities of these plants were carefully marked in the field note book. These could be easily spotted in the flowering and fruiting stages at later dates.

Immediately after cutting the twigs of about 30 cm in length or in case of small herbs the whole plant was uprooted with care, appropriate field labels were inserted with serial numbers and were kept in polythene bag to avoid further dessication and to protect from withering.

Plants were collected in the forenoon and collections were restricted to 25 numbers per day for the sake of convenience. The collected specimens were poisoned by dipping whole twig in saturated mercuric chloride in ethyl alcohol solution. Immediately the poisoned specimens were placed between blotting papers with the help of forceps. These were tied properly in iron pressors with the help of ropes. After about 24 hours the blotting papers were changed and the specimens were spread properly. Some leaves are placed facing upwards and some facing downwards to show the characters on both surfaces. These spread specimens were once again placed in other dried blotting sheets and tied again in iron pressors and this process is continued till the specimens were dried completely.

Flowers and fruits of some important and rare plants were fixed in formalin-acetic acid-alcohol in specimen tubes for further study and reference. In the case of succulents like Cacti and Euphorbias the plant parts are sliced longitudinally and transversely and placed half upwards and the other downwards. Special care was taken for the aquatic plants and small herbaceous plants.

The poisoned, pressed and dried specimens were pasted with glue on thick mount boards, measuring 42 × 28 cm and stiched with thread. Labels with all relevant information such as name of the plant, family, locality, altitude, date of collection, notes as in field notes and collector's name were affixed on the right hand bottom corner of the mounted sheets.

Provisional identifications were made with the help of Gamble's Flora of Presidency of Madras (1915–35). Later the identifications were confirmed with the help of latest monographs, Floras and comparisons with authentic specimens in the herbarium of Department of Botany, Sri Krishnadevaraya University, Anantapur. Flowering and Fruiting seasons were accurately recorded for each species together with relative degree of abundance.

After identifications the sheets were incorporated in the Herbarium of Botany Department of Sri Krishnadevaraya University, Anantapur. Of the four sets of specimens collected, one set mounted on herbarium sheets is deposited in the Herberium of Sri Krishnadevaraya University, Anantapur and the unmounted specimens are preserved for future exchange programme with other herberia in India and abroad and also for dissection purpose if necessity arises.

Plan followed in the Enumeration of Plants

The species have been enumerated systematically. The families are arranged according to Bentham and Hooker's (1862–83) system of classification with certain exceptions to accommodate recent changes in the arrangement of the families. All the genera in a family and all the species in a genus are given in Alphabetical order. Dichotomous keys have been provided to the families, genera and species which are mostly artificial. The systematic treatment under each family includes the following. Dichotomous key to the genera, name of species with original citation according to latest International Code of Botanical nomenclature (Greuter, 1988), synonymy if necessary to connect the name with Hooker's "Flora of British India" and Gamble's "Flora of Presidency of Madras", brief description, flowering and fruiting season, vernacular names if they have and local use. At the end of the description of each species the numbers of specimens examined are given. If the plant is a new record its details are given.

Common abbreviations and acronyms used in the systematic enumeration are:

MSM	:	M. Silar Mohammed
NV	:	N. Venkatappa
KH	:	K. Hanumantappa
PRSR	:	P. Rajasekhar Reddy
SSR	:	S. Sandhya Rani
KSR	:	K. Seshagiri Rao
MRS	:	M.R. Suxena
TP	:	T. Pullaiah

ca	:	circa; about
cm	:	centimetre
ed.	:	editor, edition
emend	:	emendavit, emended
et al.	:	*et aliorum,* and others
etc.	:	*etcetera* and other things
Ex	:	published
Excl.	:	Exclusus, Excluding
fl.-fr.	:	flowering-fruiting
f	:	Filius -son
ibid	:	*ibedem*, the same, in the same place
ic	:	icone, illustration
incl	:	inclusus, including
loc. cit.	:	*loco citato,* at the place cited
m	:	metre
max	:	maximum
min	:	minimum
nom. cons	:	*nomina conservanda...* conserved name
nom. ambig.	:	*nomen amibiguum*, doubtful name
nom. illegit	:	*nomina illegitimum,* illegitimate name
op. cit.	:	*opere citato*, in the work cited
p.p	:	*pro parte,* in part
sensu lato	:	in the broad sense
sp	:	species (singular)
spp.	:	species (plural)
ssp	:	subspecies
suppl	:	supplement

BSI	:	Botanical Survey of India
DD	:	Forest research Institute Herbarium, Dehradun
FBI	:	Hooker, J.D. *et al.* Flora of British India

Gamble	:	Gamble, J.S. and Fischer, C.E.C. Flora of Presidency of Madras
FFH	:	M.S. Khan's "Forest Flora of Hyderabad State"
ICBN	:	International Code of Botanical Nomenclature
RF	:	Reserve Forest
Syn	:	*Synonymum,* synonym
t	:	*tabulae*, plate
var.	:	variety
Vern.	:	Vernacular name
viz.	:	*videlicet*, namely
vol(s)	:	volume(s).

Chapter 5

FLORISTIC ANALYSIS

A total of 698 wild and naturalized species belonging to 414 genera and 110 families have been enumerated. Of these Angiosperms constitute 694 species while pteridophytes constitute 4 species.

Familes	Number of		
	Families	Genera	Species
Dicotyledons	89	331	530
Monocotyledons	17	79	164
Pteridophytes	4	4	4
Total	110	414	698

The ratio of Monocotyledons to Dicotyledons is 1:5.23 of families (17:89) 1:4.17 of genera (79:331) and 1:3.2 of species (164:530). The ratio of genera to species in Angiosperms is 1:1.68. In British India it is 1:7. This pattern of distribution of genera and species confirms the general rule that, smaller the area smaller the genus-species ratio.

In order to get an insight into the relations of the Flora of Ranga Reddi district with neighbouring floras, a comparative analysis of ten dominant families (in order of their frequency) of the flora, of Ranga Reddi District, Flora of Presidency of Madras, and Flora of India is given in order of their frequency (Table 3).

The family Fabaceae with 109 species (comprising Papilionaceae with 74 species, Caesalpiniaceae 23 and Mimosaceae 19) is dominant in Ranga Reddi district as well as in the 'Flora of the Presidency of Madras' but hold third position in the Flora of India. The family Orchidaceae holding sixth and second position in the Flora of Presidency of Madras and Flora of India with 199 and 990 species respectively is represented by only 2 species in Ranga Reddi district.

Cyperus is the largest genus with 16 species, followed by *Fimbristylis* (14), *Cassia* (10), *Eragrostis* (9), *Euphorbia* (9), *Indigofera* (9), *Ipomoea* (9), *Crotalaria* (7), *Grewia* (7) and *Phyllanthus* (7).

Table 3
Dominant families of Ranga Reddi district — A Comparative Statement

Rank	Ranga Reddi district (Present study)	No. of species	Madras Presidency Gamble	No. of species	Flora of India (BSI)	No. of species
1	Leguminosae	109	Leguminosae	432	Poaceae	1225
2	Poaceae	70	Poaceae	388	Orchidaceae	990
3	Cyperaceae	56	Rubiaceae	226	Leguminosae	775
4	Euphorbiaceae	33	Acanthaceae	202	Asteraceae	754
5	Asteraceae	32	Euphorbiaceae	199	Rubiaceae	495
6	Acanthaceae	24	Orchidaceae	199	Cyperaceae	449
7	Rubiaceae	19	Asteraceae	189	Euphorbiaceae	419
8	Convolvulaceae	17	Cyperaceae	172	Lamiaceae	393
9	Lamiaceae	17	Lamiaceae	128	Acanthaceae	379
10	Asclepiadaceae	14	Asclepiadaceae	93	Scrophulariaceae	356

1. New Records for Andhra Pradesh
   1. *Fimbristylis alboviridis* Clarke
   2. *Fimbristylis dichotoma* (L.) Vahl. ssp. *podocarpa*
   3. *Alysicarpus ovalifolius* (Schum.) J. Leon.
2. Second report for Andhra Pradesh and subsequent report
   - *Cyperus cuspidatus* Kunth
   - *Cyperus tenuiculmis* Boeck
   - *Fimbristylis tetragona* R. Br.
   - *Neanotis lancifolia* (Hook. f.) Lewis
   - *Rynchospora wightiana* Steud.
   - *Chrysanthellum americanum* (L.) Vatke
3. Subsequent Report for South India
   - *Dactyloctenium aristatum* Link
4. New Records for Telangana Region
   - *Heliotropium strigosum* Willd.
   - *Asystasia dalzelliana* Sant.
   - *Grewia orientalis* L.
   - *Tylophora fasciculata* Buch.-Ham.
   - *Leucas nutans* Spreng.
   - *Mariscus cyperinus* (Retz.) Vahl.
   - *Thelepogon elegans* Roem. & Schutt.
   - *Ziziphus glabrata* Roth
5. Rare Taxa for Andhra Pradesh
   - *Crotalaria hirta* Willd.
   - *Indigofera nummularifolia* (L.) Alston
   - *Utricularia stellaris* L.f.
   - *Drosera burmannii* Vahl

Drosera indica L.
Arisaema leschenaultii Bl.
Kyllinga melanosperma Nees

ARTIFICIAL KEY TO FAMILIES

Note: The number of times a family keyed out is indicated by the Roman numerals in brackets after family name.
In case of exceptions, the name for the genus/species is given in brackets at the end of the concerned lead.

KEY TO MAIN GROUPS

1. Plants producing seeds .... ***Angiosperms***
  2. Persistent tap-root system. Stem consisting of concentric ring of open vascular bundles with a central pith (Eu-stele), leaves with reticulate venation, stipulate or exstipulate, flowers 2-polymerous, cotyledons 2 .... ***Dicotyledons***
  2. Adventitious root system, stem consisting of scattered, closed vascular bundles without a central pith (Atactostele), leaves with parallel venation, exstipulate leaves, flowers 3-merous .... ***Monocotyledons***
1. Plants not producing seeds, containing sporangia and propogated by spores .... ***Pteridophytes***

DICOTYLEDONS

1. Reticulate venation, flowers 4–5-merous, cotyledons 2:
  2. Perianth 2-whorled, distinguished into sepals and petals:
    3. Petals free:
      4. Calyx segments distinct:
        5. Stamens arising directly from the receptacles:
          6. Plants aquatic:
            7. Leaves floating on water .... ***Nymphaeaceae***
            7. Leaves not floating, raised above water .... ***Nelumbonaceae***
          6. Plants terrestrial:
            8. Carpels free in ripe condition:
              9. Climbers........ ***Menispermaceae***
              9. Trees:
                10. Armed, fruits not edible .... ***Bombacaceae***
                10. Unarmed, fruits edible .... ***Annonaceae***
            8. Carpels not free in ripe condition:
              11. Leaves opposite .... ***Clusiaceae***
              11. Leaves alternate or whorled:
                12. Sepal 2 or 3:
                  13. Leaves spiny .... ***Papaveraceae***
                  13. Leaves not spiny .... ***Portulacaceae***
                12. Sepals 4 or more:

14. Placentation axile:
 15. Stamens monadelphous .... *Malvaceae*
 15. Stamens di or polyadelphous:
  16. Filaments of stamens free .... *Tiliaceae*
  16. Filaments of stamens united .... *Sterculiaceae*
14. Placentation other than axile:
 17. Plants insectivorous .... *Droseraceae*
 17. Plants not insectivorus:
  18. Petals 3 or 4:
   19. Plants herbs:
    20. Petals 4:
     21. Fruit a siliqua .... *Brassicaceae*
     21. Fruit a capsule .... *Cleomaceae*
    20. Petals 3 .... *Polygalaceae*
   19. Plants shrubs or stragglers:
    22. Stragglers .... *Capparaceae*
    22. Shrubs .... *Flacourtiaceae*
  18. Petals 5 or more:
   23. Flowers solitary:
    24. Flowers irregular, anther connective extends beyond the cells .... *Violaceae*
    24. Flowers regular, anther connective not extends beyond cells .... *Turneraceae*
   23. Flowers in racemes or cymes .... *Caryophyllaceae*

5. Stamens not arising directly from the receptacles:
 25. Plants herbs:
  26. Leaves compound:
   27. Flowers solitary .... *Zygophyllaceae*
   27. Flowers in umbels .... *Oxalidaceae*
  26. Leaves simple .... *Balsaminaceae*
 25. Plants trees, shrubs or twiners:
  28. Leaves gland dotted:
   29. Fruit capsule .... *Flindersiaceae*
   29. Fruit berry (Hesperidium) .... *Rutaceae*
  28. Leaves not gland dotted:
   30. Leaves compound:
    31. Placentation parietal, fruits elongated .... *Moringaceae*
    31. Placentation axile or axile pendulous, fruits not elongated:
     32. Fruits capsules or drupes, winged or lobed:
      33. Fruits other than samaroid .... *Sapindaceae*
      33. Fruits samaroid .... *Simaroubaceae*
     32. Fruits berries or drupes, not winged or lobed:
      34. Under shrubs .... *Leeaceae*
      34. Trees:
       35. Stamens monadelphous .... *Meliaceae*
       35. Stamens free .... *Burseraceae*
   30. Leaves simple:
    36. Ovule 1 per locule:
     37. Petals not exceeding calyx .... *Rhamnaceae*
     37. Petals exceeding calyx:

38. Leaves opposite .... ***Malpighiaceae***
38. Leaves alternate:
39. Stragglers or erect shrubs:
40. Flowers in cone like raceme or cyme or spike .... ***Opiliaceae***
40. Flowers in open raceme or cyme .... ***Olacaceae***
39. Trees .... ***Anacardiaceae***
36. Ovules 2 or more per locule:
41. Plants usually trees .... ***Celastraceae***
41. Plants usually stragglers .... ***Vitaceae***
4. Calyx segments united:
42. Leaf base pulvinous, fruit legume:
43. Flowers zygomorphic:
44. Corolla papilionaceous .... ***Fabaceae***
44. Corolla not papilionaceous .... ***Caesalpiniaceae***
43. Flowers actinomorphic .... ***Mimosaceae***
42. Leaf base not pulvinous, fruit other than legume:
45. Plants spiny:
46. Shrubs, stems succulent .... ***Cactaceae***
46. Trees, stems not succulent .... ***Alangiaceae***
45. Plants not spiny:
47. Climbing herbs .... ***Cucurbitaceae***
47. Erect herbs or shrubs or trees:
48. Inflorescence an umbel .... ***Apiaceae***
48. Inflorescence other than umbel:
49. Petals absent:
50. Ovules more than 3 .... ***Molluginaceae***
50. Ovules less than 3 .... ***Aizoaceae***
49. Petals present:
51. Leaves alternate, rarely clustered at the tip of branches:
52. Fruit berry .... ***Lecythidaceae***
52. Fruit capsule .... ***Onagraceae***
51. Leaves opposite decussate, rarely alternate:
53. Bark thin, papery, fruits indehiscent .... ***Myrtaceae***
53. Bark absent, if present not papery, fruit dehiscent:
54. Fruits often winged .... ***Combretaceae***
54. Fruits not winged .... ***Lythraceae***
3. Petals united:
55. Plants carnivorous .... ***Lentibulariaceae***
55. Plants not carnivorous:
56. Ovary inferior:
57. Stipules interpetiolar .... ***Rubiaceae***
57. Stipules absent:
58. Flowers in heads, ovary unilocular .... ***Asteraceae***
58. Flowers solitary, ovary 2–3 locular .... ***Lobeliaceae***
56. Ovary superior:
59. Plants leafless parasites .... ***Cuscutaceae***
59. Plants leafy autotrophs:
60. Trees:
61. Stamens alternate with petals:
62. Leaves compound .... ***Bignoniaceae***
62. Leaves simple:
63. Fruit a berry .... ***Loganiaceae***
63. Fruit a drupe .... ***Verbenaceae*** (I)

61. Stamens opposite to the petals:
64. Flowers unisexual, stamens inserted on receptacle .... ***Ebenaceae***
64. Flowers bisexual, stamens inserted on corolla .... ***Sapotaceae***
60. Shrubs or herbs:
65. Ovary 1-celled .... ***Plumbaginaceae***
65. Ovary 2-many celled:
66. Branchlets strigose or scabrous .... ***Boraginaceae***
66. Branchlets other than strigose or scabrous:
67. Inflorescence verticellaster (except *Ocimum*); style gynobasic ....***Lamiaceae***
67. Inflorescence other than verticellaster; style terminal
68. Corolla regular:
69. Leaves opposite:
70. Stamens 2:
71. Fruit berry .... ***Oleaceae***
71. Fruit capsule .... ***Nyctanthaceae***
70. Stamens 4 or more:
72. Carpels free, styles united:
73. Pollinia present .... ***Asclepiadaceae***
73. Pollinia absent .... ***Apocynaceae***
72. Carpels and styles united .... ***Gentianaceae***
69. Leaves alternate:
74. Ovary cells 2-ovuled, rarely 4-ovuled:
75. Plants twiners or herbs or shrubs, fruit capsule, not pulpy .... ***Convolvulaceae***
75. Trees, fruit drupe, pulpy .... ***Cordiaceae***
74. Ovary cells many-ovuled .... ***Solanaceae***
68. Corolla irregular:
76. Carpels 2-many-ovuled:
77. Fruits opening elastically .... ***Acanthaceae***
77. Fruits not opening elastically:
78. Plants viscid, seeds exalbuminous .... ***Pedaliaceae***
78. Plants not viscid, seeds not albuminous........................... .... ***Scrophulariaceae***
76. Carpels 1-ovuled .... ***Verbenaceae***(II)
2. Perianth 1-whorled or absent:
79. Plants submerged. Leaves dissected .... ***Ceratophyllaceae***
79. Plants not submerged. Leaves entire:
80. Flowers bisexual, anthers opening by valves:
81. Plants parasitic:
82. Plants leafy .... ***Loranthaceae***
82. Plants leafless:
83. Stems wiry .... ***Lauraceae***
83. Stems flat .... ***Viscaceae***
81. Plants not parasitic:
84. Ovary inferior .... ***Aristolochiaceae***
84. Ovary superior:
85. Stipules ochreate .... ***Polygonaceae***
85. Stipules not ochreate:
86. Perianth scarious .... ***Amaranthaceae***
86. Perianth not scarious .... ***Nyctaginaceae***
80. Flowers unisexual, anthers not opening by valves:
87. Fruits samaroid .... ***Hernandiaceae***
87. Fruits not samaroid:
88. Ovary 1-loculed:

89. Herbs, latex absent .... ***Urticaceae***
89. Trees, latex present:
90. Latex milky, fruit hypanthodium .... ***Moraceae***
90. Latex not milky, fruit other than hypanthodium .... ***Ulmaceae***
88. Ovary 2 or more-loculed .... ***Euphorbiaceae***

MONOCOTYLEDONS

1. Leaves parallel veined. Flowers 3-merous. Cotyledon 1:
91. Plants aquatic, free floating or submerged:
92. Plants free floating, stoloniferous (*Pistia*) .... ***Araceae*** (I)
92. Plants not free floating, non-stoloniferous:
93. Perianth 2-seriate:
94. Flowers in spathes .... ***Hydrocharitaceae***
94. Flowers in spikes or racemes .... ***Pontederiaceae***
93. Perianth 1-seriate .... ***Aponogetonaceae***
91. Plants terrestrial:
95. Flowers in spikelets;
96. Stems hollow, stigmas feathery, styles 2 .... ***Poaceae***
96. Stems solid, stigmas entire, style 1 .... ***Cyperaceae***
95. Flowers not in spikelets:
97. Climbers .... ***Dioscoreaceae***
97. Non-climbers (Except *Gloriosa* and *Asparagus*)
98. Inflorescence a pappose brown spike .... ***Typhaceae***
98. Inflorescence otherwise:
99. Leaves fan like .... ***Arecaceae***
99. Leaves simple, not fan like:
100. Perianth absent (*Arisaema, Pilea*) .... ***Araceae*** (II)
100. Perianth present:
101. Leaves succulent, tip spiny .... ***Agavaceae***
101. Leaves not as above:
102. Leaves radical .... ***Eriocaulaceae***
102. Leaves not radical:
103. Flowers in whorls on long peduncles ... ***Alismataceae***
103. Flowers not as above:
104. Flowers actinomorphic; perfect stamens 3–6:
105. Bracts present, 4 cm in length .... ***Amaryllidaceae***
105. Bracts absent, even if present not as above:
106. Perianth segments corolline:
107. Flowers yellow .... ***Hypoxidaceae***
107. Flowers otherwise .... ***Liliaceae***
106. Outer perianth segments calycine ... . ***Commelinaceae***
104. Flowers zygomorphic, perfect stamens 1–2 .... ***Orchidaceae***

PTERIDOPHYTES

1. Aquatic herbs:
   2. Rhizomatous herbs .... ***Marsileaceae***
   2. Non-rhizomatous herb .... ***Azollaceae***
1. Terestrial herbs:
   3. Leaves heterophyllous .... ***Selaginellaceae***
   3. 'Leaves homophyllous .... ***Actinopteridaceae***

Chapter 6

SYSTEMATIC ENUMERATION

ANNONACEAE

1. Leaves glabrous, fruits syncarpic, not clustered .... ***Annonna***
1. Leaves tomentose, fruits not syncarpic, clustered .... ***Miliusa***

ANNONA Linnaeus

Annona squamosa L. Sp. Pl. 537. 1753; FBI 1: 78; Gamble 20(14); Khan in FFH 52.

A shrub, up to 6 m tall. Leaves oblong-lanceolate, 3–8 × 2.5–3 cm, base cuneate, margin entire, tip subacute, glabrescent above, glaucous below. Flowers greenish-yellow, terminal, solitary. Syncarp up to 8 cm long, glaucous, 1-seeded. Seeds brown, smooth.

Common. Fl. & Fr.: May–Sept. *Vern.:* Sitaphalam. Mohammadabad beat, *MSM & KH* 10206.

MILIUSA Leschenault ex A.P. de Candolle

Miliusa tomentosa (Roxb.) Sinclair in Gard. Bull. Straits Settlem. 14: 378. 1955. *Uvaria tomentosa* Roxb. Pl. Cor. t. 35. 1795. *Saccopetalum tomentosum* (Roxb.) Hook. f. & Thoms. Fl. Ind. 152. 1855; FBI 1: 88; Gamble 22(16); Khan in FFH 53.

A deciduous tree, up to 20 m tall, branchlets densely woolly. Leaves ovate, 4–8 × 2.5–5 cm, base obtuse, margin entire, tip acute, tomentose above, woolly below. Flowers purple in axillary fascicles. Berries 6–8 in a cluster, 2 cm long, pubescent, 3–4-seeded. Seeds brown.

More or less common in forests. Fl. & Fr.; Mar.–May. *Vern.:* Chilkadudi, Mohammadabad beat, *MSM & KH* 10213.

Polyalthia longifolia (Sonn.) Thw. is planted as an avenue tree.

MENISPERMACEAE

1. Leaves peltate .... *Cissampelos*
1. Leaves other than peltate:
   2. A twining shrub, leaves villous .... *Cocculus*
   2. A climbing shrub, leaves glabrous .... *Tinospora*

CISSAMPELOS Linnaeus

Cissampelos pareira L. Sp. Pl. 1031. 1753. var. **hirsuta** (Buch.-Ham. ex DC.) Forman in Kew Bull.22: 356. 1968. *C. hirsuta* Buch-Ham. ex DC. Syst. Nat. 1: 535. 1817. *C. pareira* L. Sp. Pl. 1031. 1753. p.p; FBI 1: 103; Gamble 30(21); Khan in FFH 55.

A twining shrub, branchlets wiry tomentose. Leaves ovate-reniform, 4–6 × 5–6.5 cm, base cordate, margin entire, tip mucronate. Flowers greenish-yellow, unisexual, male in sub-corymbs, female in thyrsoid clusters. Drupes 4 mm long, hairy, 1-seeded. Seeds brown.

Common in hedges and on trees in forest. Fl. & Fr.: Aug.–Oct. Anantagiri RF, *MSM & TP* 11060.

COCCULUS A.P. de Candolle

Cocculus hirsutus (L.) Diels in Engl. Pflanzenr. IV 94. 236. 1910. *Menispermum hirsutum* L. Sp. Pl. 341. 1753. *Cocculus villosus* DC. Prodr. 1: 98. 1824; FBI 1: 101; Gamble 29(21); Khan in FFH 55.

A bushy twiner, up to 6 m, branchlets tomentose. Leaves ovate-cordate, 2–5 × 1–3 cm, base cordate, margin entire, tip mucronate, pubescent above, villous below. Flowers white, unisexual, male in axillary cymose panicles, female in axillary clusters. Drupes 0.5 cm long, glabrous, 1-seeded. Seeds brown.

Common twiner on bushes and trees. Fl. & Fr.: Mar.–May. *Vern.:* Dusara tiga. Mohammadabad beat, *MSM & KH* 10241; Kanmankalva beat, *MSM & KH* 10306.

TINOSPORA Miers *nom. cons.*

Tinospora cordifolia (Willd.) Hook. f. & Thoms Fl. Ind. 1: 184. 1855; FBI 1: 97; Gamble 26(19); Khan in FFH 55. *Menispermum cordifolium* Willd. Sp. Pl. 4: 826. 1806.

A woody climbing shrub, branchlets glabrous, bark papery in young one and corky in mature one. Leaves cordate, 4–8 × 5–8.5 cm, base cordate, margin entire, tip acuminate, glabrous on both sides. Flowers green in axillary pseudo-racemes. Male in fascicles, female solitary. Drupes globose, 0.5 cm long, 1-seeded. Seeds black.

Common in hedges. Fl. & Fr.: All seasons. *Vern.:* Tippa teega. Naskal, *MSM & PRSR* 15122.

NELUMBONACEAE

NELUMBO Adanson

Nelumbo nucifera Gaertn. Fruct. 1: 73. 1788; Subramanyam Aquat. Angiosp. 8.f.4. 1962. *Nelumbium speciosum* Willd. Sp. Pl. 2: 1258. 1799; FBI 1: 116; Gamble 34 (24).

A perennial aquatic herb. Leaves rotund, 30–50 cm across, glaucous below, margin entire. Flowers white with light rose tinge. Carpels many, free, ripe carpels nut like. Ovules pendulous.

Occasional in ponds (Seshagiri Rao, 1986).. Fl. & Fr.: All seasons. *Vern.:* Tamara.

NYMPHAEACEAE

NYMPHAEA Linnaeus

Nymphaea nouchalii Burm. f. Fl. Ind. 120. 1768. *N. stellata* Willd. Sp. Pl. 2: 1153. 1799; Gamble 34(24); Khan in FFH 57.

A large perennial aquatic herb, up to 70 cm tall. Rhizome submerged. Leaves floating, peltate, 10–18 × 9–16 cm, petiole up to 40 cm, base cordate, margin entire, tip rounded, glabrous on both sides. Flowers white, terminal, solitary. Fruit globose, 6.5 cm long, glabrous. Seeds striate.

Common in ponds. Fl. & Fr.: Sept.–Jan. *Vern.:* Kaluva. Kanmankalva beat, *MSM* 10469.

PAPAVERACEAE

ARGEMONE Linnaeus

Argemone mexicana L. Sp. Pl. 508. 1753; FBI 1: 117; Gamble 35(25); Khan in FFH 58.

An annual erect prickly herb, up to 45 cm tall. Stems with yellow sap. Leaves oblong, 6–8 × 2–4 cm, base sub amplexicaul, margin spinulose-dentate, glabrous on both sides. Flowers yellow, terminal, solitary. Capsules septicidal, 4.5 cm long, spinous, many-seeded. Seeds black.

Common on roadsides. Fl. & Fr.: May–Aug. *Vern.*: Yerri Kusuma. Mohammadabad beat, *MSM & KH* 10204.

BRASSICACEAE

BRASSICA Linnaeus

Brassica juncea (L.) Czern. & Coss in Czern. Consp. Pl. Chark. 8: 5.1859; FBI 1: 157. 1872; Gamble 38(28). *Sinapsis juncea* L. Sp. Pl. 668. 1753.

An annual erect herb up to 60 cm tall. Leaves pinnatifid, simple, oblanceolate-obovate, base attenuate, margin dentate, tip sub acute. Flowers yellow in terminal racemes. Siliqua sub terete, glabrous. Seeds ovoid.

Common in waste lands as an escape. Fl. & Fr.: Jan.–Apr. *Vern.:* Avalu. Lingampalle, *MSM* 19220.

CAPPARACEAE

1. Unarmed shrubs. Stamens 4 or 6, inserted unilaterally on the slender base of the gynophore .... ***Cadaba***
1. Armed shrubs or trees. Stamens numerous, inserted on the torus at the base of the gynophore .... ***Capparis***

CADABA Forsskal

Cadaba fruticosa (L.) Druce Bot. Exch. Club. Soc. Brit. Isles. 3: 415. 1914. *Cleome fruticosa* L. Sp. Pl. 671. 1753. *Cadaba indica* Lam. Encycl. 1: 544. 1785; FBI 1: 172; Gamble 43(31).

A shrub up to 1 m tall. Leaves oblong, lanceolate, 2.5–3.8 × 0.5–2.5 cm, base cuneate, margin entire, tip obtuse-apiculate. Flowers green in terminal corymbs. Berries torulose, many-seeded. Seeds small.

Occasional in scrub jungle (Seshagiri Rao, 1986). Fl. & Fr.: Apr.–Aug.

CAPPARIS Linnaeus

1. Trees unarmed, bark yellow, leaves not mucronate .... ***C. grandis***
1. Shrubs armed, bark other than yellow, leaves mucronate:
    2. Flowers in corymbose umbels. Fruits globose, 0.8 cm across, single-seeded .... ***C. sepiaria***
    2. Flowers in clusters. Fruits ellipsoid, 2.5 cm across, many-seeded .... ***C. zeylanica***

Capparis grandis L.f. Suppl. Pl. 263. 1781; FBI 1: 176; Gamble 46(33); Khan in FFH 62.

A deciduous tree up to 7 m tall, branchlets tomentose, bark yellow. Leaves elliptic, 2–4 × 1.5–2 cm, base cuneate, margin entire, tip obtuse, glabrous on both sides. Flowers cream in axillary or terminal umbels. Berries long, glabrous, 3–6–seeded.

Common in forests. Fl. & Fr.: Mar.–July. Dharur RF, *MSM & TP* 11095.

Capparis sepiaria L. Syst. Nat. ed. 10, 1571. 1759; FBI 1: 177; Gamble 46(32); Khan in FFH 63.

An armed straggling shrub, up to 3 m, branchlets glabrous, spines recurved. Leaves elliptic-ovate, 1–4.5 × 1–3.5 cm, base cuneate, margin entire, tip obtuse, glabrous on both sides. Flowers white in terminal umbels. Berries 0.8 cm long, glabrous, black when ripe, 1-seeded.

Occasional in forests. Fl. & Fr.: Apr.–Sept. Mohammadabad beat, *MSM* 10481.

Capparis zeylanica L. Sp. Pl. (ed. 2) 720. 1762; Gamble 46(33); Khan in FFH 62. *Capparis horrida* L.f. Suppl. Pl. 264. 1781.

A prickly climbing shrub, up to 6 m, prickles recurved. Leaves ovate, 4.5–6.5 × 2.5–4 cm, base cuneate, margin entire, tip mucronate, glabrous on both sides. Flowers white, in axillary, serial clusters on fresh shoots. Berries 4 cm long, pubescent, many-seeded. Seeds brown.

Common in forests. Fl. & Fr.: Feb.–June. *Vern.:* Adavi nimma. Mohammadabad beat, *MSM & KH* 10215.

CLEOMACEAE

CLEOME Linnaeus

Note: Based on presence or absence of disk and gynophore and the number of stamens, the genus *Cleome* has been variously split into a number of taxa or sections. A.P. De Cadolle (Prodr. 1: 242. 1824) treated *Cleome* L., *Gynandropsis* DC. and *Polanisia* Rafin. as distinct genera. Schultes f. (Syst. 1829) merged all the three genera into one genus *Cleome* L. distinguishing four sections. Based on morphological characters Iltis (Brittonia 10: 1958 & 12: 1960) and Jocobs (Fl. Males. 6: 100. 1960) have treated *Gynandropsis* under *Cleome* L. as *C. gynandra*. In view of good evidence from various disciplines, the conception of Iltis (loc. cit.) and Jacobs (loc. cit.) in treating *Gynandropsis* under *Cleome* is adopted in the present work.

1. Leaves simple .... ***C. monophylla***
1. Leaves 3–5-foliolate
    2. Stamens 6
        3. Flowers 2.5 cm across, white or pink, stamens well exerted from the petals .... ***C. gynandra***
        3. Flowers 3 mm across, yellow, stamens inserted within the flower... ***C. aspera***
    2. Stamens more than 6:
        4. Capsules glandular pubescent .... ***C. viscosa***
        4. Capsules smooth .... ***C. chelidonii***

Cleome aspera J. Koenig ex DC. Prodr. 1: 241. 1824; FBI 1: 169; Gamble 41(29).

An annual prostrate herb, branchlets glandular. Leaves 3-foliolate, leaflets obovate-oblanceolate, 1–2 × 0.3–0.5 cm, base cuneate, margin ciliate, tip obtuse. Flowers yellow, axillary, solitary. Capsules cylindrical, glabrous, many-seeded. Seeds brown.

Occasional in waste lands (Seshagiri Rao, 1986) . Fl. & Fr.: July–Jan.

Cleome chelidonii L.f. Suppl. Pl. 300. 1781; FBI 1: 170; Gamble 41(29).

An annual erect herb, up to 50 cm tall, branchlets glandular-pubescent. Leaves 3-foliolate apically and 5–7-foliolate basally, leaflets obovate, 1.5–3 × 0.5–1.3 cm, base cuneate, margin enitre, tip obtuse. Flowers rosy in terminal and axillary racemes. Capsules cylindrical, glabrous, striate, many-seeded. Seeds yellowish brown.

Occasional in forest (Seshagiri Rao, 1986). Fl. & Fr.: Sept.–Feb.

Cleome gynandra L. Sp. Pl. 671. 1753. *Gynandropsis pentaphylla* DC. Prodr. 1: 238. 1824; FBI 1: 172; Gamble 40(29).

An annual erect herb up to 50 cm tall, branchlets glandular pubescent. Leaves 3–5-foliolate, leaflets obovate, 2.5–5 × 1–2.5 cm, base cuneate, margin entire, tip acute. Flowers white in terminal corymbs. Capsules cylindrical, many-seeded.

Common in waste lands (Seshagiri Rao, 1986). Fl. & Fr.: July–Feb. *Vern.:* Vamitna.

Cleome monophylla L. Sp. Pl. 672. 1753; FBI 1: 168; Gamble 41(29).

An annual erect herb, up to 70 cm tall, branchlets glandular-pubescent. Leaves simple, lanceolate, 3–8 × 1–1.5 cm, base truncate, margin ciliate, tip acute, glandular hairy on both sides. Flowers pink in terminal racemes. Capsules septicidal, 10 cm long, pubescent, many-seeded. Seeds black.

Occasional in forests. Fl. & Fr.: May–Aug. Malkalcheruvu, *MSM & KH* 10280; Venkat Reddipalli, *MSM & KH* 10307.

Cleome viscosa L. Sp. Pl. 672. 1753; FBI 1: 170; Gamble 41(29). *Polanisia icosandra* Wight & Arn. Prodr. Fl. Ind. Orient. 22. 1834.

An annual erect herb, up to 60 cm tall, branchlets viscid-glandular. Leaves 3–5-foliolate, leaflets obovate, 1–1.5 × 0.5–0.75 cm, base cuneate, margin ciliate, tip acute, glandular hairy on both sides. Flowers yellow in axillary racemes. Capsules septicidal, 8 cm long, glandular hairy, many-seeded. Seeds black.

Common in waste places. Fl. & Fr.: Apr.–Dec. *Vern.:* Kukka Vaminta. Dharmapuram, *MSM & KH* 10285.

VIOLACEAE

HYBANTHUS N.J. Jacquin *nom. cons.*

Hybanthus enneaspermus (L.) F. Muell. Fragm. 10: 81. 1876. *Viola enneasperma* L. Sp. Pl. 937. 1753. *V. suffruticosa* L. Sp. Pl. 937. 1753. *Ionidium suffruticosum* (L.) Ging ex DC. Prodr. 1: 311. 1834; FBI 1: 185; Gamble 49(35).

A perennial erect herb, up to 35 cm tall. Leaves linear-lanceolate, 3–6 × 0.5 cm, base attenuate, margin entire, tip acute, glabrous above, glaucous below. Flowers pink, axillary, solitary. Capsules 0.6 cm long, globose, pilose. Seeds ovoid, striate.

Common in forest undergrowth. Fl. & Fr.: May–Jan. *Vern.:* Ratnapurusa. Rangammagudem beat, *MSM & KH* 10318; Kusumasamudram beat, *MSM & KH* 10345; Lingampalli, *MSM & PRSR* 12763.

FLACOURTIACEAE

FLACOURTIA Commerson ex L. Heritier De Brutelle

Flacourtia indica (Burm. f.) Merr. Interpr. Rump. Herb. Ambion. 377. 1917. *Gmelina indica* Burm. f. Fl. Indica. 132. t. 39. f. 5. 1768. *Flacourtia sepiaria* Roxb. Pl. Cor. t. 68. 1796; FBI 1: 194; Gamble 54(39); Khan in FFH 65.

A small tree up to 3 m tall. Leaves clustered at the apex, obovate, 1.5–3.5 × 1–1.5 cm, base attenuate, margin entire, tip acute, glabrous, on both sides. Flowers white in axillary racemes. Berries fleshy, globose red when ripe.

Occasional in Central University campus (Seshagiri Rao, 1986). Fl. & Fr.: Oct.–May.

POLYGALACEAE

POLYGALA Linnaeus

1. Inflorescence extra axillary, 2.5 cm long. Seeds two .... ***P. arvensis***
1. Inflorescence lateral, leaf opposed, 7.5 cm long. Seed solitary .... ***P. elongata***

Polygala arvensis Willd. Sp. Pl. 3: 876. 1802. *P. chinensis* auct. non. L.; FBI 1: 204; Gamble 58(42).

An annual erect herb, up to 25 cm tall, branchlets pubescent. Leaves oblong-lanceolate, 2.5–4 × 0.5–1 cm, base subacute, margin entire, tip apiculate, pubescent above, glaucous below. Flowers cream in axillary racemes. Capsules 0.4 cm long, margin pubescent, 2-seeded. Seeds hairy.

Common in scrub jungles. Fl. & Fr.: June–Dec. Kusumasamudram beat, *MSM & NV* 10371; Gadirayal beat, *MSM* 10520.

Polygala elongata Klein ex Willd. Sp. Pl. 3: 879. 1802; FBI 1: 203; Gamble 58(42).

An annual erect herb, up to 20 cm tall, branchlets pubescent. Leaves oblong-lanceolate, 2–3 × 0.5–0.8 cm, base attenuate, margin entire, tip apiculate, glabrous above, glaucous below. Flowers cream in lateral racemes. Capsules 0.4 cm long, ciliate, 1-seeded. Seeds white pilose.

Occasional in open forests. Fl. & Fr.: June–Dec. Kusumasamudram beat, *MSM & NV* 10370.

CARYOPHYLLACEAE

POLYCARPAEA Lamarck *nom. cons.*

Polycarpaea corymbosa (L.) Lamk. Tabl. Encycl. 2: 129. 1792; FBI 1: 245; Gamble 65(46). *Achyranthes corymbosa* L. Sp. Pl. 205. 1753.

An annual erect herb, up to 20 cm tall, branchlets pubescent. Leaves whorled, linear, 1–3 × 0.1 cm, base cuneate, margin entire, tip acute, stipules membranous, fimbriate, glabrous on both sides. Flowers white in terminal cymes. Capsules 0.1 cm long, shiny, many-seeded. Seeds granular.

Common in open forests, roadsides. Fl. & Fr.: June–Sept. Kusumasamudram beat, *MSM & NV* 10344; Gadirayal, *MSM* 10521.

PORTULACACEAE

PORTULACA Linnaeus

1. Nodes without hairs. Petals 5 .... *P. oleracea*
1. Nodes with hairs. Petals 4:
   2. Leaves decussate, glabrous. Flowers yellow. Stamens ca. 8 .... *P. quadrifida*
   2. Leaves alternate, puberulous. Flowers pink. Stamens ca. 16 .... *P. suffruticosa*

Portulaca oleracea L. Sp. Pl. 445. 1753; FBI 1: 246; Gamble 66(47).

A prostrate herb up to 50 cm, branchlets succulent-green-purplish. Leaves obovate-spathulate, 0.5–1.8 × 0.5–1.5 cm, base attenuate, margin entire, tip obtuse, glabrous on both sides. Flowers yellow in terminal clusters. Capsules ovoid, glabrous, many-seeded. Seeds tubercled.

Abundant in waste lands. Fl. & Fr.: All seasons. Central University, *MSM* 19201.

Portulaca quadrifida L. Mant. Pl. 1: 73. 1767; FBI 1: 247; Gamble 66(47).

An annual prostrate succulent herb up to 50 cm, branchlets creeping, pinkish. Leaves ovate-lanceolate, 4–7 × 1–3 mm, base cuneate, margin entire, tip acute, succulent, glabrous on both sides. Flowers yellow, terminal, solitary. Capsules oblong, seeds 10, tubercled.

Occasional weed in cultivated fields (Seshagiri Rao, 1986). Fl. & Fr.: All seasons.

Portulaca suffruticosa Wight & Arn. Prodr. Fl. Ind. Orient. 356. 1834; FBI 1: 247; Gamble 66(48).

An annual erect herb, up to 12 cm tall, branchlets from a stumpy base, nodal hairs setose, sparse. Leaves linear, succulent, 0.5–2.5 × 0.1 cm, base cuneate, margin entire, tip subacute, glabrous on both sides. Flowers pink in axillary cymes. Capsules 0.5 cm long, glabrous, shiny, many-seeded.

Occasional in marshy places of open forests. Fl. & Fr.: June–Aug. Kusumasamudram beat, *MSM* & *NV* 10376.

Portulaca grandiflora Hook. (Gaddiroja) is planted as a pot herb for its attractive flowers.

CLUSIACEAE *(nom. alter.* GUTTIFERAE)

CALOPHYLLUM Linnaeus

Calophyllum inophyllum L. Sp. Pl. 513. 1753; FBI 1: 273; Gamble 76(55).

A tree up to 8 m tall, latex milky. Leaves oblong-obovate, 6–9 × 4–6 cm, base round, margin entire, tip rounded. Flowers white in axillary racemes. Drupes globose. 1-seeded.

Occasional (Seshagiri Rao, 1986). Fl. & Fr.: All seasons.

MALVACEAE

1. Herbs or shrubs:
   2. Fruit capsule:
      3. Epicalyx persistent. Calyx campaunlate .... ***Hibiscus***
      3. Epicalyx deciduous. Calyx spathaceous .... ***Abelmoschus***
   2. Fruit a schizocarp:
      4. Epicalyx absent:
         5. Mericarps 1-seeded, awned .... ***Sida***
         5. Mericarps 2–3-seeded, beaked .... ***Abutilon***
      4. Epicalyx present:
         6. Leaves simple or entire. Staminal column truncate at apex ..... ***Malvastrum***
         6. Leaves lobed. Staminal column dentate at apex:
            7. Staminal column antheriferous throughout. Mericarps winged ..... ***Pavonia***
            7. Staminal column antheriferous in the upper half only. Mericarps not winged.. ***Urena***
1. Trees .... ***Thespesia***

ABELMOSCHUS Medikus

Note: Hochreutiner (1924) erected *Abelmoschus* Med. as a distinct genus (formerly treated under *Hibiscus* L.) based on the calyx character, which is shortly 5-toothed, spathaceous and adnate to corolla and falls with the latter after flowering. These characters are specific to *Abelmoschus* Med. and no where found in any Malvaceae member (Borssum Walkes, 1966).

Abelmoschus ficulneus (L.) Wight & Arn. Prodr. 53. 1834; Borssum Walkes in Blumea 14: 101. 1966. *Hibiscus ficulneus* L. Sp. Pl. 695. 1753; FBI 1: 340; Gamble 97(70).

A spreading herb up to 40 cm long, branchlets hispid. Leaves 3-lobed, orbicular, lobes obovate-spathulate, 1–2.3 × 1–2.5 cm, base cordate, margin crenate-serrate, tip obtuse, hispid. Flowers white, axillary, solitary. Capsule oblong-ovate, tomentose. many-seeded.

Occasional in cultivated fields. Fl. & Fr.: Jan.–Mar. Pargi, *MSM & PRSR* 15104.

ABUTILON P. Miller

Abutilon indicum (L.) Sweet Hort. Brit. ed. 1. 54. 1826; FBI 1: 326; Gamble 91(65). ssp. ***indicum***; Borssum Walkes in Blumea 14: 171. 1966. *Sida indica* L. Cent. Pl. 2. 96. 1756.

A sub shrub up to 1 m tall, branchlets stellate-tomentose. Leaves cordate, 2–8 × 2–6.5 cm, base cordate, margin dentate, tip acute, stellate hairy above, glaucous below. Flowers yellow, axillary, solitary. Schizocarps globose, mericarps many, stellate hairy. Seeds ovoid.

Common on road sides. Fl. & Fr. July–Jan. *Vern.:* Thuthura benda. Central University, *MSM* 19204.

HIBISCUS Linnaeus *nom. cons.*

Hibiscus ovalifolius (Forssk.) Vahl Symb. Bot. 1: 50. 1790. *Urena ovalifolia* Forssk. Fl. Aegypt-Arab. 124. 1775. *Hibiscus micranthus* L.f. Suppl. Pl. 308. 1781; FBI 335; Gamble 97(70).

A sub-shrub, up to 1 m tall, branchlets stellate-tomentose. Leaves ovate, 1–2 × 1–1.5 cm, base sub-cordate, margin serrate, tip acute, stellately tomentose on both sides. Flowers pink, axillary, solitary. Capsules 0.6 cm long, puberulous, many-seeded. Seeds hairy.

Occasional in forest undergrowth. Fl. & Fr.: Mar.–May. *Vern.:* Nitya malli. Mohammadabad beat, *MSM & KH* 10221.

MALVASTRUM A. Gray *nom. cons.*

Malvastrum coromandelianum (L.) Garcke, Bonplandia 5: 295. 1857; Gamble 88(64). *Malva coromandeliana* L. Sp. Pl. 687. 1753. *M. tricuspidata* R. Br. in Aiton f. Hortus Kew ed. 2. 4: 210. 1812. *Malvastrum tricuspidatum* (R. Br.) A. Gray Pl. Wright 1: 16. 1852; FBI 1: 321.

A shrub up to 90 cm tall, branchlets stellate tomentose. Leaves ovate lanceolate, 2–4 × 1.5–3 cm, base truncate, margin crenate-serrate, tip acute. Flowers yellow, axillary, solitary or in terminal clusters, schizocarp not exceeding calyx, mericarps reniform. Seed 1 per mericarp, reniform, compressed, glabrous.

Common in waste lands. Fl. & Fr. All seasons. Central University, *TP & MSM*, 19248.

PAVONIA Cavanilles *nom. cons.*

1. Upper leaves lobed, mericarps winged .... ***P. zeylanica***
1. Upper leaves not lobed, mericarps not winged .... ***P. odorata***

Pavonia odorata Willd. Sp. Pl. 3: 837. 1800; FBI 1: 331; Gamble 93(67).

An erect herb, up to 75 cm tall, branchlets viscid. Leaves ovate-sub orbicular, 3–7 × 1–2.5 cm, base cordate, margin dentate to sub entire, tip acute, stellate tomentose above, dense stellate tomentose below. Flowers pink, axillary, solitary. Schizocarps 0.7 cm long, pubescent. Seed 1 per mericarp, pubescent.

Occasional in scrub jungles. Fl. & Fr.: June–Sept. Nayanigutta, *MSM & NV* 10327.

Pavonia zeylanica (L.) Cav. Diss. 3: 134. t. 48. f.2. 1787; FBI 1: 331. Gamble 93(67).

An erect herb up to 80 cm tall, branchlets viscid. Leaves ovate-suborbicular, 3–6 × 1–2.5 cm, base cordate, margin dentate, tip acute, stellate tomentose above and dense tomentose below. Flowers pinkish-white, axillary, solitary. Schizocarp 5 mm long, pubescent. Seed 1 per mericarp pubescent.

Occasional in open forest undergrowth. Fl. & Fr.: June–Sept. Central University, *MSM & TP* 15140.

SIDA Linnaeus

1. Leaves palmi-nerved, mericarps 5 .... ***S. cordata***
1. Leaves penni-nerved, mericarps 7–10:
    2. Leaf base cuneate. Mericarps beaked .... ***S. rhombifolia***
    2. Leaf base other than cuneate, Mericarps awned:

3. Leaves lanceolate, carpels 5 .... *S. acuta*
3. Leaves cordate, carpels 10 .... *S. cordifolia*

Sida acuta Burm. f. Fl. Ind. 147. 1768, emend. K. Schum. in C. Martius. Fl. Bros. 12. 3: 326. 1891; Gamble 90(64); *S. carpinifolia* sensu Mast. in Hook. f. Fl. Brit. India 1: 323 (non L.f. 1781).

An annual erect, much branched sub shrub, up to 50 cm tall, branchlets glabrous. Leaves 1.5–6 × 0.4–1.2 cm, base cuneate, margin serrate, tip acute-acuminate, glabrous on both sides. Flowers yellow, axillary, solitary. Schizocarps 0.5 cm long, mericarps 6, glabrous, many-seeded, 3-gonous. Seeds dark brown.

Common in open forests. Fl. & Fr.: Sept.–Jan. Mohammadabad beat, *MSM* 10436.

Sida cordata (Burm. f.) Borssum Walkes in Blumea. 14: 182. 1966. *Melochia cordata* Burm. f. Fl. Ind. 143. 1768. *Sida veronicaefolia* Lam. Encycl. 1: 5. 1783; Gamble 89(64). *S. humilis* Cav. Diss. 5: 277 t. 134. f. 2. 1788; FBI 1: 322.

An annual erect herb, up to 70 cm tall, branchlets glabrous. Leaves ovate, 1.5–8 × 1–4.5 cm, base subcordate, margin serrate, tip acuminate, stellate hairy on both sides. Flowers yellow in axillary lax racemes. Schizocarps 0.4 cm long, not exceeding calyx, mericarps 5, glabrous. Seeds 3-gonous.

Occasional in forests. Fl. & Fr.: Sept.–Dec. Rangammagudem beat, *MSM* 10476.

Sida cordifolia L. Sp. Pl. 684. 1753; FBI 1: 324; Gamble 89(64).

A perennial erect sub shrub, up to 1 m tall, branchlets densely tomentose, intermingled with long simple hairs. Leaves ovate, 1–5.5 × 0.5–3 cm, base cordate, margin serrate-crenate, tip acute, grey tomentose above, villous below. Flowers yellow, in axillary and terminal, solitary or 5–7-flowered cymes. Schizocarps 0.6 cm long, exceeding calyx, mericarps 8–10, scabrid. Seeds 3-gonous.

Common in open forests. Fl. & Fr.: All seasons. Kanmankalva beat, *MSM* 10459; Mohammadabad beat, *MSM & KH* 10557; Central University, *MSM* 15165.

Sida rhombifolia L. Sp. Pl. 684. 1753; emend Masters in Hook f. Fl. Brit. India 1: 323; Gamble 90(65).

An erect sub shrub, up to 1 m tall, branchlets glabrous. Leaves ovate, 1.5–3 × 0.5–2.5 cm, base cuneate, margin serrate, tip obtuse, glabrous above, glaucous below. Flowers yellow in axillary and terminal solitary or 5–7-flowered cymes. Schizocarps 0.4 cm long, not exceeding calyx, mericarps 7–10, tomentose at apex. Seeds ovoid.

Common in rocky places. Fl. & Fr.: All seasons. Anantasagar beat, *MSM* 10512.

THESPESIA Solander ex Correa *nom. cons.*

Thespesia populnea (L.) Solander ex Correa in Ann. Mus. Natl. Hist. Nat. 9: 290. t. 8. f. 1. 1807; FBI 1: 345; Gamble 101(72); Khan in FFH 72. Borssum Walkes in Blumea 14: 106. 1966. *Hibiscus populneus* L. Sp. Pl. 694. 1753.

An evergreen tree, up to 8 m tall. Leaves ovate, 5–10 × 7–10 cm, base cordate, margin entire, tip acuminate. Flowers yellow, axillary, solitary. Capsules globose. 2–4-seeded.

Occasional on roadsides. Fl. & Fr.: All seasons. *Vern.:* Ganga ravi. Central University, *MSM* 19208.

URENA Linnaeus

Urena lobata L. Sp. Pl. 692. 1753; FBI 1: 329; Gamble 92(66); ssp. ***sinuata*** (L.) Borssum Walkes in Blumea 14: 142. 1966. *U. sinuata* L. Sp. Pl. 692. 1753.

A shrub, up to 1 m tall, branchlets hairy. Leaves deeply palmilobed, ovate, 3–5 × 3–4 cm, base cuneate, margin crenate, dentate, tip obtuse, stellate tomentose above, woolly below. Flowers pink, axillary, solitary. Schizocarps globose, mericarps 5, 3-gonous, 0.8 cm long, glochidiate, stellate-tomentose. Seeds reniform.

Occasional in forest under growth. Fl. & Fr.: Dec.–Mar. Mohammadabad beat, *MSM & KH* 10549.

Several species of *Gossypium* (*Vern.:* Pathi) are cultivated for cotton. *Abelmoschus esculentus* (L.) Moench (*Vern.:* Benda) and *Hibiscus cannabinus* L. (*Vern.:* Gongura) are cultivated as fruit and leafy vegetables respectively. *Hibiscus rosa-sinensis* L., (*Vern.:* Mandara), *H. schizopetalus* (Mast.) Hook. f. (*Vern.:* Mandara), *Alcea rosea* L. (=*Althaea rosea* (L.) Cav.) are grown as ornamentals.

BOMBACACEAE

BOMBAX Linnaeus *nom. cons.*

Bombax ceiba L. Sp. Pl. 511. 1753. *B. malabaricum* DC. Prodr. 1: 479. 1824; FBI 1: 349; Gamble 99(71); Khan in FFH 73.

A deciduous armed tree, up to 15 m tall, branchlets prickled, bark grey. Leaves 5–7-foliolate, leaflets elliptic, 8–12 × 4–6.5 cm, base cuneate, margin entire, tip acuminate, glabrous on both sides. Flowers red. Capsules 12 cm long, oblong, many-seeded.

Occasional in deciduous forest. Fl. & Fr.: Dec.–May. *Vern.:* Buruga. Mohammadabad RF, *MSM* 19228.

Ceiba pentandra (L.) Gaertn. (*Vern.:* Burga) is commonly grown for silk cotton.

STERCULIACEAE

1. Herbs or shrubs:
   2. Calyx shortly lobed, flowers pink in capitate terminal cymes .... ***Melochia***
   2. Calyx deeply lobed, flowers yellow in axillary clusters .... ***Waltheria***
1. Trees:
   3. Leaves lobed. Petals absent .... ***Sterculia***
   3. Leaves simple. Petals 5:
      4. Petals appendaged, Crimson. Fruits globose capsules .... ***Guazuma***
      4. Petals clawed, yellow. Fruits twisted follicles .... ***Helicteres***

GUAZUMA P. Miller

Guazuma ulmifolia Lam. Encycl. 3: 52. 1789. *Theobroma guazuma* L. Sp. Pl. 782. 1753; *Guazuma tomentosa* Kunth Nov. Gen. Sp. 5: 320. 1823; FBI 1: 375; Gamble 111(79).

A deciduous tree, up to 10 m tall, branchlets stellate tomentose. Leaves lanceolate, 6–12 × 2–5 cm, base subcordate, margin serrulate, tip acuminate, scabrous above, tomentose below. Flowers yellow in axillary and terminal cymes. Capsules globose, woody, many-seeded.

Common in villages. Fl. & Fr.: Aug.–Sept. Central University, *MSM* 17207.

HELICTERES Linnaeus

Helicteres isora L. Sp. Pl. 763. 1753; FBI 1: 365; Gamble 107(77); Khan in FFH 77.

A deciduous tree, up to 5 m tall, branchlets clothed with stellate hairs. Leaves obovate-orbicular, 4–10 × 3–5.5 cm, base subcordate, margin serrate, tip acute. Flowers red in axillary clusters. Follicles 5, 7 cm long, spirally twisted, tomentose, many-seeded.

Common in forest. Fl. & Fr.: Dec.–May. *Vern.:* Nuttudda. Central University, *MSM & TP* 15132; *MSM* 15165.

MELOCHIA Linnaeus

Melochia corchorifolia L. Sp. Pl. 675. 1753; FBI 1: 374; Gamble 110(79).

An erect sub shrub, up to 60 cm tall, branchlets glabrous, reddish. Leaves ovate, 1.5–5 × 0.5–2 cm, base subcordate or truncate, margin serrate, tip acute, glabrous on both sides. Flowers pink in terminal cymes. Capsules loculicidal, 0.2 cm long, 5-seeded.

Common on the banks of ponds and rivers. Fl. & Fr.: Sept.–Dec. Dharampur, *MSM* 10486.

STERCULIA Linnaeus

Sterculia urens Roxb. Pl. Cor. t. 24. 1795; FBI 1: 355; Gamble 106(75); Khan in FFH 75.

A deciduous tree, up to 10 m tall, branchlets smooth, glabrous, bark smooth, grey, exfoliating in thin flakes. Leaves crowded at apices of branches, palmately 3–5-lobed, 15–40 × 10–20 cm, base cordate, margin entire, tip acuminate, scabrid above, stellate tomentose below. Flowers yellowish-green in terminal viscid panicles. Follicles 4–5 in a cluster, 4 cm long, yellowish-brown, stiff hairy, beaked, 3–6-seeded. Seeds black.

Common in forests. Fl. & Fr.: Dec.–Mar. *Vern.*: Tapsy. *Uses*: It yields gum. Anantasagar beat, *MSM & KH* 11014.

WALTHERIA Linnaeus

Waltheria. indica L. Sp. Pl. 673. 1753; FBI 1: 374; Gamble 111(79). *W. americana* L. Sp. Pl. 673. 1753.

An erect sub shrub, up to 80 cm tall, branchlets stellate tomentose. Leaves elliptic-ovate, 1.5–7 × 1–3.5 cm, base obtuse, margin serrate, tip obtuse, stellate tomentose above, densely stellate tomentose below. Flowers yellow in axillary clusters. Capsules 0.2 cm long, enclosed in calyx, 1-seeded.

Common in open forest and scrub jungle. Fl. & Fr.: All seasons. Kanmankalva beat, *MSM* 10463.

TILIACEAE

1. Trees or shrubs, fruits drupes .... *Grewia*
1. Sub shrubs or herbs, fruits capsules:
   2. Leaves lobed. Petals glandular at base, capsules prickled .... *Triumfetta*
   2. Leaves not lobed. Petals eglandular at base, capsules unarmed .... *Corchorus*

CORCHORUS Linnaeus

1. Capsule beak divided:
   2. Capsules not winged .... *C. tridens*
   2. Capsules winged .... *C. aestuans*
1. Capsule beak undivided:
   3. Capsules angled:

4. Leaves strigose, dentate, capsules strigose .... *C. urticifolius*
4. Leaves glabrous, crenate-serrate, capsules scabrid .... *C. trilocularis*
3. Capsules terete:
5. Leaf tip acuminate, capsule glabrous .... *C. olitorius*
5. Leaf tip obtuse, capsule strigose .... *C. fascicularis*

Corchorus aestuans L. Syst. Nat. 1079. 1759. *C. acutangulus*. auct. non. Forssk. 1775; FBI 1: 398; Gamble 121(86).

An annual erect herb, up to 60 cm tall, branchlets pubescent. Leaves oblong-lanceolate, 2–3.5 × 0.5–2 cm, base obtuse, margin serrate, tip obtuse, pubescent on both sides. Flowers yellow, leaf-opposed, solitary. Capsules triangular, winged, strigose, beaked, many-seeded. Seeds truncate, brown.

Abundant in waste lands. Fl. & Fr.: May–Dec. Central University, *TP & MSM* 19247.

Corochorus fascicularis Lam. Encycl. 2: 104. 1786; FBI 1: 398; Gamble 121(87).

An annual erect herb, up to 15 cm tall, branchlets glabrous. Leaves oblong-lanceolate, 1–3 × 0.4–1 cm, base acute, margin serrate, tip acute, glabrous on both sides. Flowers yellow in leaf-opposed cymes. Capsules 1 cm long, beaked, 3-valved, densely pubescent, many-seeded. Seeds 3-gonous, black.

Common on the banks of ponds. Fl. & Fr.: Sept.–Dec. Rangammagudem beat, *MSM & KH* 10574.

Corchorus olitorius L. Sp. Pl. 529. 1753; FBI 1: 397; Gamble 122(87).

An erect herb, up to 50 cm tall, branchlets glabrous. Leaves oblong-lanceolate, 5–8 × 2–4 cm, base obtuse, margin serrulate, tip acuminate, glabrescent on both sides. Flowers yellow in leaf opposed cymes. Capsules glabrous, cylindrical. Seeds brown.

Common on roadsides and streams. Fl. & Fr.: Aug.–Dec. *Vern.:* Goni. Central University, Naskal *MSM* 15117, 15129.

Corchorus tridens L. Mant. Pl. 566. 1771; FBI 1: 398; Gamble 121(87).

An erect sub-shrub, up to 70 cm, branchlets glabrous. Leaves oblong-lanceolate, 3–8 × 0.5–2 cm, base cuneate, margin serrate, tip acuminate. Flowers yellow, leaf opposed, solitary. Capsule not winged, many-seeded. Seeds truncate.

Common in waste lands and forests (Seshagiri Rao, 1986) Fl. & Fr.: May–Dec.

Corchorus trilocularis L. Mant. Pl. 77. 1767; FBI 1: 397; Gamble 122(87).

An annual erect herb, up to 50 cm tall, branchlets glabrous. Leaves oblong-lanceolate, 2–4 × 1–1.3 cm, base cuneate, margin serrate, tip acute, glabrous on both sides. Flowers yellow, axillary, solitary. Capsules 6 cm long, stellate hairy when young, scabrid when old, many-seeded. Seeds black.

Common in waste lands. Fl. & Fr.: June–Sept. Kusumasamudram beat, *MSM & NV* 10372.

Corchorus urticifolius Wight & Arn. Prodr. Fl. Ind. Orient. 73. 1834; FBI 1: 397; Gamble 122(87).

An erect sub-shrub, up to 1 m tall, branchlets pubescent. Leaves obovate-lanceolate, 5–7 × 2–4 cm, base obtuse, margin dentate, tip acute. Flowers in leaf-opposed, 2–4-flowered cymes. Capsules falcate, strigose, many-seeded. 3-gonous, wrinkled.

Occasional in forest (Seshagiri Rao, 1986). Fl. & Fr.: May–Dec.

GREWIA Linnaeus

1. Leaves glaucous below:
    2. Leaves 3-nerved from base, drupes 2–4-lobed .... ***G. orientalis***
    2. Leaves 5-nerved from base, drupes 1–2-lobed .... ***G. orbiculata***
1. Leaves not glaucous below:
    3. Flowers polygamous .... ***G. hirsuta***
    3. Flowers bisexual:
        4. Flowers white .... ***G. abutilifolia***
        4. Flowers yellow and dull yellow:
            5. Drupes 4-lobed, extra axillary .... ***G. villosa***
            5. Drupes 2-lobed, axillary or leaf opposed:
                6. Leaves 3-nerved at base. Drupes scabrid .... ***G. flavescens***
                6. Leaves 5-nerved at base. Drupes glabrous .... ***G. tiliifolia***

Grewia abutilifolia Vent ex A.L. Juss. Ann. Mus. Nat. Hist. Nat. 4. 92.1804; FBI 1: 390. *G. aspera* Roxb. Fl. Ind. 2: 591. 1832; Gamble 119(85).

A shrub, up to 3 m tall, branchlets tomentose. Leaves ovate, 5–10 × 4–8 cm, base rotund, margin serrate, tip acute, acuminate, stellate tomentose above and velvety below. Flowers white in axillary umbellate cymes. Drupes subglobose, 4-lobed, stellate tomentose.

Occasional in forest (Seshagiri Rao, 1986). Fl. & Fr.: July–Dec.

Grewia flavescens A.L. Juss. Ann. Must. Nat. Hist. Nat. 4: 91. 1804; Gamble 119(85). *G. carpinifolia* auct. non A.L. Juss; Roxb. Fl. Ind. 2: 587. 1832; FBI 1: 387.

A climbing shrub, up to 6 m, branchlets stellate tomentose. Leaves oblong, 7–10 × 4–5 cm, stellate tomentose above, dense stellate tomentose below. Flowers yellow in axillary cymes, bisexual. Drupes densely stellate tomentose.

Occasional in forests. Fl. & Fr.: July–Sept. Mohammadabad beat, *MSM & NV* 10427.

Grewia hirsuta Vahl, Symb. Bot. 1: 34. 1790; FBI 1: 391; Gamble 119(85).

A much branched shrub, up to 2 m, branchlets rusty tomentose. Leaves oblong-lanceolate, 5–9 × 1.5–2.5 cm, base oblique, margin serrate-crenate, tip acute, tomentose above, grey tomentose below. Flowers in axillary cymes, polygamous. Drupes 0.8 cm long, hairy, wrinkled, slightly 4-lobed.

Common in forests. Fl. & Fr.: Apr.–Sept. Lothuvagu, *MSM* 10451; Lingampalli, *MSM* 12744.

Grewia orbiculata Rottl. in Ges. Naturf. Freunde Berlin Neue Schriften 4: 205. 1803; FBI 1: 386. *G. rotundifolia* Juss. in Ann. Mus. Natl. Hist. Nat. 4: 92. 1804; Gamble 118(84).

A tree up to 3 m tall, branchlets velvety. Leaves oblong-ovate, 1–2.5 × 0.5–1 cm, base oblique, obtuse, margin-serrate, tip obtuse, pubescent above, glaucous below. Flowers yellow in axillary and terminal cymes. Drupes subglobose, grey tomentose, seeds stony.

Occasional in forests. Fl. & Fr.: July–Dec. Vikarabad, *TP & MSM* 11094.

Grewia orientalis L. Sp. Pl. 964. 1753; Gamble 117(84). *G. columnaris* J.E. Smith in Rees Cyclop. 17: 5. 1811; FBI 1: 383.

A shrub up to 2 m tall, branchlets virgate, stellate hairy. Leaves elliptic-lanceolate, 4–8 × 2–4 cm, base slightly cordate, margin serrulate, tip acute. Flowers white in axillary umbellate cymes. Drupes globose yellow, wrinkled, 2–4-lobed.

Common in forest. Fl. & Fr.: July–Dec. Central University, *MSM* 15163

Grewia tiliifolia Vahl, Sym. Bot. 1: 35. 1790; FBI 1: 386; Gamble 48(84); Khan in FFH 79. *G. arborea* Roxb. ex Rottl. Ges. Naturf. Freunde Berlin Neue Schriften 4: 205. 1803.

A deciduous tree, up to 8 m tall. Leaves ovate, 7–13 × 4–9 cm, base cordate, margin crenate-dentate, tip sub acute, glabrescent on both sides. Flowers yellow in axillary cymes, bisexual, peduncle shorter than petiole. Drupes 0.8 cm long, sparsely hirsute.

Occasional in forests. Fl. & Fr.: May–Sept. Rangammagudem beat, *MSM & KH* 10312; Dharur RF, *MSM & TP* 11094; Central University, *MSM* 15164.

Grewia villosa Willd. in Ges. Naturf. Frenunde Berlin Neue. Schriften 4: 205. 1803; FBI 1: 388; Gamble 119(85).

A shrub up to 3 m tall. Leaves ovate, obricular 5–9 × 5–10 cm, base cordate, margin crenate-serrulate, tip obtuse-subacute, rugose above, villous below. Flowers yellow in leaf opposed cymes. Drupes globose, stellate hairy, 4-stoned.

Common in scrub jungle. Fl. & Fr.: Apr.–Sept. Central University, *MSM* 15161.

TRIUMFETTA Linnaeus

1. Leaves lobed:
   2. Prickles on capsules hispid below .... *T. pentandra*

2. Prickles on capsules glabrous throughout .... *T. rhomboidea*
1. Leaves not lobed .... *T. pilosa*

Triumfetta pentandra A. Rich in Guill. & Perr. Fl. Seneg. Tent. 1: 93. t. 19. 1831; Gamble 120(86). *T. neglecta* Wight & Arn. Prodr. Fl. Ind. Orient. 75. 1834; FBI 1: 396.

An erect sub shrub, up to 1 m tall, branchlets hairy. Lower leaves palmately lobed, 5 × 8 cm, upper leaves elliptic, 1.5–4.5 × 1–2.5 cm, base cuneate, margin serrate-crenate, tip acuminate, strigose on both sides. Flowers yellow in axillary and terminal racemes, bracts lanceolate. Capsules 0.2 cm long, hispid. Seeds beaked.

Common in waste lands and hedges. Fl. & Fr.: All seasons. Mohammadabad, *MSM* 10439; Central University, *MSM* 15156, Gacchibowli, *MSM* 12772.

Triumfetta pilosa Roth Nov. Pl. Spec. 228. 1821; FBI 1: 394; Gamble 129(86).

A sub shrub, up to 1 m tall, branchlets hispid. Leaves ovate-lanceolate, 6–10 × 2.5–5 cm, base subcordate, margin serrate, tip acuminate, tomentose above and woolly below. Flowers yellow in leaf opposed umbellate cymes. Capsule globose, hispid. Seeds ovoid.

Common in waste lands (Seshagiri Rao, 1986). Fl. & Fr.: Sept.–Dec.

Triumfetta rhomboidea Jacq. Enum. Carib. 22. 1760; FBI 1: 395; Gamble 120(86).

An erect much branched sub shrub, up to 1 m tall, branchlets stellate-tomentose. Leaves orbicular-rhomboid, 1–5.3 × 0.5–2 cm, base cuneate, margin crenate-serrate, tip obtuse. Flowers yellow in terminal racemes. Capsules globose, 1–2-seeded. Seeds ovoid, beaked.

Common in hedges and under growth of forest (Seshagiri Rao, 1986). Fl. & Fr.: Aug.–Jan. *Vern.:* Bankathuthara.

MALPIGHIACEAE

ASPIDOPTERYS A. Jussieu

Aspidopterys cordata (Heyne ex Wall.) A. Juss. in Ann. Sci. Nat. Bot. Ser. 2. 13: 267. 1840; FBI 1: 421; Gamble 130(92). *Hiraea cordata* Heyne ex Wall. Pl. As. Rar. 1: 13. 1829.

A climbing shrub, up to 2 m, branchlets grey tomentose. Leaves broadly ovate, 3–14 × 2.5–9.5 cm, base cordate, margin enitre, tip acuminate, tomentose above, densely grey tomentose below. Flowers cream in axillary and terminal lax panicles. Samaras 3 cm long, elliptic, reticulately veined, glabrous.

Rare in forest. Fl. & Fr.: June–Sept. Ramayakunta, *MSM* 10490, 10491.

ZYGOPHYLLACEAE

TRIBULUS Linnaeus

Tribulus terrestris L. Sp. Pl. 387. 1753; FBI 1: 423; Gamble 130(92).

An annual prostrate herb, branchlets glaucous. Leaves pinnate, opposite leaflets oblong, 1–1.5 × 0.5–0.7 cm, base oblique, margin entire, tip mucronate. Flowers yellow, axillary, solitary. Schizocarps globose, 1 cm long, spiny, 5-seeded.

Common on roadsides. Fl. & Fr.: All seasons. *Vern.:* Palleru. Tandur, *MSM & TP* 12702.

OXALIDACEAE

1. Leaves digitately 4-foliolate .... ***Oxalis***
1. Leaves even-pinnate .... ***Biophytum***

BIOPHYTUM A.P. de Candolle

Biophytum sensitivum (L.) DC. Prodr. 1: 690. 1824; FBI 1: 436; Gamble 133(95): *Oxalis sensitiva* L. Sp. Pl. 434. 1753; p.p.

An annual erect herb, up to 10 cm tall, branchlets glabrous. Leaves pinnate, leaflets all crowned at the tip, leaflets oblong, 0.5–15 × 0.4–0.7 cm, base obtuse, margin entire, tip apiculate, glabrous on both sides. Capsules 0.5 cm long, oblong, obovoid, many-seeded.

Occasional in forest. Fl. & Fr.: Aug.–Oct. Anantagiri RF, *MSM & TP* 11086; Pargi, *MSM* 15106.

OXALIS Linnaeus

1. A non bulbous herb, leaves cauline .... ***O. corniculata***
1. A bulbous herb, leaves radical .... ***O. latifolia***

Oxalis corniculata L. Sp. Pl. 435. 1753; FBI 1: 431; Gamble 132(74).

An annual diffuse herb up to 15 cm, branchlets creeping. Leaves 3-foliolate, leaflets obcordate, 0.5–1.2 × 0.5–1.5 cm, base cuneate, margin entire, tip acute. Flowers yellow in axillary pseudo umbels. Capsules oblong, many-seeded. Seeds ovoid, ribbed.

Occasional in shady places. Fl. & Fr.: Sept.–Dec. Central University, *MSM* & TP 15134.

Oxalis latifolia H.B.K. Nov. Gen. Sp. 5. 237. t. 467. 1822.

An annual herb with conical bulbs. Leaves radical. Flowers yellow.

Weed in Central University nursery. *MSM & TP* 15135.

BALSAMINACEAE

IMPATIENS Linnaeus

Impatiens balsamina L. Sp. Pl. 938: 1753; FBI 1: 453; Gamble 142(101).

An annual erect herb, up to 70 cm tall, branchlets glabrescent. Leaves lanceolate, 8–10 × 1.5–2 cm, base attenuate, margin serrate, tip acuminate. Flowers pink, axillary, solitary. Capsules 1 cm long, oblong, many-seeded.

Common in open forest. Fl. & Fr.: Aug.–Oct. Anantagiri RF, *MSM & TP* 11081.

RUTACEAE

1. Trees unarmed .... ***Murraya***
1. Trees armed:
   2. Petiole and rachis not winged. Stamens numerous .... ***Aegle***
   2. Petiole and rachis winged. Stamens up to 12:
      3. Flowers bisexual. Ovules 1 or 2 in each cell .... ***Naringi***
      3. Flowers polygamous. Ovules more than 2 in each cell. .... ***Limonia***

AEGLE Correa *nom. cons.*

Aegle marmelos (L.) Corr. Serr in Trans. Linn. Soc. London 5: 223. 1800; FBI 1: 516; Gamble 161(115); Khan in FFH 89. *Crataeva marmelos* L. Sp. Pl. 444. 1753.

An armed deciduous tree, up to 10 m tall, branchlets pubescent. Leaves 3-foliolate, leaflets elliptic, terminal leaflet large, lateral inequilateral, 5–8 × 1.2–4 cm, base cuneate, margin crenate, pubescent on both sides. Flowers greenish-white in axillary panicles. Berries 8 cm long, glabrous, many-seeded.

Common in forest. Fl. & Fr.: Jan.–May. *Vern.:* Maredu. D. Tanda, *MSM & KH* 10253.

LIMONIA Linnaeus

Note: Swingle (in J. Wash. Acad. Sci. 4: 328. 1914) is of the view that the generic name *Limonia* L. (1764) is illegitimate 'being a mere variant of *Limonium* Miller (1754)' of Plumbaginaceae. He therefore took the next available name *Feronia* Corr. for the genus which has been subsequently followed by many. Airy Shaw (in Kew Bull. 1939: 293. 1939) states that *Limonia* is not an orthographic variant of *Limonium* of Plumbaginaceae and the legitimate name *Limonia* L. must be retained.

Limonia acidissima L. Sp. Pl. ed. 2. 554. 1762; B.C. Stone & D.H. Nicolson in Taxon. 27: 551–552. 1978. *Feronia limonia* (L.) Swingle in J. Wash. Acad. Sci. 4: 428. 1914. *Schinus limonia* L. Sp. Pl. 389. 1753. *Limonia* *elephantum* (Corr.) Panigr. in Taxon. 26: 577. 1977. *Feronia elephantum* Corr. in Trans. Linn. Soc. London 5: 224. 1800; FBI 1: 516. Gamble 160(114).

An armed tree up to 15 m tall, branchlets tomentose. Leaves imparipinnate, leaflets oblong-obovate, 2–4 × 1–2 cm, base cuneate, margin entire, tip obtuse. Flowers cream in axillary and terminal panicles. Berries globose, woody, many-seeded.

Common in villages, open and deciduous forest. Fl. & Fr.: Aug.–Mar. *Vern.*: Velaga. Central University, *MSM* 19215.

Note: Panigrahi (loc. cit.) has argued for the rejection of the name *Limonia acidissima* L. as Nomen ambiguum. However, Stone and Nicolson (loc. cit.) have argued for retaining the original name. The concept of Stone and Nicolson is followed here.

MURRAYA J.S. Koenig ex Linnaeus *nom. cons.*

Murraya koenigii (L.) Spreng. Syst. Veg. 2: 315. 1826; FBI 1: 503; Gamble 156(111). *Bergera koenigii* L. Mant. Pl. 565, 1771.

A small tree up to 5 m tall. Leaves imparipinnate, leaflets oblong-lanceolate, 2–5 × 1–2 cm, base obtuse, margin entire, tip acute. Flowers white in corymbs. Berries globose, black when ripe, 1–2-seeded.

Common in villages and occasional in forest. Fl.-Fr.: All seasons. Mohammadabad, *MSM* 19229.

NARINGI Adanson

Naringi crenulata (Roxb.) Nicolson in Saldanha & Nicolson Fl. Hassan 387. 1976. *Limonia crenulata* Roxb. Pl. Cor. t. 86. 1798; Gamble 157(112). *L. acidissima* auct. non. L. 1762; FBI 1: 507; Khan in FFH 86.

An armed deciduous tree, up to 6 m tall, branchlets glabrous. Leaves clustered. Leaflets oblong, 2.5–5 × 1–2 cm, base sub acute, margin crenulate, tip sub acute, glabrescent on both sides. Flowers white in axillary racemes. Berries globose, 3–4-seeded. Seeds ovoid.

Occasional in forest. Fl. & Fr.: Apr.–May. Anantagiri RF, *MSM & TP* 11056.

Murraya koenigii (L.) Spreng. is cultivated for its leaves, used as a condiment. *Citrus aurantifolia* (Christm.) Swingle (Country lime, *Vern.*: Nimma) and *C. sinensis* (L.) Osbeck (Sweet orange, *Vern.*: Bathai) are cultivated for fruits.

SIMAROUBACEAE

AILANTHUS Desfontaines *nom. cons.*

Ailanthus excelsa Roxb. Pl. Cor. 24. t. 23. 1795; FBI 1: 581; Gamble 163(116); Khan in FFH 91.

A deciduous tree, up to 15 m tall, bark rough, light grey. Leaves imparipinnate, leaflets 16–24, elliptic-lanceolate, 10–25 × 4–8 cm, base truncate, margin bluntly dentate, tip acuminate, puberulous above, tomentose below. Flowers yellow in large panicles. Samaras lanceolate, flat, 1-seeded.

Common in villages. Fl. & Fr.: Jan.–May. *Vern.:* Peddamanu., Central University, *MSM* 19230.

BURSERACEAE

1. Bark peeling off into flakes, flowers white .... ***Boswellia***
1. Bark not peeling off into flakes, flowers yellow .... ***Garuga***

BOSWELLIA Roxburgh

Boswellia serrata Roxb. ex. Colebr. in Asiat. Res. 9: 379. t. 5; FBI 1: 528. *B. glabrata* Roxb. Pl. Cor. t. 207. 1811; Gamble 168(120).

A deciduous tree, up to 15 m tall, branchlets smooth, bark yellow, peeling off into flakes. Leaves ovate-lanceolate, 2–6 × 0.5–3.5 cm, base oblique, margin serrate, tip rounded. Flowers white in axillary racemes. Drupes trigonous, 2 cm long, bony.

Common in deciduous forest. Fl. & Fr.: Mar.–June. *Vern.:* Anduk. Anantagiri RF, *MSM & TP* 11071.

GARUGA Roxburgh

Garuga pinnata Roxb. Pl. Cor. t. 208. 1811; FBI 1: 528; Gamble 169(121).

A deciduous tree, up to 10 m tall, branchlets pubescent. Leaves compound, leaflets 6–8 pairs, ovate-obovate, 2.5–6 × 1.5–2 cm, base cuneate, oblique, margin crenate, tip acute. Flowers yellow in axillary panicles. Drupes oblong, pyrenes 2–3. Seed 1 per pyrene.

Occasional in forest (Seshagiri Rao, 1986). Fl. & Fr.: Feb.–Mar.

Bursera penicellata (DC) Engl. is grown by forest department.

MELIACEAE

1. Leaflets 4.5–7 cm long. Fruits drupes:
    2. Leaves 1-pinnate. Flowers white. Drupes oblong-ovoid .... ***Azadirachta***
    2. Leaves 2–3 pinnate. Flowers blue or lilac. Drupes globose .... ***Melia***
1. Leaflets 8.5–13 cm long. Fruits capsules .... ***Soymida***

AZADIRACHTA A.H.L. Jussieu

Azadirachta indica A. Juss. in Mem. Mus. Hist. Nat. 19: 221, t, 2, f, 5, 1830; Gamble 177(127). *Melia azadirachta* L. Sp. Pl. 385. 1753; FBI 1: 544; Khan in FFH 96.

An evergreen tree, up to 20 m tall, branchlets glabrous, bark brown. Leaves pinnate, leaflets asymmetrically lanceolate, 6–8.5 × 2–2.5 cm, base cuneate, margin serrate, tip acuminate. Flowers white in axillary and terminal panicles. Drupes oblong, 1.5 cm long, glabrous.

Common in villages. Fl. & Fr.: Jan.–June. *Vern.*: Vepa. Central University, *MSM*, 19231.

MELIA Linnaeus

Melia azedarach L. Sp. Pl. 384. 1753; FBI 1: 544; Gamble 176(126); Khan in FFH 97.

An evergreen tree, up to 10 m tall, branchlets glabrous. Leaves bipinnate, leaflets ovate-lanceolate, 2–5 × 1–1.5 cm, base oblique, margin serrate, tip acuminate. Flowers blue or lilac in axillary panicles. Drupes globose, 1 cm long, glabrous.

Common in villages. Fl. & Fr.: Dec.–Feb. *Vern.*: Turaka vepa. Naskal, *MSM & PRSR* 15113.

SOYMIDA A.H.L. Jussieu

Soymida febrifuga (Roxb.) A. Juss. Mem. Mus. Hist. Nat. 19. 251. 1830; FBI 1: 567; Gamble 185(133); Khan in FFH 97. *Swietenia febrifuga* Roxb. Pl. Cor. t. 17. 1795.

A deciduous tree, up to 8 m tall, branchlets glabrous, bark brown. Leaves pinnate, leaflets ovate, 5–10 × 5–7 cm, base oblique-truncate, margin entire, tip obtuse. Flowers white in branched compact panicles. Capsules obovoid, 7 cm long, woody, many-seeded.

Common in deciduous forest. Fl. & Fr.: Feb.–May. *Vern.*: Somi. Central University, *MSM* 19232.

FLINDERSIACEAE

CHLOROXYLON A.P. de Candolle *nom. cons.*

Chloroxylon swietenia DC. Prodr. 1: 625. 1824; FBI 1: 569; Gamble 152(109); Khan in FFH 99. *Swietenia chloroxylon* Roxb. Pl. Cor. t. 64. 1796.

A deciduous tree, up to 8 m tall, branchlets tomentose, bark corky-rough, deeply furrowed, yellowish. Leaves paripinnate, leaflets oblong, margin entire, tip obtuse, glabrous above, glaucous below. Flowers cream in axillary panicles. Capsules 3 cm long, glabrous, 12–15-seeded. Seeds brown.

Common in forests. Fl. & Fr.: Feb.–May. *Vern.:* Billu. Gadirayal beat, *MSM & KH* 10257.

OLACACEAE

1. Leaves oblong, up to 9 cm long. Staminodes ca. 6; fruit up to 1.5 cm long ... ***Olax***
1. Leaves ovate, up to 3 cm long. Staminodes 0, fruit up to 3 cm long .... ***Ximenia***

OLAX Linnaeus

Olax scandens Roxb. Pl. Cor. t. 102. 1799; FBI 1: 575; Gamble 190(36); Khan in FFH 101.

A climbing shrub, up to 5 m, branchlets sparsely pubescent. Leaves oblong, 5–9 × 2–4 cm, base cuneate, margin entire, tip acute, glabrous on both sides. Flowers cream in axillary racemes. Drupes half enclosed by accrescent calyx, 1.5 cm long, apiculate, 1-seeded.

Common in forests. Fl. & Fr.: Mar.–July. *Vern.* Harduli. Kusumasamudram beat, *MSM & NV* 10353, 10369; Anantagiri RF, *MSM* 11067.

XIMENIA Linnaeus

Ximenia americana L. Sp. Pl. 1193. 1793; FBI 1: 574; Gamble 189(135); Khan in FFH 100.

An armed straggler, up to 5 m. Leaves ovate, 3–5 × 2.5–4 cm, base obtuse, margin entire, tip obtuse, glabrous on both sides. Flowers cream in axillary racemes. Drupes 3 cm long, glabrous, 1-seeded.

Common in forests. Fl. & Fr.: Feb.–May. *Vern.:* Nakira. Mohammadabad beat, *MSM & KH* 10217; Gadirayal beat, *MSM & KH* 10256.

OPILIACEAE

OPILIA Roxburgh

Opilia amentacea Roxb. Pl. Cor. t. 158. 1802 & Fl. Ind. 2: 87. 1832; FBI 1: 583; Gamble 192(138).

A climbing shrub, up to 10 m, branchlets pubescent. Leaves ovate-lanceolate, 2–5 × 1–1.5 cm, base obtuse, margin entire, tip acuminate, glabrous on both sides. Flowers greenish yellow in axillary racemes. Drupes 1 cm long with fleshy pericarp, 1-seeded. Seeds yellow.

Occasional in forests. Fl. & Fr.: Apr.–July. Anantasagar beat, *MSM& NV* 10412.

CELASTRACEAE

1. Unarmed scandant shrub, flowers polygamous, greenish in pendulous panicles or racemes .... ***Celastrus***
1. Armed erect shrub, flowers bisexual, white in erect cymes .... ***Maytenus***

CELASTRUS Linnaeus

Celastrus paniculatus Willd. Sp. Pl. 1: 1125. 1798; FBI 1: 617; Gamble 208(150); Khan in FFH 102.

A climbing shrub, up to 8 m. Leaves ovate, 4–12 × 2.5–7.5 cm, base obtuse, margin crenate-serrulate, tip acuminate, glabrous above, pubescent below. Flowers white in terminal panicles. Capsules 1 cm long, 6-seeded.

Common in forests. Fl. & Fr.: Mar.–May. *Vern.:* Manertiga-Mulliak. Mohammadabad beat, *MSM & KH* 10245, 10246; Dharmapur, *MSM & KH* 10293.

MAYTENUS Molina

Maytenus emarginata (Willd.) Ding Hou in Steenis, Fl. Males. 1. 6: 241. 1962. *Celastrus emarginatus* Willd. Sp. Pl. 1: 1128. 1798. *Gymnosporia emarginata* (Willd.) Thw. Enum. Pl. Zey. 409. 1864; FBI 1. 621; Gamble 210(151). *Celastrus montanus* Roth in Roem. & Schult. in L. Syst. Veg. 5: 427. 1819. *Gymnosporia montana* (Roth) Benth. Fl. Austral. 1: 400. 1863; FBI 1: 621; Gamble 209(150). *Celastrus senegalensis* Lam. Encyl. 1: 661. 1783; Khan in FFH 103.

An erect armed shrub, up to 3 m tall. Leaves oblong, 3–10 × 2–4.5 cm, base attenuate, margin crenate-serrate, tip rounded, glaucous on both sides. Flowers white in axillary dichotomous cymes. Capsules 1 cm long. Seeds brown.

Common in scrub jungles. Fl. & Fr.: May–Sept. *Vern.:* Danti. Gadirayal beat, *MSM & KH* 10255.

Note: *Maytenus senegalensis* (Lam.) Exell. which was once considered as the valid name for the Indian taxon is now found out to be a different African plant and hence the next available name, *M. emarginata* is now used for the species.

RHAMNACEAE

1. Trees or stragglers, leaves palminerved, drupe not winged .... ***Ziziphus***
1. Lianes, leaves penninerved, drupe winged .... ***Ventilago***

VENTILAGO J. Gaertner

Ventilago denticulata Willd. in Ges. Natürf. Freunde Berlin Neue Schriften 3: 417. 1811. *V. calyculata* Tul in Ann. Sci. Nat. Bot. Ser. 4. 8: 124. 1857; FBI 1: 631; Gamble 218(516); Khan in FFH 104.

An unarmed liane, branchlets striate, tomentose. Leaves elliptic-ovate, 4–10 x 2.5–7.5, base rounded, margin dentate, tip sub acute. Flowers green in terminal panicles. Samaras globose, 0.8 cm, wings 3 cms.

Common in forests. Fl. & Fr. Dec.–Mar. Anantagiri RF, *MSM & PRSR* 12779.

ZIZIPHUS P. Miller

1. Stragglers, drupes small, black when ripe .... ***Z. oenoplia***
1. Trees, drupes other than black when ripe:
   2. Unarmed .... ***Z. glabrata***
   2. Armed:
      3. Styles 2, connate. Drupes glabrous .... ***Z. mauritiana***
      3. Styles 3, free. Drupes white tomentose .... ***Z. xylopyrus***

Ziziphus glabrata Heyne ex Roth, Nov. Pl. Spec. 159. 1821; FBI 1: 633. *Z. trinervia* Roxb. Pl. Ind. 1: 606. 1832. non Poir. Gamble 220(158).

An unarmed tree up to 15 m tall, branchlets tomentose. Leaves elliptic-ovate, 4–5 × 2–3.5 cm, base cuneate, margin serrate, tip obtsue. Flowers green in axillary cymes. Drupes globose, 1 cm long, glabrous, many-seeded.

Common in forest. Fl. & Fr.: Aug.–Oct. Rudraram RF, *MSM & TP* 12714.

Ziziphus mauritiana Lam. Encyl. 3: 319. 1789. var. ***mauritiana.*** *Z. jujuba* (L.) Gaertn. Fruct. 1: 203. 1788; FBI 1: 632; Gamble 219(157); Khan in FFH 105. *Rhamnus jujuba* L. Sp. Pl. 194. 1753.

An armed deciduous tree, up to 8 m tall, branchlets tomentose. Leaves orbicular-rotund, 1–2 × 0.8–1 cm, base oblique, margin glandular-serrate, tip rotund, glabrous above, tomentose below. Flowers greenish in axillary cymes. Drupes 1 cm long, glabrous, 1-seeded. Seeds brown.

Common in scrub jungles. Fl. & Fr.: May–Aug. *Vern.:* Regu. Mancherla, *MSM & KH* 10270; Rudraram, *MSM & TP* 12713.

Ziziphus oenoplia (L.) Miller, Gard. Dict. (ed.8).n.3. 1768; FBI 1: 634; Gamble 220(158); Khan in FFH 106. *Rhamnus oenoplia* L. Sp. Pl. 194. 1753.

Prickly straggling shrub, up to 8 m, branchlets densely tomentose. Leaves ovate, 2.5–4.5 × 1.5–2.5 cm, base oblique-sub acute, margin denticulate, tip acute, pubescent above, sericeous below. Flowers creamish green in axillary cymes. Drupes 1 cm long, shining, pulp scanty, 1-seeded.

Common in scrub jungles and hedges. Fl. & Fr.: July–Nov. *Vern.:* Pariki. Anantasagar beat, *MSM & NV* 10401.

Ziziphus xylopyrus (Retz.) Willd. Sp. Pl. 1: 1104. 1798; FBI 1: 634; Gamble 220(158); Khan in FFH 106. *Rhamnus xylopyrus* Retz. Obs. Bot. 2: 11. 1781.

A sparsely armed tree, up to 4 m tall, branchlets rusty tomentose. Leaves orbicular-rotund, 3–6 × 1.8–4 cm, base subcordate, oblique, margin serrulate, tip rounded, glabrescent above, white tomentose below. Flowers green in axillary cymes. Drupes 2 cm long, white tomentose, many-seeded. Seeds brown.

Common in forests. Fl. & Fr.: May–July. *Vern.*; Goti. Kothapalli beat, *MSM & KH* 10301; Kusumasamudram beat, *MSM & NV* 10336.

VITACEAE

1. Tendrils on peduncle .... ***Ampelocissus***
1. Tendrils on stem:
   2. Leaves simple, lobed, berries 1-seeded .... ***Cissus***
   2. Leaves 3–5-foliolate, berries 2–4 seeded .... ***Cayratia***

AMPELOCISSUS J.E. Planchon *nom. cons.*

Ampelocissus latifolia (Roxb.) Planch. in J. Vigne. Am. 1884. 8: 374. 1884; Gamble 230(165). *Cissus latifolia* Lam. Encycl. 1: 30. 1783. *Vitis latifolia* Roxb. Fl. Ind. ed. 2. 1: 661. 1832; FBI 1: 652; Khan in FFH 107.

A climbing shrub, branchlets weak, finely striate, glabrescent. Leaves 3–5-lobed, cordate, 20–25 × 20–22 cm, base cordate, margin dentate-serrate, tip acute, glabrous on both sides. Flowers deep reddish brown in leaf-opposed cymes, peduncles bearing tendrils below the cymes. Berries 0.8 cm long, glabrous, 2–3-seeded. Seeds smooth, polished.

Common in forests. Fl. & Fr.: June–Sept. *Vern.:* Pulatiga, Dobatiga. Anantasagar beat, *MSM & NV* 10402.

Remarks: The specific epithet *latifolia* was validly published by Lamarck in Dict. Encycl. Bot. 1: 30. 1783. Subsequent authors including Planchon have credited the specific epithet to Roxburgh; but Roxburgh had just made a new combination under *Vitis* Linn. de Candolle (Prodr. 1: 6346. 1824) and Cooke (Fl. Pres. Bombay 1: 270. rep. ed. 1958) have given Vahl as the author for the species, Vahl's name however dates from 1794.

CAYRATIA A.L. Jussieu *nom. cons.*

1. Leaves 3-foliolate, stipules triangular, 4 mm long .... *C. trifolia*
1. Leaves 5–7-foliolate, stipules falcate, 1.5 cm long .... *C. auriculata*

Cayratia auriculata (Wall.) Gamble Fl. Pres. Madras 1: 237 (170). 1918. *Cissus auriculata* Wall. in Roxb. Fl. Ind. 430. 1820. *Vitis auriculata* (Wall.) Wight & Arn. Prodr. 129. 1834; FBI 1: 658.

A climbing shrub, branchlets succulent, striate, sparsely pubescent, tendrils leaf-opposed. Leaves 5-foliolate, stipules auriculate, leaflets obovate, 6–12 × 6.5–15 cm, base cuneate, margin crenate-serrate, tip acuminate, pubescent along nerves above, dense tomentose below. Flowers greenish-white in axillary umbels. Berries 1.5 cm long, glabrescent, 1-seeded. Seeds rugose.

Common in hedges. Fl. & Fr.: May–Nov. *Vern.:* Aidu Akula pundukura. Kothapalli beat, *MSM & KH* 10300.

Cayratia trifolia (L.) Domin. in Biblioth. Bot. 89. 371. 1927. *Vitis trifolia* L. Sp. Pl. 203. 1753. *Cissus carnosa* Lam. Encycl. 1: 31. 1783. *Vitis carnosa* Wall. ex Wight & Arn., Prodr. 127. 1834; FBI 1: 654. *Cayratia carnosa* (Lam.) Gagnepain in Lecomte Notul. Syst. (Paris) 1: 347. 1911; Gamble 237(169).

A climbing shrub, branchlets pubescent. Leaves trifoliolate, leaflets ovate, 4–5 × 2–2.5 cm, base obtuse, margin dentate, tip acute. Flowers greenish-white in axillary cymes. Berries obovoid, 0.5 cm long, 2-seeded.

Common on hedges. Fl. & Fr.: Aug.–Oct. Anantagiri RF, *MSM & TP* 11075.

CISSUS Linnaeus

1. Stem 4-angular, 4-winged .... *C. quadrangularis*
1. Stem terete, not winged:
   2. Branchlets densely pubescent. Flowers cream .... *C. vitiginea*
   2. Branchlets glabrescent. Flowers white .... *C. pallida*

Cissus pallida (Wight & Arn.) Planch. in A. & C. DC. Mongr. Phan. 5: 477. 1887; Gamble 234(167). *Vitis pallida* Wight & Arn. Prodr. 125. 1834; FBI 1: 647; Khan in FFH 108.

An erect shrub, up to 1.5 m tall, branchlets terete, striate, tendrils leaf-opposed. Leaves ovate-cordate, 23–25 × 20–21 cm, base cordate, margin distantly glandular serrulate, tip shortly acuminate, glabrous on both sides. Flowers white in leaf-opposed cymes. Berries 0.8 cm long, 1–2-seeded. Seeds smooth.

Common in forest. Fl. & Fr.: June–Oct. *Vern.*: Kurdandi. Kusumasamudram beat, *MSM & NV* 10368.

Cissus quadrangularis L. Mant. Pl. 39. 1767; Gamble 228(167). *Vitis quadrangularis* (L.) Wall. ex Wight & Arn. Prodr. 125. 1834; FBI 1: 645; Khan in FFH 107.

A climbing shrub, stem 4-angled, fleshy, jointed at nodes. Leaves ovate, 3–4 × 2–3 cm, base truncate, margin serrate, tip acute. Flowers red in axillary umbellate cymes. Berries globose, 0.5 cm long, 1-seeded.

Occasional on hedges. Fl. & Fr.: All seasons. Central University, *MSM* 19209.

Cissus vitiginea L. Sp. Pl. 117. 1753; Gamble 234(167). *Vitis linnaei* Wall. ex Wight & Arn. Prodr. 126. 1834; FBI 1: 649.

A climbing shrub, up to 4 m, branchlets densely pubescent. Leaves cordate, 5–10 × 4–7.5 cm, base cordate, margin dentate, tip acuminate, glandular pubescent above, grey dense tomentose below. Flowers cream in leaf-opposed cymes. Berries 1 cm long, glaucous, apiculate, 1-seeded. Seeds rugulose.

Common in forest. Fl. & Fr.: June–Nov. *Vern.:* Adavi draksha. Kusumasamudram beat, *MSM & NV* 10366.

Vitis vinifera L. (Grapes, *Vern.:* Draksha) is extensively cultivated for its fruits.

LEEACEAE

LEEA D. Van Royen ex Linnaeus

Leea asiatica (L.) Ridsdale in Manilal Bot. Hist. Hort. Malab. 189. 1980. *Phytolacca asiatica* L. Sp. Pl. 474. 1753. *Leea crispa* L. Syst. Nat. ed. 12. 2. 627. 1767; FBI 1: 665. 1875; Gamble 240(171). *L. aspera* Edgew in Trans. Linn. Soc. London 20: 36. 1846. non Wall. ex G. Don. 1831; FBI 1: 665. 1875; Gamble 240(171).

A shrub, up to 2 m tall, branches 8–10-angled, grooved. Leaves pinnate, leaflets 5–7, elliptic ovate or oblong, 10–15 × 3–6 cm, base cordate, margin serrate-crenate, tip rounded. Flowers pale green in axillary and terminal cymes. Berries depressed, 6-lobed. Seeds wedge-shaped.

Rare in forest (Seshagiri Rao, 1986). Fl. & Fr. Oct.–Jan.

SAPINDACEAE

1. Climbers .... ***Cardiospermum***
1. Shrubs or trees:
   2. Leaves simple, glabrous beneath .... ***Dodonaea***
   2. Leaves odd pinnate, pubescent beneath:
      3. Leaves 3-foliolate, margin serrate, berries glabrous, small .... ***Allophylus***
      3. Leaves 3–5-foliolate, margin entire, drupes tomentose, large .... ***Sapindus***

ALLOPHYLUS Linnaeus

Allophylus serratus (Roxb.) Kurz in J. Asiat. Soc. Bengal, Pt. 3, Nat. Hist. 44: 185; Gamble 246(175). *Ornitrophe serrata* Roxb. Pl. Cor. t. 61. 1796. *Schmidella serrata* DC. Prodr. 1: 610. 1824. *Allophylus cobbe* auct. non Raeusch; FBI 1: 673.

A shrub, up to 8 m tall, branchlets densely pubescent. Leaves 3-foliolate, leaflets ovate, terminal one large, laterals inequilateral, 6–9 × 3–4.5 cm, base acute, margin crenate-dentate, tip obtuse, sparsely hirsute above, villous below. Flowers greenish-white in axillary spikes. Berries 0.8 cm long, glabrous, 1-seeded. Seeds arillate.

Occasional in forests. Fl. & Fr.: July–Sept. Anantasagar beat, *MSM & NV* 10398.

CARDIOSPERMUM Linnaeus

1. Fruits sharply 3-lobed, not bloated .... *C.* ***halicacabum*** var. ***luridum***
1. Fruits less clearly 3-lobed, bloated .... *C.* ***halicacabum*** var. ***microcarpum***

Cardiospermum halicacabum L. Sp. Pl. 366. 1753; FBI 1: 670; Gamble 244(175); var. ***luridum*** (Bl.) Adelb. in Blumea 6: 322. 1948. *C. luridum* Bl. in Rumphia 3: 184. 1847.

A climbing tendril bearing herb, branchlets sparsely pubescent. Leaves biternate, leaflets ovate-lanceolate, terminal one large, lateral inequilateral, 2.5–4 × 1.5–2.5 cm, base obtuse, margin irregularly deeply incised crenate-serrate, tip mucronate, sparsely pubescent above, dense tomentose below. Flowers white in axillary racemes. Capsules loculicidal, bloated, 3.5 cm long, 3-lobed, not winged, tomentose. Seeds black.

Common in forest. Fl. & Fr.: July–Sept. Anantasagar beat, *MSM & NV* 10414.

Cardiospermum halicacabum L. var. **microcarpum** (Kunth) Bl. in Rumphia 3: 185. 1847. *C. microcarpum* Kunth Nov. Gen. Sp. 5: 104. 1821.

A climbing tendril bearing herb, branchlets glabrous. Leaves biternate, bipinnate, leaflets ovate-lanceolate, 3–4 × 1–1.5 cm, base attenuate, margin incised, tip acuminate, glabrous above, glaucous below. Flowers white in axillary racemes. Capsules sharply 3-lobed, not bloated, 1.5 cm long, pubescent, winged at angles. Seeds black.

Occasional in forest. Fl. & Fr.: Sept.–Dec. Rangammagudem beat, *MSM & KH* 10572.

DODONAEA P. Miller

Dodonaea viscosa (L.) Jacq. Enum. Pl. Carib 19. 1760; FBI 1: 697; Gamble 253(181). *Ptelea viscosa* L. Sp. Pl. 118. 1753; Khan in FFH 111.

An evergreen tree, up to 3 m tall, branchlets brown, bark fissured. Leaves oblong-spathulate, 2–8 × 0.5–2 cm, base attenuate, margin entire, tip obtuse, glabrous on both sides. Flowers whitish-cream in axillary panicles. Capsules 1.5 cm long, 2–4-winged, 1–3-seeded. Seeds black.

Common in forests. Fl. & Fr.: June–Jan. *Vern.*: Bandaru. Mohammadabad beat, *MSM & KH* 10233; Anantasagar beat, *MSM & NV* 10389.

SAPINDUS Linnaeus

Sapindus emarginatus Vahl, Symb. Bot. 3: 54. 1794; Gamble 250(178). *S. trifoliatus* auct. non L; FBI 1: 682.

A deciduous tree up, to 8 m tall, branchlets tomentose. Leaves 3–5-foliolate, leaflets elliptic-oblong, 5–10 × 2.5–5 cm, base cuneate, margin entire, tip emarginate, glabrous above and pubescent below. Flowers greenish-yellow in terminal racemes. Drupes ovoid-tomentose, wrinkled when ripe. Seeds smooth, black.

Occasional in forests. Fl. & Fr.: Oct.–Feb. *Vern.:* Kunkudu. *Uses*: Fruits used as a substitute for soap. Central University, *MSM* 19214.

ANACARDIACEAE

1. Leaves simple.
   2. Drupe with a basal hypocarp .... ***Semecarpus***
   2. Drupe without a basal hypocarp.
      3. Drupe oblong-fleshy. Secondary nerves horizontal towards margins .... ***Mangifera***
      3. Drupe lenticular, not fleshy. Secondary nerves curved upwards... ***Buchanania***
1. Leaves compound:
   4. Leaflets 3. Sepals, petals and stamens 5 each. Drupe to 7 mm long .... ***Rhus***
   4. Leaflets 7–9. Sepals, petals and stamens 4 each. Drupe to 1 cm long..... ***Lannea***

BUCHANANIA K.P.J. Sprengal

Buchanania lanzan Spreng. in J. Bot. (Schrader) 2: 234. 1801; Gamble 258(184). *B. latifolia* Roxb. Fl. Ind. 2: 385. 1832; FBI 2: 23; Khan in FFH 115.

A deciduous tree, up to 8 m tall, branchlets glabrous. Leaves ovate, 12–20 × 5–10 cm, base cuneate, margin entire, tip retuse, glabrous above, densely tomentose below. Flowers cream in axillary and terminal panicles. Drupes 1.3 cm long, obliquely lentiform, black, stone hard, 2-valved. Seed gibbous, acute at one end.

Common in forest. Fl. & Fr.: Dec.–Mar. *Vern.*: Chironjii, Morli. Anantasagar beat, *MSM & KH* 11012.

LANNEA A. Richard *nom. cons.*

Lannea coromandelica (Houtt.) Merr. in J. Arnold Arbor. 19: 353. 1938. *Dialium coromandelicum* Houtt. Handl. Pl. Kruidk. 2: 39. t. 5. f. 2. 1774. *Odina wodier* Roxb. Fl. Ind. 2: 293. 1832; FBI 2: 29; Gamble 263(187); Khan in FFH 115.

A deciduous tree, up to 12 m, branchlets rusty tomentose, bark ash-coloured. Leaves imparipinnate, leaflets ovate-oblong, 3–8 × 1.5–3.5 cm, base obtuse-oblique, margin entire, tip acuminate, sparsely pubescent, glaucous above, tomentose below. Flowers purplish in terminal racemes. Drupes 1.5 cm long, compressed.

Common in forests. Fl. & Fr.: Feb.–May. *Vern.*: Gumpana. Lothuvagu, *MSM & KH* 10295.

MANGIFERA Linnaeus

Mangifera indica L. Sp. Pl. 200. 1753; FBI 2:13; Gamble 259(185); Khan in FFH 112.

A tree, up to 20 m tall, branchlets glabrous. Leaves elliptic, 10–15 × 4–6.5 cm, base oblique, margin entire, tip acuminate. Flowers greenish-yellow in terminal panicles. Drupes ovoid, 10 cm long.

Common in villages. Fl. & Fr.: Dec.–Mar. Central University, *MSM* 19233.

RHUS Linnaeus

Rhus mysorensis G. Don, Gen. Syst. 2: 74. 1832; FBI 2: 91; Gamble 264(188).

A small tree, up to 3 m tall, branchlets woolly at apices. Leaves 3-foliolate, leaflets oblanceolate, 1–2.5 × 0.3–0.8 cm, base cuneate, margin sinuate-toothed, tip obtuse. Flowers greenish-yellow in axillary panicles. Drupes globose, brown.

Common on roadside. Fl. & Fr.: Sept.–May. Gacchibowli, *MSM* 12775.

SEMECARPUS Linnaeus f.

Semecarpus anacardium L.f. Suppl. Pl. 182. 1781; FBI 2; 30; Gamble 266(190); Khan in FFH 116.

A deciduous tree, up to 8 m tall, young branchlets rusty tomentose, latex milky. Leaves oblong-elliptic, 9–20 × 4–10 cm, base obtuse, margin entire, tip rounded, sparsely pubescent above, grey tomentose below. Flowers greenish-white in terminal panicles. Drupes 2.5 cm long, reniform, glabrous. Seeds thick.

Common in forests. Fl. & Fr.: May–Jan. *Vern.:* Jeedi, Bhilawa. Dharmapur *MSM & KH* 10289.

Anacardium occidentale L. (*Vern.:* Jeedi mamidi) is cultivated for its edible nuts.

MORINGACEAE

MORINGA Adanson

Moringa oleifera Lam. Encycl. 1: 398. 1785; Gamble 270(192). *M. pterygosperma* Gaertn. Fruct. 2: 314. 1791; FBI 2: 45.

A medium tree, up to 6 m tall, trunk grey white. Leaves 3-more-pinnate terminal odd leaflet obovate-elliptic, 1.4–2.3 × 1–1.5 cm. Flowers white in terminal panicles. Capsules loculicidal. Seeds winged.

Cultivated for its fruits which are used as vegetable. *Vern.:* Munaga. Fl. & Fr. Jan.–July. Mohammadabad, *MSM* 19234.

FABACEAE

1. Trees:
    2. Leaves 3-foliolate:
        3. Branchlets prickled:
            4. Leaflet apex acute, sub-acute or acuminate. Flowers scarlet ...... ***Erythrina***
            4. Leaflet apex obtuse, retuse. Flowers yellow ...... ***Sesbania***
        3. Branchlets not prickled ...... ***Butea***
    2. Leaves more than 3-foliolate:
        5. Leaflets opposite:
            6. Pods 1–2 seeded, woody ...... ***Pongamia***
            6. Pods more than 3-seeded, not woody ...... ***Gliricida***
        5. Leaflets alternate:
            7. Leaflets large, 7–9 × 4.5–8.5 cm. Pods winged, rounded ...... ***Pterocarpus***
            7. Leaflets small, 2–5 × 1.5–4.5 cm. Pods not winged, oblong ...... ***Dalbergia***
1. Shrubs or herbs:
    8. Climbers (except *Rhynchosia suaveolens*):
        9. Leaves 3-foliolate
            10. Corolla orange-scarlet, velvety ...... ***Butea (superba)***
            10. Corolla not as above:
                11. Pods with stinging hairs, brown ...... ***Mucuna***
                11. Pods not as above:
                    12. Pods compressed, thickened at sutures ...... ***Canavalia***
                    12. Pods not as above:
                        13. Leaflets and calyx gland dotted:
                            14. Pods enclosed in membranous calyx ...... ***Paracalyx***
                            14. Pods not as above:

15. Pods 2-seeded .... ***Rhynchosia***
15. Pods 3-many-seeded .... ***Atylosia***
13. Leaflets and calyx not as above:
16. Flowers yellow .... ***Dolichos***
16. Flowers other than yellow:
17. Pods glabrous .... ***Teramnus***
17. Pods viscid .... ***Pseudarthria***
9. Leaves 5-many-folioalate.
18. Seeds scarlet with a black spot, shining .... ***Abrus***
18. Seeds not as above:
19. Flowers white or blue, solitary, pod dehiscent .... ***Clitoria***
19. Flowers whitish-pink, panicled, pod indehiscent .... ***Derris***
8. Non-climbers:
20. Leaves 2-foliolate .... ***Zornia***
20. Leaves simple or 3-many foliolate:
21. Pods moniliform:
22. Pod joints turgid .... ***Alysicarpus***
22. Pod joints flattened:
23. Leaves simple or 3-foliolate .... ***Desmodium***
23. Leaves many-foliolate .... ***Aeschynomene***
21. Pods non-moniliform:
24. Leaves or leaflets lobed .... ***Vigna***
24. Leaves or leaflets not lobed:
25. Flowers yellow:
26. Anthers uniform:
27. Pods spine tipped, many-seeded .... ***Rothia***
27. Pods beaked, 2–3-seeded .... ***Stylosanthes***
26. Anthers dimorphic:
28. Pods turgid .... ***Crotalaria***
28. Pods flattened .... ***Goniogyna***
25. Flowers other than yellow:
29. Leaves or leaflets margin dentate:
30. Leaves simple .... ***Psoralea***
30. Leaves 3-foliolate .... ***Melilotus***
29. Leaves or leaflets margin entire:
31. Pods boat-shaped .... ***Eleotis***
31. Pods not as above:
32. Vegetative parts hairy .... ***Indigofera***
32. Vegetative parts not as above .... ***Tephrosia***

ABRUS Adanson

Abrus precatorius L. Syst. Nat (ed. 12) 2: 472. 1767; FBI 2: 175; Gamble 349(247); Khan in FFH 127; ssp. **precatorius**; Verdc. in Kew Bull. 24: 241. 1970.

A climbing herb, up to 5 m. Leaves paripinnate, leaflets linear-oblong, 1–2.5 × 0.5–0.8 cm, base obtuse, margin entire, tip apiculate, sparsely pubescent on both sides. Flowers pink in axillary racemes. Pods 4.5 cm long, pilose, 5-seeded. Seeds scarlet with a black spot, shining.

Common in scrub jungles. Fl. & Fr.: July–Dec. *Vern.*: Gurivinda. Kusumasamudram beat, *MSM & NV* 10361.

AESCHYNOMENE Linnaeus

Aeschynomene indica L. Sp. Pl. 713. 1753; FBI 2: 151; Gamble 331(234).

An annual erect herb, up to 60 cm tall, branchlets striate, glabrescent. Leaves imparipinnate, leaflets oblong, 15–25 pairs, 0.5–0.75 × 0.1–0.2 cm; base oblique, margin entire, tip obtuse, glabrous on both sides. Flowers yellow, axillary, solitary. Pods 4 cm long, 6–10-seeded, jointed, glabrous. Seeds shining.

Common in moist places on banks of ponds. Fl. & Fr.: July–Dec. Kusumasamudram beat, *MSM & NV* 10332.

ALYSICARPUS Desvaux *nom. cons.*

1. Leaves upper 3-foliolate, lower 1-foliate .... ***A. racemosus***
1. Leaves all 1-foliate.
   2. Calyx not exceeding first joint of pod:
      3. Pseudoracemes lax-flowered, mostly longer than 4 cm, pods without septa inside .... ***A. ovalifolius***
      3. Pseudoracemes densely-flowered, mostly shorter than 4 cm, pods with septa inside:
         4. Pod not moniliform. Joints tetragonous, strongly veined .... ***A. vaginalis***
         4. Pod moniliform. Joints subglobose, faint veined .... ***A. monilifer***
   2. Calyx exceeding or much longer than the first joint of pod:
      5. Pods longer than calyx. Joints straw-coloured, unveined ...... ***A. bupleurifolius***
      5. Pods shorter than calyx or as long as calyx. Joints green-black, veined......... .... ***A. heyneanus***

Alysicarpus bupleurifolius (L.) DC. Prodr. 2: 352. 1825; FBI 2: 158; Gamble 338(239). *Hedysarum bupleurifolium* L. Sp. Pl. 745. 1753.

A perennial prostrate herb, up to 50 cm, branchlets glabrous. Leaves oblong, 2–3.5 × 0.5–0.8 cm, base rounded, margin entire, tip acute, glabrous on both sides. Flowers orange in terminal lax racemes. Pods cylindrical, 1 cm long, glabrous, 2–4-seeded.

Common in moist places. Fl. & Fr.: Sept.–Jan. Dharmapur beat, *MSM* 10487.

Alysicarpus heyneanus Wight & Arn. Prodr. 234. 1834. *A. rugosus* (Willd.) DC. Prodr. 50. 2. 353. 1825; FBI 2: 159; Gamble 338(239). *Hedysarum rugosum* Willd. Sp. Pl. 3: 1172. *Alysicarpus wallichii* Wight & Arn. Prodr. 234. 1834.

An annual sub-erect herb, up to 20 cm, branchlets pilose. Leaves oblong-lanceolate, 1.5–3 × 0.5–1 cm, base subcordate, margin entire, tip obtuse. Flowers purple in axillary racemes. Pods moniliform, articles 3–6, pubescent. Seeds rugose.

Common in waste lands and scrub jungle. Fl. & Fr.: Jan.–Mar. Pargi, *MSM & PRSR* 12797; Central University, *MSM* 15155.

Alysicarpus monilifer (L.) DC. Prodr. 2: 353. 1825; FBI 2: 157; Gamble 338(238). *Hedysarum moniliferum* L. Mant. Pl. 102. 1767.

A perennial diffuse herb, branchlets hairy. Leaves elliptic-oblong, 1.5–3.5 × 0.5–2 cm, base sub-cordate, margin entire, tip apiculate, glabrous above, pubescent along nerves below. Flowers pink in axillary racemes. Pods moniliform, 2 cm long, hairy, 4–7-seeded.

Common in forest in moist places. Fl. & Fr.: May–Sept. Mohammadabad beat, *MSM* 10494.

Alysicarpus ovalifolius (Schum.) J. Leon. in Bull. Jard. Bot. Estat. Brux. 24: 88. 1954. *Hedysarum ovalifolium* Schum. Beskr. Guin. Pl. 359. 1827. *Desmodium ovalifolium* (Schum.) Walp. Rep. 1: 737. 1842.

An erect or spreading annual herb. Leaves usually dimorphic, elliptic and oblong to narrowly lanceolate. Pseudoracemes lax flowered. Pods puberulous with hooked hairs, articles 4–6.

Rare in open places. Fl. & Fr.: July–Nov. Kanman Kalva, *MSM* 10468.

Alysicarpus racemosus Benth. in Linneaa 24: 162. 1851; Gamble 339(240). *A. belgaumensis Wight var. racemosus* (Benth.) Baker in Hook. f. Fl. Brit. India 1: 160. 1876.

An annual diffuse herb, up to 40 cm tall, branchlets hairy. Leaves 1-foliate below, 3-foliolate above, leaflets obovate, to 2 × 1.5 cm, base round, margin entire, tip obtuse. Flowers purple turning blue in terminal racemes. Pod moniliform, 2-5-jointed, glabrous.

Rare in waste lands (Seshagiri Rao, 1986). Fl. & Fr.: July–Sept.

Alysicarpus vaginalis (L.) DC. Prodr. 2: 353. 1825 var. **vaginalis**. FBI 2: 158; Gamble 338(239). *Hedysarum vaginalis* L. Sp. Pl. 746. 1753. *Alysicarpus vaginalis* (L.) DC. var. *nummularifolius* Miq. Fl. Ind. Bot. 1: 232. 1855; FBI 2: 158; Gamble 338(238).

A prostrate herb, up to 1 m, branchlets sparsely pubescent. Leaves lanceolate, 2–8.5 × 0.5–2.5 cm, base truncate or cuneate, margin entire, tip acute, glabrous above, sparsely hairy along midrib and nerves below. Flowers purple in lateral or axillary racemes. Pods 2.2 cm long, downy pubescent, articles 5–7, 4-gonous, strongly nerved, 4–5-seeded.

Common in forests and moist places. Fl. & Fr.: May–Sept. Kanmankalva beat, *MSM* 10468.

ATYLOSIA R. Wight et Arnott

Atylosia scarabaeoides (L.) Benth. in Miq. Pl. Jungh 243. 1852; FBI 2: 215; Gamble 369(261). *Dolichos scarabaeoides* L. Sp. Pl. 726. 1753.

A climbing herb, branchlets pubescent. Leaves 3-foliolate, leaflets ovate-lanceolate, 3–5 × 1–1.5 cm, terminal leaflet larger than laterals, base cuneate, margin entire, tip obtuse, velvety pubescent on both sides. Flowers yellow in axillary 2–6-flowered clusters. Pods 1.5 cm long, densely pubescent with transverse depressions, 4–6-seeded. Seeds oblong.

Common on hedges. Fl. & Fr.: June–Dec. Dharampur, *MSM & KH* 10543; Lingampalle, *MSM* 12759.

BUTEA J. Koenig ex Roxburgh

1. Trees, lower calyx lobes much shorter than the side ones, base of the pod round .... ***B. monosperma***
1. Climbing shrubs, lower calyx lobes as long as laterals, base of the pod narrowed .... ***B. superba***

Butea monosperma (Lam.) Taub. in Engl. Nat. Pflanzanf. 3(3): 366. 1894. *Erythrina monosperma* Lam. Encycl. 1: 391. 1785. *Butea frondosa* J. Koening ex Roxb. Asiat. Res. 3: 469. 1792; FBI 2: 194; Gamble 357(252); Khan in FFH 129.

A deciduous tree, up to 10 m tall, branchlets densely tomentose, bark rough. Leaves 3-foliolate, leaflets broadly obovate, 5–12.5 × 4–14 cm, base cuneate, margin entire, tip obtuse, glabrescent above, silky tomentose below. Flowers flame coloured, fascicled on rigid terminal racemes. Pods 20 cm long, silky tomentose. Seeds compressed.

Common in forests. Fl. & Fr.: Mar.–May. *Vern.:* Modugu, Mohammadabad beat, *MSM & KH* 10242.

Butea superba Roxb. Pl. Cor. 1: 23. t. 22. 1795; FBI 2: 195; Gamble 358(252); Khan in FFH 129.

A deciduous woody climbing shrub, branchlets tomentose. Leaves 2-foliolate, leaflets rhomboid-unequally ovate, 10–20 × 10–12 cm, base cuneate, margin entire, tip rounded, tomentose when young. Flowers orange-scarlet in axillary and terminal panicles. Pods oblong, velvety, 1-seeded.

Common in forests. Fl. & Fr.: Feb.–June. *Vern.:* Teega moduga. Mohammadabad, *MSM* 19235.

CANAVALIA A.P. De Candolle

Canavalia gladiata (Jacq.) DC. Podr. 2: 404. 1825. *Dolichos gladiatus* Jacq. Coll. Bot. 2: 276. 1788. *Canavalia ensiformis* sensu Baker in Hook. f. FBI 2: 195. 1876 non DC., Gamble 359(254).

A climbing shrub, branchlets pubescent. Leaves 3-foliolate, leaflets ovate, 5–8 × 4–5.5 cm, terminal larger than laterals, base cuneate, margin entire, tip abruptly acuminate, glabrous on both sides. Flowers purplish in axillary racemes. Pods 12 cm long, compressed, thickened along sutures, 10–15-seeded. Seeds pinkish brown.

Occasional in forests. Fl. & Fr.: Sept.–Dec. Dharampur, *MSM & KH* 10545.

CLITORIA Linnaeus

Clitoria ternatea L. Sp. Pl. 753. 173; FBI 2: 208; Gamble 365(258).

A perennial climbing herb, up to 6 m, branchlets sparsely pubescent. Leaves imparipinnate, leaflets obovate, 5–7 pairs, 2.5–4 × 1.5–2.5 cm, base obtuse, margin entire, tip retuse, sparsely appressed pubescent on both sides. Flowers white, axillary, solitary. Pods 9 cm long, sparsely appressed pubescent, 6–10-seeded. Seeds yellow-brownish, smooth.

Common in forest clearings, roadsides and scrub jungles. Fl. & Fr.: June–Jan. Nayanigutta, *MSM & NV* 10329.

CROTALARIA Linnaeus

1. Leaves 3-foliolate:
   2. Diffuse herb. Racemes 6–9-flowered, pods 2-seeded .... ***C. medicaginea***
   2. Shrubs or subshrubs. Racemes many-flowered, pods many-seeded:
      3. Leaflets obovate. Stipe of the pod < 2 mm .... ***C. pallida***
      3. Leaflets ovate-elliptic. Stipe of the pod > 2 cm .... ***C. laburnifolia***
1. Leaves simple:
   4. Stipules foliaceous, branchlets angled .... ***C. verrucosa***
   4. Stipules not foliaceous, branchlets terete:
      5. Pods glabrous .... ***C. mysorensis***
      5. Pods hairy or pubescent;
         6. Erect herbs. Standard petal yellow
            7. Plants to 15 cm tall. Pods villous. .... ***C. pusilla***
            7. Plants to 30–75 cm tall. Pods hirsute .... ***C. hirsuta***
         6. Diffuse herbs. Standard petal blue within .... ***C. bifaria***

Crotalaria bifaria L.f. Suppl. Pl. 322. 1781; FBI 2: 169. 1876; Gamble 293(207).

A branched herb. Leaves simple, ovate-orbicular, 2–4 × 1–2 cm, base cuneate, margin entire, tip obtuse, glabrescent above and puberulous below. Flowers yellow without, blue within, in lateral racemes. Pod oblong, puberulous, 10-seeded.

Occasional in open forest. Fl. & Fr.: Nov.–Dec. Lingampalle, *MSM* 12756.

Crotalaria hirsuta Willd. Sp. Pl. 3: 978. 1803; FBI 2: 68; Gamble 293(207).

An annual erect herb up to 50 cm tall, branchlets hairy. Leaves ovate, 3–5 × 2–2.5 cm, base cuneate, margin entire, tip acute. Flowers yellow in axillary racemes. Pods oblong, hirsute, 8–10-seeded.

Occasional in open forest. Fl. & Fr.: Sept.–Nov. Lingampalle, *MSM* 12749.

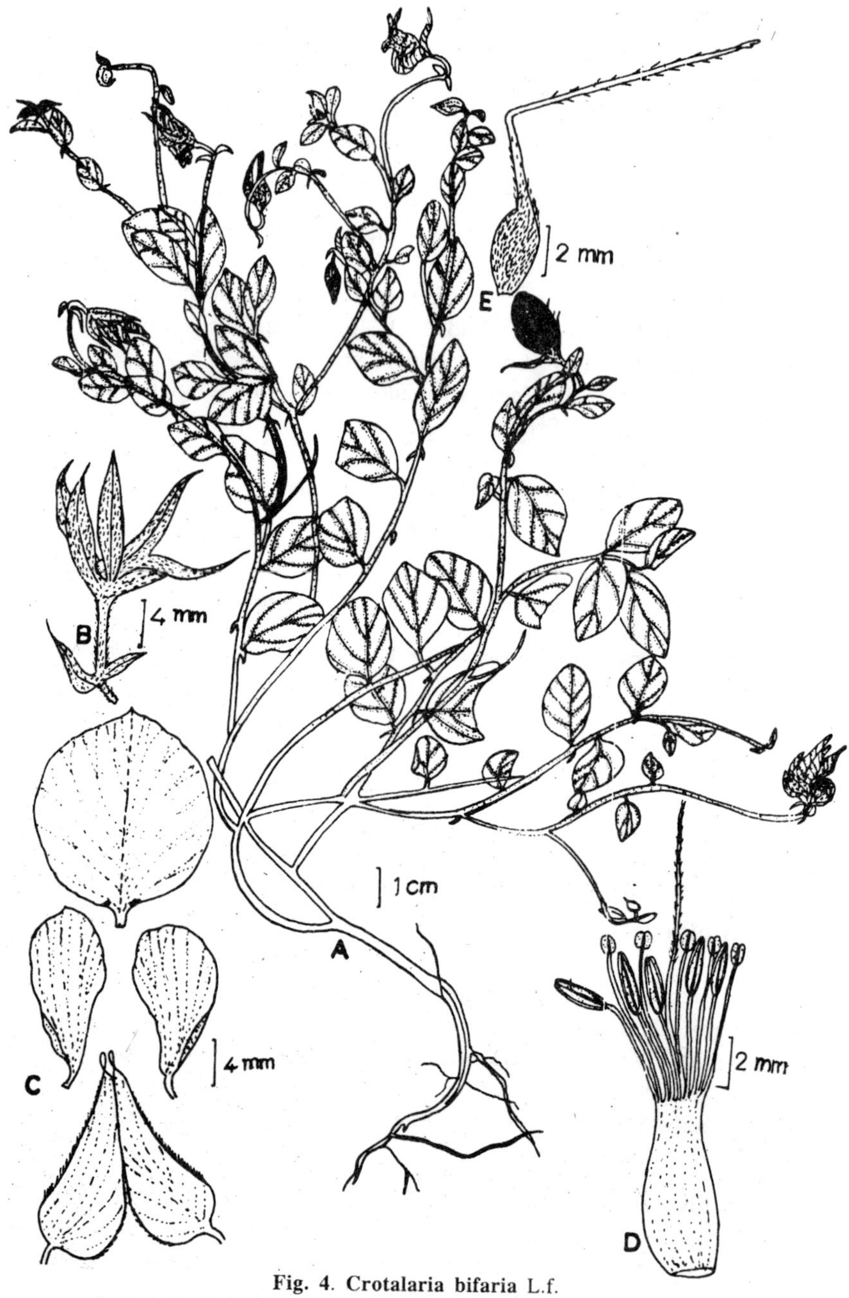

Fig. 4. Crotalaria bifaria L.f.
A. Twig B. Calyx C. Corolla D. Staminal Column with Pistil E. Pistil

Crotalaria laburnifolia L. Sp. Pl. 715. 1753; FBI 2: 84; Gamble 301(213).

An erect herb, up to 1.5 m tall, branchlets glabrous. Leaves 3-foliolate, leaflets ovate, 2.5–6 × 2.3–5 cm, base acute, margin entire, tip obtuse, sparsely appressed pubescent on both sides. Flowers yellow, falcate, in lateral lax racemes. Pods 5 cm long, glabrous, 20–30-seeded.

Occasional on roadsides. Fl. & Fr.: July–Oct. Nancherla gate, *MSM & NV* 10393.

Crotalaria medicaginea Lam. Encycl. 2: 201. 1786; FBI 2: 81; Gamble 299(211). *C. neglecta* Wight & Arn. Prodr. 192. 1834.

A perennial prostrate herb, up to 50 cm, branchlets appressed pubescent. Leaves 3-foliolate, terminal leaflet larger than laterals, obovate, 0.5–2.5 × 0.3–1 cm, base attenuate, margin entire, tip rounded, glabrous above, appressed pubescent below. Flowers in axillary and terminal, 2–4-flowered racemes. Pods 0.5 cm long, densely pubescent, beaked, 2-seeded.

Common in forests. Fl. & Fr.: All seasons. Anantasagar beat, *MSM* 10506.

Crotalaria mysorensis Roth, Nov. Pl. Sp. 338. 1821; FBI 2: 70. 1876; Gamble 295(208).

Annual erect herb, up to 1 m tall, branchlets tomentose. Leaves simple, sessile, linear-elliptic, oblong, 3–6 × 1–2 cm, base cuneate, margin entire, tip apiculate, densely pilose on both sides. Flowers yellow with maroon streaks in terminal racemes. Pods obovoid, bloated, glabrous, many-seeded.

Common in forest under growth. Fl. & Fr.: Sept.–Nov. Lingampalle, *MSM* 12753.

Crotalaria pallida Aiton, Hort. Kew. 3: 20. 1789. *C. mucronata* Desv. J. Bot. (Desvana) 3: 76. 1814. *C. striata* DC. Prodr. 2: 131. 1825; FBI 2: 84; Gamble 301(212).

A subshrub, up to 1 m tall, branchlets puberulous. Leaves 3-foliolate, leaflets obovate, 3–6 × 2–4 cm, base cuneate, margin entire, tip obtuse-rotund. Flowers yellow in terminal racemes. Pods oblong, many-seeded.

Occasional in forest under growth (Seshagiri Rao, 1986). Fl. & Fr.: All seasons.

Crotalaria pusilla Heyne ex Roth Nov. Pl. Spec. 335. 1821; FBI 2: 70; Gamble 296(209).

An annual diffuse herb, up to 15 cm long, branchlets hairy. Leaves oblong, 0.5–2 × 0.2 cm, base obtuse, margin entire, tip acute, appressed hairy on both sides. Flowers yellow in terminal or lateral racemes. Pods oblong, 0.5 cm long, villous, 4–8-seeded.

Common in forest. Fl. & Fr.: All seasons. Mohammadabad beat, *MSM* 10482.

Crotalaria verrucosa L. Sp. Pl. 715. 1753; FBI 2: 77; Gamble 297(210).

A subshrub, up to 1 m tall, branchlets glabrous. Leaves ovate-rhomboid, 1.5–4 × 1–3 cm, base attenuate, margin entire, tip acute. Flowers white-blue. Pods oblong, bloated, sparsely pubescent, 16–20-seeded.

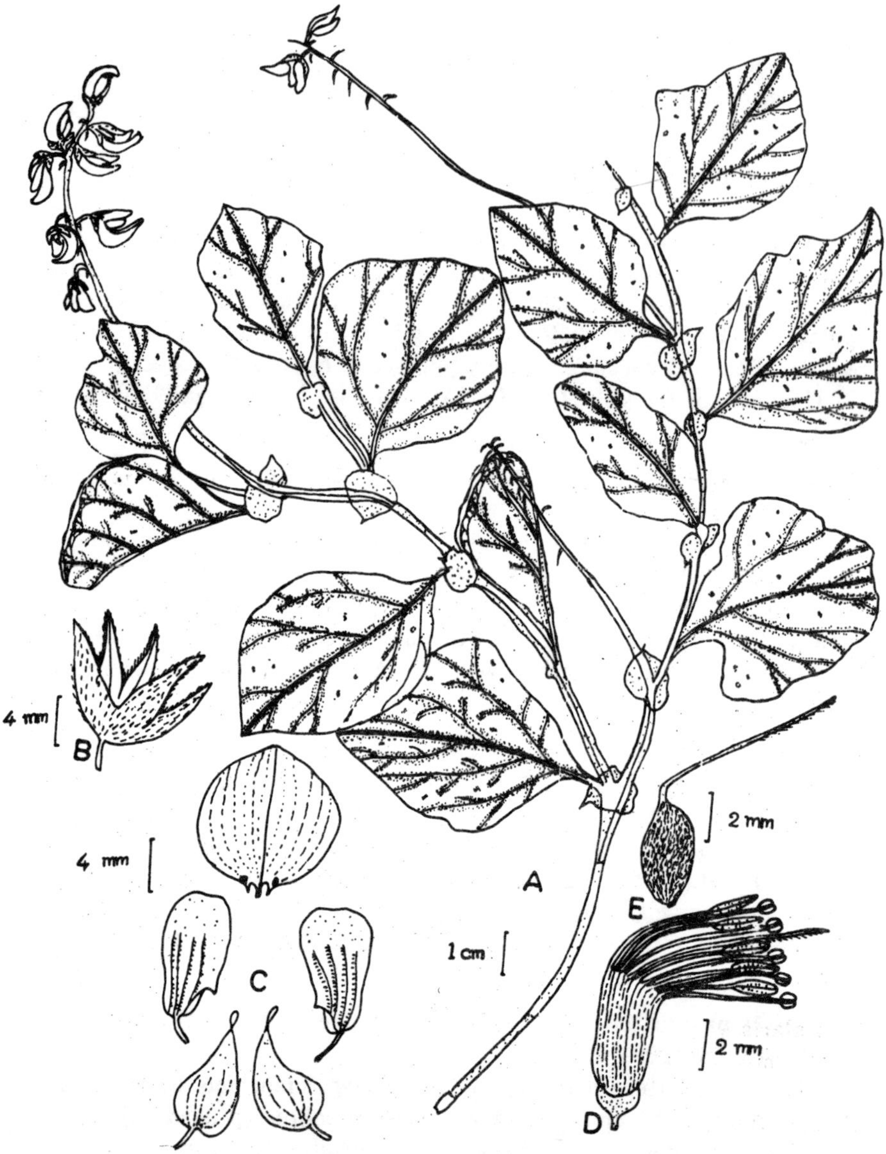

Fig. 5. Crotalaria verrucosa L.
A. Twig B. Calyx C. Corolla D. Staminal column with Pistil E. Pistil.

Common weed of waste land. Fl. & Fr.: Sept.–Nov. Central University, *MSM & TP* 15148.

DALBERGIA Linnaeus f.

1. Leaflets 9-above. Stamens 5+5:
   2. Flowers pink, pods drying grey .... ***D. lanceolaria***
   2. Flowers white, pods drying black .... ***D. paniculata***
1. Leaflets 3–7. Stamens 9:
   3. Pod 1 cm across, 1 or 2-seeded. Plains .... ***A. sissoo***
   3. Pod 2 cm across, 1–3-seeded. Hills .... ***D. latifolia***

Dalbergia lanceolaria L.f. Suppl. 316. 1781; FBI 2: 235; Gamble 383(270).

A deciduous tree, up to 10 m tall, young branchlets pubescent. Leaves imparipinnate, 5–7-pairs, leaflets obovate, 1–2 × 0.5–2 cm, base obtuse, margin entire, tip obtuse, glabrous on both sides. Flowers pinkish-white in axillary panicles. Pods 1 cm long, lanceolate, 1–3-seeded.

Occasional in forest. Fl. & Fr.: Aug.–Dec. Anantagiri RF, *SSR* 18513.

Dalbergia latifolia Roxb. Pl. Cor. t. 113. 1799; FBI 2: 231; Gamble 383(270).

A deciduous tree, up to 10 m tall, branchlets glabrous. Leaves imparipinnate, leaflets 3–4-pairs, obovate-orbicular, 3–5 × 3–4.5 cm, base subacute, margin entire, tip retuse, glabrous above, puberulous below. Flowers white in axillary panicles. Pods 9 cm long, oblong, 1–3-seeded.

Occasional in forest. Fl. & Fr.: Aug.–Dec. Dharur RF, *MSM & TP* 11096.

Dalbergia paniculata Roxb. Pl. Cor. t. 114. 1799; FBI 2: 236; Gamble 383(270); Khan in FFH 136.

A deciduous tree, up to 12 m tall, branchlets rusty tomentose. Leaves imparipinnate, 5–6-pairs, leaflets oblong-obovate, 1–3.5 × 1–2.5 cm, base rounded, margin entire, tip retuse, pubescent above, tomentose below. Flowers white in terminal branched panicles. Pods 7 cm long, glabrescent, 1–2-seeded.

Common in forests. Fl. & Fr.: May–July. *Vern.:* Sophera. Mohammadabad beat, *MSM & KH* 10243.

Dalbergia sissoo Roxb. Fl. Ind. 3: 223; 1832; FBI 2: 231; Gamble 283(270); Khan in FFH 134.

A deciduous tree, up to .10 m tall, branchlets tomentose. Leaves imparipinnate, leaflets obovate, 3–7 × 2–6 cm, base cuneate, margin entire, tip acuminate, pubescent above, densely pubescent below. Flowers cream in terminal panicles. Pods 7.5 cm long, glabrescent, 1–2-seeded.

Common on banks of rivers. Fl. & Fr.: Jan.–May. *Vern.:* Sissoo. Kanmankalva beat, *MSM & KH* 10311.

DERRIS Loureiro *nom. cons.*

Derris scandens (Roxb.) Benth. in J. Linn. Soc. Bot. 4(Suppl.) 103. 1860; FBI 2: 240; Gamble 387(273); Khan in FFH 138. *Dalbergia scandens* Roxb. Pl. Cor. t. 192. 1805.

A woody climbing shrub, up to 4 m, branchlets glabrous. Leaves imparipinnate, leaflets oblong-elliptic, 2–8 × 1–3.5 cm, base truncate-cuneate, margin entire, tip obtuse, glabrous on both sides. Flowers whitish-pink in clusters, in axillary and terminal panicles. Pods 6.5 cm long, appressed pubescent, winged, 5-seeded.

Common in deciduous forests. Fl. & Fr.: Apr.–Sept. Anantasagar beat, *MSM* 10504; Dharur, *MSM* 11098.

DESMODIUM Desvaux *nom. cons.*

1. Leaves 3-foliolate:
   2. Herb, pods 3–6-jointed .... ***D. triflorum***
   2. Subshrub, pods 2-jointed .... ***D. biarticulatum***
1. Leaves simple .... ***D. gangeticum***

Desmodium biarticulatum (L.) F.V. Muell. Fragm. Phyt. Austr. 2. 121. 1861; FBI 2; 163. 1876; Gamble 344(243). *Hedysarum biarticulatum* L. Sp. Pl. 747. 1753. *Dicerma biarticulatum* (L.) DC. Mem. Leg. 7: 328. 1825 & Prodr. 2: 339. 1825.

A slender under shrub, up to 60 cm tall. Leaves 3-foliolate, leaflets oblong 0.5–15 × 0.3–0.5 cm, base cuneate, margin entire, tip obtuse. Flowers red in terminal lax racemes. Pod with 1–2 articles, pubescent. Seeds black.

Occasional in forest undergrowth. Fl. & Fr.: Apr.–Oct. Central University, *MSM* 15158; Rudraram RF, *MSM & TP* 12728.

Desmodium gangeticum (L.) DC. Prodr. 2: 327. 1825; FBI 2: 168; Gamble 345(244). *Hedysarum gangeticum* L. Sp. Pl. 746. 1753.

A diffuse herb, up to 1 m, branchlets pubescent. Leaves elliptic-ovate, 4–10 × 2–5 cm, base truncate, margin entire, tip acute, pubescent above, densely pubescent below. Flowers violet in terminal racemes. Pods moniliform, 3 cm long, glabrous, articles 4–6.

Common in forest in moist places. Fl. & Fr.: May–Sept. Anantasagar, *MSM* 10507.

Desmodium triflorum (L.) DC. Prodr. 2: 334. 1825; FBI 2: 173; Gamble 347(245). *Hedysarum triflorum* L. Sp. Pl. 749. 1753.

A prostrate herb, up to 50 cm, branchlets pilose. Leaves 3-foliolate, leaflets obovate, 4–5 × 3–5 mm, base cuneate, margin entire, tip obtuse, glabrous above, pubescent below. Flowers pink, axillary, solitary. Pods 1 cm long, articles 4–6, hooked pubescent. Seeds oblong.

Common weed in cultivated fields. Fl. & Fr.: Sept.–Dec. Lingampalle, *MSM & TP* 15138.

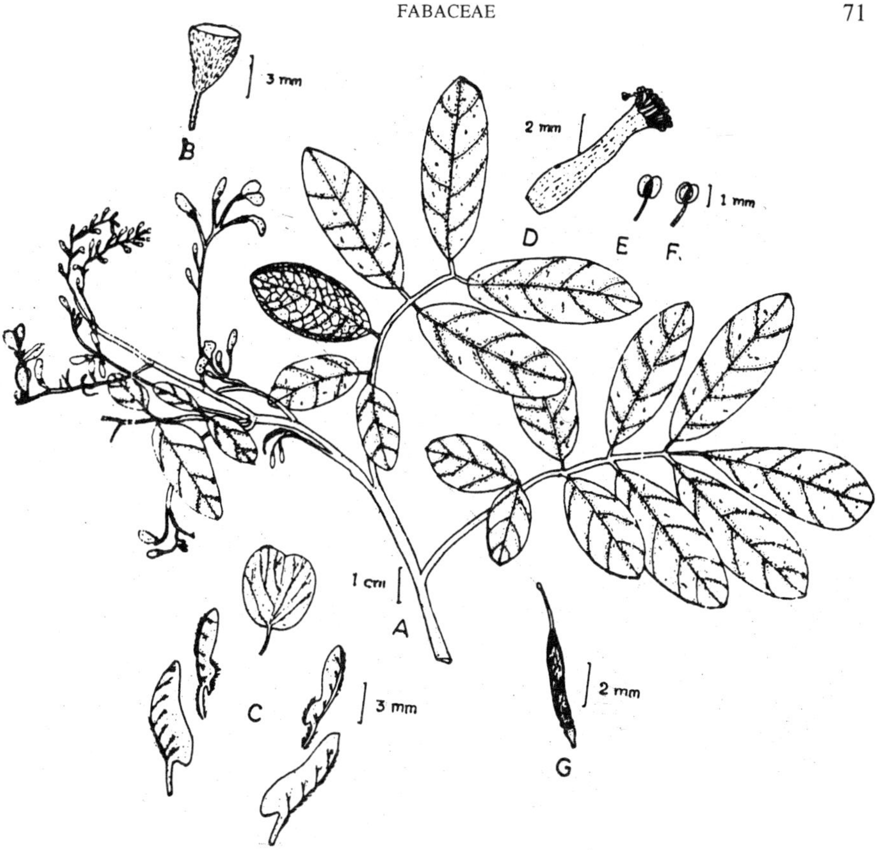

Fig. 6. Derris scandens (Roxb.) Benth.
A. Twig B. Calyx C. Corolla D. Staminal column E. & F. Anthers G. Pistil

DOLICHOS Linnaeus

Dolichos trilobus L. Sp. Pl. 736. 1753. *D. falcatus* Klein ex Willd. Sp. Pl. 3: 1047; 1802; FBI 2: 211; Gamble 366(259).

A perennial climbing herb, branchlets hairy. Leaves 3-foliolate, 2–6.5 × 2–5 cm, base truncate, margin entire, tip acute, sparsely pubescent on both sides. Flowers yellow in long axillary lax racemes. Pods linear, 7 cm long, glabrous, 4–5-seeded.

Common in forests and moist places. Fl. & Fr.: May–Sept. Kanmankalva beat, *MSM* 10461; Mohammadabad beat, *MSM & KH* 10560.

ELEIOTIS A.P. de Candolle

Eleiotis monophylla (Burm. f.) DC. Mem. Legum. 350. 1826. *Glycine monophylla* Burm. f. Fl. Indica. 161. t. 50. f. 2. 1768. *Eleiotis sororia* DC. Prodr. 2: 348. 1825; FBI 2: 153; Gamble 333(235).

An annual prostrate herb up to 30 cm, branchlets glabrous. Leaves 1-foliate, obovate-orbicular, 0.5–1.5 × 0.6–2 cm, base cordate, margin entire, tip rotund. Flowers red in lax racemes. Pods boat-shaped, 1-seeded. Seeds oblong.

Rare in cultivated fields. Fl. & Fr.: Sept.–Dec. Lingampalle, *MSM & TP* 15121.

ERYTHRINA Linnaeus

Erythrina stricta Roxb. Fl. Ind. 3: 251. 1832; FBI 2: 189; Gamble 354(250). Khan in FFH.

A deciduous armed tree, up to 10 m tall, branchlets prickled. Leaves 3-foliolate, leaflets ovate, 8–12 × 11–14 cm, base truncate, margin entire, tip acuminate, glabrescent on both sides. Flowers scarlet in terminal long racemes. Pods linear oblong, glabrous, 5–6-seeded.

Abundant in villages. Fl. & Fr. Jan.–May. *Vern.:* Tella varjam. Mokarlabad, *MSM & KH* 10261.

GLIRICIDIA Humboldt, Bonpland et Kunth

Gliricidia sepium (Jacq.) Kunth ex Walp. Repert. 1: 679. 1849. *Robinia sepium* Jacq. Enum. Syst. Pl. 28. 1760.

A deciduous tree, up to 9 m tall, branchlets pubescent. Leaves imparipinnate, leaflets 13–17 pairs, elliptic-lanceolate. 2.5–6 × 1–2.5 cm, base obtuse, margin entire, tip acuminate, glabrous on both sides. Flowers pink, in long racemes. Pods oblong, glabrous, 8–10-seeded.

Cultivated. Fl. & Fr.: July–Jan.

GONIOGYNA A.P. de Candolle

Goniogyna hirta (Willd.) Ali in Taxon 16. 463. 1967. *Hallia hirta* Willd. Sp. Pl. 3(2): 1169. 1803. *Heylandia latebrosa* sensu FBI 2: 65. 1876; non. (L.) DC. 1825; Gamble 280(199).

An annual prostrate herb, branchlets with silky hair. Leaves cordate-ovate, 1–1.5 × 0.5–0.7 cm, base truncate, margin entire, tip obtuse. Flowers yellow, axillary, solitary. Pods oblong, compressed, 1-seeded.

Common in open forest. Fl. & Fr.: Aug.–Oct. Anantagiri, RF, *MSM & TP* 11087.

INDIGOFERA Linnaeus

1. Leaves simple:
   2. Leaf blade linear .... ***I. linifolia***
   2. Leaf blade other than linear:

3. Leaves ovate-cordate. Pod 2-seeded .... ***I. cordifolia***
3. Leaves obovate, cuneate. Pod 1-seeded .... ***I. nummularifolia***

1. Leaves 3-many-foliolate:
    4. Leaves 3-foliolate:
        5. Leaflets gland dotted below .... ***I. glandulosa***
        5. Leaflets not gland dotted below .... ***I. trita***
    4. Leaves 5-many foliolate:
        6. Herbs:
            7. Prostrate herbs. Pod 2-seeded .... ***I. linnaei***
            7. Erect herbs. Pods to 12-seeded .... ***I. glabra***
        6. Shrubs.
            8. Leaves not exceeding 8 cm in length .... ***I. tinctoria***
            8. Leaves exceeding 8 cm in length:
                9. Pods hirsute .... ***I. hirsuta***
                9. Pods glabrous .... ***I. cassioides***

Indigofera cassioides Rottler ex DC. Prodr. 2: 225. 1825. *I. pulchella* Roxb. Fl. Ind. 3: 382. 1832; FBI 2: 101; Gamble 313(221); Khan in FFH 122.

An erect shrub, up to 1.5 m tall, branchlets pubescent. Leaves imparipinnate, leaflets elliptic-oblong, 1.5–2 × 0.5–1 cm, base obtuse, margin entire, tip truncate-mucronate. Flowers purple in axillary long racemes. Pods cylindrical, 6 cm long, glabrous, 8–10-seeded. Seeds black, globose.

Occasional in forest. Fl. & Fr.: Oct.–Dec. Anantagiri RF, *MSM & PRSR* 12776.

Indigofera cordifolia Heyne ex Roth, Nov. Pl. Spec. 357. 1821; FBI 2: 93; Gamble 309(218).

An annual prostrate herb, up to 50 cm, branchlets pubescent. Leaves ovate, 0.5–1 × 0.5–0.8 cm, base cordate, margin entire, tip acute, pilose on both sides. Flowers red in axillary racemes. Pods linear, 2.5 cm long, pubescent, 2–4-seeded.

Common in forests. Fl. & Fr.: Sept.–Dec. Rangammagudem, *MSM* 10471; Gadirayal, *MSM* 10516.

Indigofera glabra L. Sp. Pl. 751. 1753; Gamble 309(219). *I. pentaphylla* Murr. Syst. Veg. ed. 13. 564. 1774; FBI 2: 95.

An erect herb, up to 40 cm tall, branchlets brown, terete. Leaves imparipinnate, leaflets 5–7-foliolate, obovate, 0.5–2 cm, base obtuse, margin entire, tip acute. Flowers red in axillary racemes. Pods 2 cm long, linear, pointed, 8–10-seeded. Seeds black.

Occasional in forest. Fl. & Fr.: Oct.–Dec. Lingampalle, *MSM* 12742 & 12760.

Indigofera glandulosa Roxb. ex Willd. Sp. Pl. 3: 1227. 1800; FBI 2: 94; Gamble 309(218).

An erect herb, up to 40 cm tall, branchlets pubescent. Leaves 3-foliolate, leaflets oblanceolate, 1.5–3 × 0.5–0.6 cm, base cuneate, margin entire, tip retuse. Flowers red in axillary racemes. Pods oblong, 5 mm long, deflexed, winged, 2-seeded. Seeds ovoid.

Fig. 7. Indigofera cassioides DC.
A. Twig B. Calyx C. Corolla D. Staminal column E. Pistil.

Common in fields. Fl. & Fr.: Sept.–Dec. Naskal, *MSM & PRSR* 15118.

Indigofera hirsuta L. Sp. Pl. 751. 1753; FBI 2: 98; Gamble 312(221).

An annual erect herb, up to 70 cm tall, branchlets white pilose. Leaves imparipinnate, leaflets elliptic-oblanceolate, 2–3.5 × 1–1.12 cm, base acute, margin entire, tip obtuse, densely hairy on both sides. Flowers pink in dense axillary racemes. Pods straight in clusters, 4-gonous, 2 cm long, hairy, 5-seeded. Seeds black.

Occasional in forest undergrowth. Fl. & Fr.: June–Dec. Kusumasamudram beat, *MSM & KH* 11030.

Indigofera linifolia (L.f.) Retz. Obs. Bot. 4: 29. 1786; FBI 2: 92; Gamble 309(218). *Hedysarum linifolium* L.f. suppl. Pl. 331. 1781.

An annual diffuse herb, up to 40 cm, branchlets silvery canescent. Leaves linear, 1–2 × 0.2–0.4 cm, base cuneate, margin entire, tip mucronate. Flowers red, axillary, solitary. Pods globose, 0.2 cm long, pubescent, 1-seeded.

Occasional in forest. Fl. & Fr.: Oct.–Dec. Anantagiri RF, *MSM* 11078; Lingampalle, *MSM* 12762.

Indigofera linnaei Ali in Bot. Not. 111: 549. 1958. *Hedysarum prostratum* L. Mant. Pl. 102. 1767. *Indigofera enneaphylla* L. Mant. Pl. 272, 571. 1771, *nom. superfl*, FBI 2: 94; Gamble 309(218).

An annual prostrate herb, branchlets appressed white hirsute. Leaves imparipinnate, leaflets oblanceolate, 0.3–0.6 × 0.2–0.4 cm, base cuneate, margin entire, tip retuse, appressed densely hirsute on both sides. Flowers red in axillary racemes. Pods 0.4 cm long, white hirsute, 2-seeded. Seeds globose.

Occasional weed in forests and wastelands. Fl. & Fr.: May–Jan. Malkalcheruvu, *MSM & KH* 10281.

Note: The earliest valid name for this plant is *Hedysarum prostratum* L. (Mant. 1: 102. 1767). According to articles 55 and 70 of the Intern. Code. Bot. Nomencl. the correct name should be *I. prostrata*. As Linnaeus did not make this combination his new name *I. enneaphylla* is illegitimate, nor can the earliest epithet be reinstated now, as it would be a later homonym of Willdenow 1808. Hence Ali has chosen the new name "*Indigofera linnaei*."

Indigofera nummularifolia (L.) Liv. ex Altson in Trimen Fl. Ceylon (Suppl.) 72. 1931. *Hedysarum nummularifolium* L. Sp. Pl. 746. 1753. *Indigofera echinata* Willd. Sp. Pl. 3: 1222. 1805; FBI 2: 92; Gamble 308(218).

An annual prostrate herb, branchlets tomentose. Leaves 1-foliate, obovate, 1–1.5 × 0.5–1 cm, base cuneate, margin entire, tip rotund, glabrous on both sides. Flowers red in 5–10-flowered axillary racemes. Pods reniform, 0.8 cm long, hooked spiny, 1-seeded. Seed brown, oblong-subreniform.

Occasional in ponds. Fl. & Fr.: June–Dec. Anantasagar beat, *MSM & KH* 11018.

Indigofera tinctoria L. Sp. Pl. 751. 1753; FBI 2: 99; Gamble 320(220).

An erect herb, up to 1.5 m tall, branchlets appressed pubescent. Leaves imparipinnate, 9–13-foliolate, leaflets oblong-obovate, 1–2.5 × 0.5–1 cm, base acute, margin entire, tip apiculate, glabrescent above, appressedly pubescent below. Flowers pink in axillary simple racemes. Pods 3 cm long, sparsely pubescent, slightly curved, apiculate, 8–12-seeded, seeds brown,

Occasional in wastelands and open forests. Fl. & Fr.: July–Jan. Anantasagar beat, *MSM & NV* 10385.

Indigofera trita L.f. Suppl. Pl. 335. 1781; FBI 2: 96; Gamble 310(219).

A subshrub up to 1 m tall, branchlets pubescent. Leaves 3-foliolate, leaflets obovate, 1–2 × 0.5–1.5 cm, base cuneate, margin entire, tip emarginate. Flowers pink in long raceme Pods linear, densely pubescent, 5–8-seeded.

Common in forest. Fl. & Fr.: Mar.–Sept. Anantagiri RF, *MSM & TP* 11064.

MELILOTUS P. Miller

Melilotus indica (L.) All. Fl. Pedam. 1: 308. 1875; Gamble 303 (215). *Trifolium indicum* L. Sp. Pl. 765. 1753. *Melilotus parviflora* Desf. Fl. Atlant. 2: 192. 1799; FBI 2: 89. 1876.

An annual small herb, up to 15 cm tall. Leaves 3-foliolate, leaflets obovate-oblong, 1.5–5 × 0.3–0.7cm, base cuneate, margin serrate, tip mucronate. Flowers cream in axillary racemes. Pods oblong, 1-seeded.

A common weed in fields (Seshagiri Rao, 1986). Fl. & Fr.: July–Nov.

MUCUNA Adanson

Mucuna pruriens (L.) DC. Prodr. 2: 405. 1825; FBI 2: 187. *Dolichos pruriens* L. Herb. Amb. 23. 1754. *Stizolobium pruriens* (L.) Medik. Vorles. Churpfalz. Phys. Ocon. Ges. 2: 339. 1787. *Mucuna prurita* Hook. J. Bot. Kew Gard. Misc. 2: 348. t. 13. 1830; *nom. superfl*; Gamble 356(251); Khan in FFH 127.

A slender climbing herb, branchlets downy pubescent. Leaves 3-foliolate, leaflets ovate-rhomboid, 6–9 × 5–7 cm, base obtuse-subacute, margin entire, tip acute, glabrescent above, grey pubescent below. Flowers purple in racemes. Pods curved at base and apex, 8 cm long, brown-bristly, 6-seeded. Seeds black.

Rare in forests. Fl. & Fr.: Sept.–Jan. *Vern.:* Duldamma, Kusumasamudram beat, *MSM & KH* 11034.

PARACALYX S.I. Ali

Paracalyx scariosus (Roxb.) Ali in Univ. Stud. (Karachi) 5(3): 95. 1968. *Cylista scariosa* Roxb. Pl. Cor. t. 92. 1798; FBI 2: 219; Gamble 371(262).

A climbing subshrub, up to 50 cm, branchlets villous, brown. Leaves 3-foliolate, ovate, terminal leaflet deltoid, lateral leaflets ovate, inequilateral, 3–4.5 × 2–3 cm, base obtuse, margin entire, tip acuminate, pubescent above, silky tomentose below. Flowers yellow-orange in terminal racemes, stamens diadelphous. Pods ellipsoid, 0.8 cm long, flat, pubescent, 1-seeded. Seed reniform, estrophiolate.

Occasional in forest undergrowth. Fl. & Fr.: Dec.-Mar. Mohammadabad, *MSM & KH* 10561.

PONGAMIA Ventenat *nom. cons.*

Pongamia pinnata (L.) Pierre. Fl. Forest Cochinch Sub. t. 385. 1899. *Cytisus pinnatus* L. Sp. Pl. 741. 1753. *Pongamia glabra* Vent. Jard. Malm. 28. 1803. *nom illegit.* FBI 2: 240; Gamble 385(272); Khan in FFH 137.

An evergreen tree, up to 5 m tall. Leaves imparipinnate, leaflets ovate-elliptic, 5–12 × 6–7 cm, base truncate, margin entire, tip acuminate. Flowers purplish-white in densely arranged, long panicles. Pods oblong, glabrous, 1-seeded.

A common avenue tree often as an escape. Fl. & Fr.: All seasons. *Vern.* Kanuga. Central University, *MSM* 19216.

PSEUDARTHRIA Wight & Arnott

Pseudarthria viscida (L.) Wight & Arn. Prodr. 209. 1834; FBI 2: 154; Gamble 334(236). *Hedysarum viscidum* L. Sp. Pl. 747. 1753.

An annual climbing herb, branchlets pubescent. Leaves 3-foliolate, leaflets ovoid-rhomboid, terminal leaflet large, lateral leaflets small, 4–6.5 × 2–5 cm, glabrous on both sides. Flowers pink in terminal lax racemes. Pods linear, 1.5 cm long, viscid, 4–6-seeded.

Occasional in forest and common in wet places. Fl. & Fr.: Sept.–Jan. Anantasagar beat, *MSM* 10501.

PSORALEA Linnaeus

Psoralea corylifolia L. Sp. Pl. 764. 1753; FBI 2: 103; Gamble 315(222).

An annual erect herb, up to 70 cm tall, branchlets glandular. Leaves 1-foliate, ovate, 4–6 × 3–4 cm, base cuneate, margin dentate, tip apiculate. Flowers purple in axillary racemes. Pods oblong, 1-seeded. Seeds reniform.

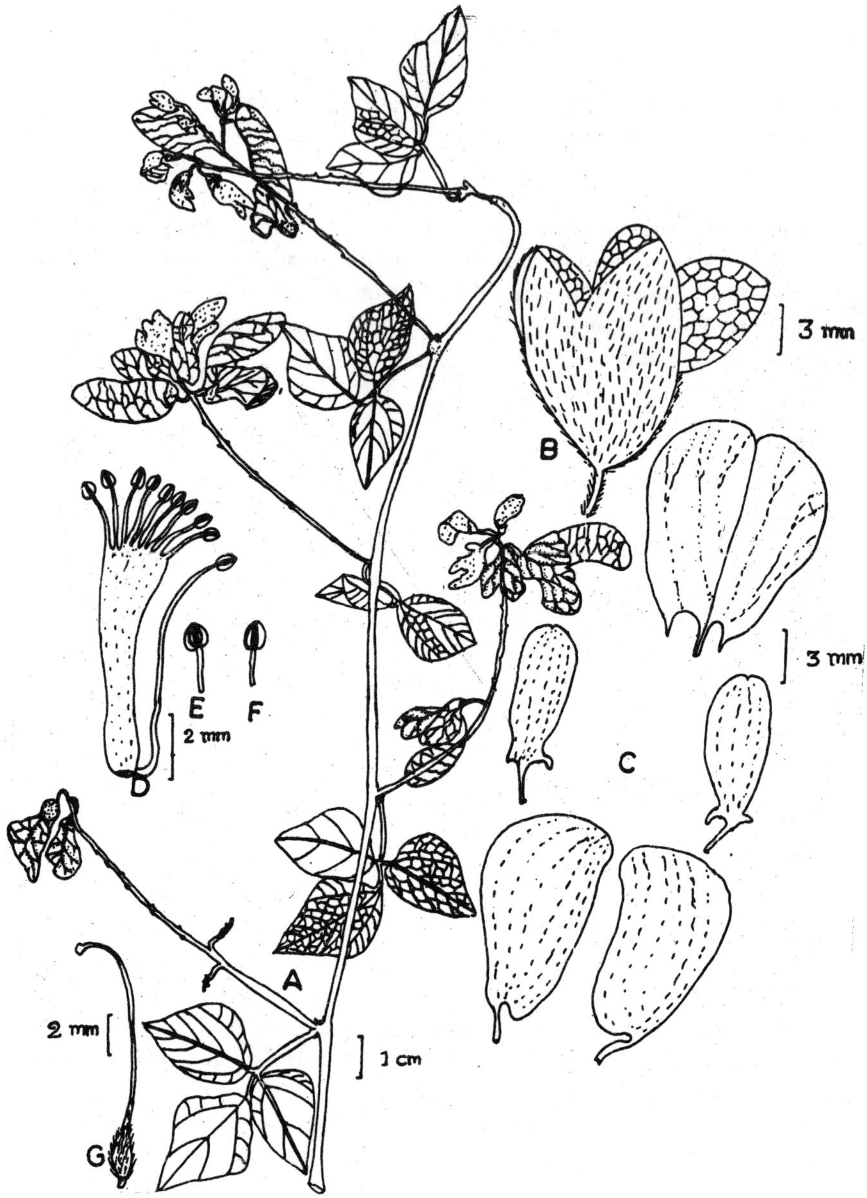

Fig. 8. Paracalyx scariosus (Roxb) Ali
A. Twig B. Calyx C. Corolla D. Staminal column E & F. Anthers G. Pistil.

Common in fields. Fl. & Fr.: Sept.–Dec. *Vern.*: Bhanchalam. Naskal, *MSM & PRSR* 15110.

PTEROCARPUS N.J. Jacquin *nom. cons.*

Pterocarpus marsupium Roxb. Pl. Cor. t. 116. 1799; FBI 2: 239; Gamble 385(271); Khan in FFH 136.

A deciduous tree, up to 15 m tall, bark grey. Leaves imparipinnate, leaflets 5–7, elliptic-oblong, 6–13.5 × 5–7.5 cm, base cuneate, margin entire, tip emarginate, glabrous above, pubescent below. Flowers greenish-yellow in axillary and terminal long panicles. Pods orbicular, winged, 1-seeded.

Occasional in forest. Fl. & Fr.: June–Dec. *Vern.*: Pedda Eegi. Naskal, *MSM & PRSR* 12800.

RHYNCHOSIA Loureiro *nom. cons.*

1. Flowers yellow with maroon streaks. Pod 1-seeded .... ***R. rufescens***
1. Flowers yellow without maroon streaks. Pod 2-seeded:
   2. Woody undershrub. Pods septate between seeds .... ***R. suaveolens***
   2. Woody stragglers. Pods not septate between seeds:
      3. Corolla yellow .... ***R. minima***
      3. Corolla purple .... ***R. viscosa***

Rhynchosia minima (L.) DC. Prodr. 2: 385. 1825; FBI 2: 223; Gamble 375(264). *Dolichos minimus* L. Sp. Pl. 726. 1753.

An annual trailing herb, stem pubescent. Leaves 3-foliolate, leaflets obovate-rhomboid, 1–1.5 × 1–2 cm, base obtuse, margin entire, tip truncate-rounded. Flowers yellow in axillary long racemes. Pods oblong, pubescent. 2-seeded.

Common in hedges. Fl. & Fr.: Aug.–Dec. Mekavarpalli, *MSM & TP* 11043.

Rhynchosia rufescens (Willd.) DC. Prodr. 2: 387. 1825; FBI 2: 220; Gamble 373(263). *Glycine rufescens* Willd. in Ges. Naturf. Freunde Berlin Neue Schriften 4: 222. 1803.

A twining shrub. Leaves 3-foliolate, leaflets elliptic-ovate, 2–4.5 × 1.5–3 cm, base rounded, margin entire, tip acute, pubescent on both sides. Flowers yellow with maroom streaks in long racemes. Pods ovoid, pointed at apex, 1-seeded.

Common in forest. Fl. & Fr.: July–Jan. Anantagiri RF, *MSM & PRSR* 12778.

Rhynchosia suaveolens (L.f.) DC. Prodr. 2: 387. 1825; FBI 2: 221; Gamble 374(264). *Glycine suaveolens* L.f. Suppl. Pl. 326. 1781.

An annual subshrub, up to 70 cm tall, branchlets viscid. Leaves 3-foliolate, leaflets ovate, 3–5 × 2–4 cm, base truncate, margin entire, tip acuminate. Flowers yellow, axillary, solitary. Pods oblong, pubescent, 2-seeded.

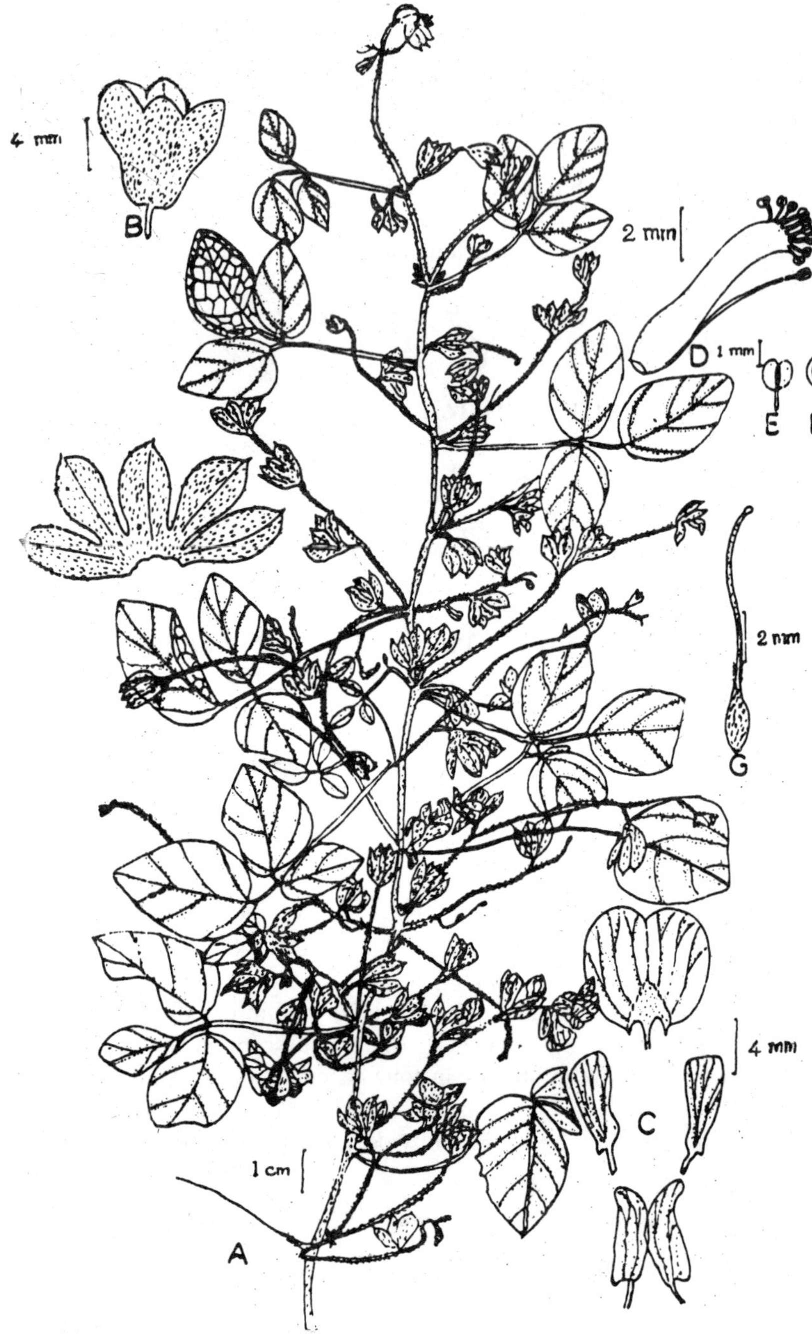

Fig. 9. **Rhynchosia rufescens** (Willd.) DC.
A. Twig B. Calyx C. Corolla D. Staminal column E & F. Anthers G. Pistil.

Occasional in forest. Fl. & Fr.: Aug.–Jan. *Vern.*: Karra ulava. Central University, *MSM* 18160.

Rhynchosia viscosa DC. Prodr. 2: 387. 1825; FBI 2: 225; Gamble 375(265).

An annual climbing shrub. Stem quadrangular, pubescent. Leaves 3-foliolate, leaflets ovate-deltoid, 5–9 × 4.5–8 cm, base sub-cordate, margin entire, tip obtuse-acuminate, pubescent above, so densely below. Flowers purple in axillary, long racemes. Pods oblong, pubescent, 2-seeded.

Occasional in hedges. Fl. & Fr. Nov.–Apr. Central University, *MSM* 15130.

ROTHIA Persoon

Rothia indica (L.) Druce in Bot. Exch. Club. Soc. Brit. Isles. 3: 423. 1914. *Trigonella indica* L. Sp. Pl. 778. 1753. *Rothia trifoliata* (Pers.) DC. Prodr. 2: 382. 1825; FBI 2: 63; Gamble 279(166). *Dilwynia trifoliata* Pers. Syn. Pl. 2: 302. 1807.

An annual diffuse herb, branchlets hirsute. Leaves 3-foliolate, leaflets oblanceolate, 2–2.5 × 0.5–0.8 cm, base attenuate, margin entire, tip acute, hispid on both sides. Flowers yellow, solitary. Pods linear, hairy, spine-tipped, many-seeded.

Occasional weed of cultivated lands. Central University, *MSM & TP* 15136.

SESBANIA Scopoli

Sesbania bispinosa (Jacq.) W. Wight in U.S.D.A. Bur. Pl. Industr. Bull. no. 137: 15. 1909. *Aeschynomene bispinosa* Jacq. Icon. Pl. Rar. 3: 13. t. 564. 179. *Sesbania aculeata* (Willd.) Poir. in Lam. Encycl. 7: 128. 1806; *nom. illegit*; FBI 2: 114; Gamble 323(228). *Coronilla aculeata* Willd. Sp. Pl. 3: 1147. 1802.

An armed tree, up to 4 cm tall, branchlets prickled, glabrous, latex watery. Leaves paripinnate, leaflets 7–10 pairs, oblong, 0.5–1.2 × 0.3–0.5 cm, base cuneate, margin entire, tip obtuse, glabrous on both sides. Flowers yellow in axillary racemes. Pods 28 cm long, torulose, many-seeded. Seeds oblong, brown.

Occasional in forest. Fl. & Fr.: Sept.–Jan. Anantasagar beat, *MSM & KH* 11013.

STYLOSANTHES O. Swartz

Stylosanthes fruticosa (Retz.) Alston in Trimen. Fl. Ceylon 6 (Suppl.) 77. 1931. *Arachis fruticosa* Retz. Obs. Bot. 5: 26. 1788. *Stylosanthes mucronata* Willd. Sp. Pl. 3: 1166. 1802 *nom. illegit.* FBI 2: 148; Gamble 326(230).

A prostrate herb up to 60 cm long, branchlets tomentose. Leaves 3-foliolate, leaflets oblong, elliptic, 0.5–1.5 × 0.2–0.5 cm, base subacute, margin ciliate, tip obtuse. Flowers yellow in axillary and terminal racemes. Pods oblong, beaked, 2–3-seeded. Seeds reniform.

Common in wastelands and fields. Fl. & Fr.: Oct.–Feb. Gacchibowli, *MSM* 12768.

TEPHROSIA Persoon *nom. cons.*

1. Prostrate herbs:
   2. Flowers white .... ***T. pumila***
   2. Flowers pink .... ***T. maxima***
1. Erect herbs:
   3. Inflorescence extra axillary .... ***T. purpurea***
   3. Inflorescence axillary:
      4. Branchlets velvety. Style pubescent .... ***T. tinctoria***
      4. Branchlets sericeous. Style glabrous .... ***T. villosa***

Tephrosia maxima Pers. Syn. 2: 329. 1807; Gamble 1: 329(224). *T. purpurea* (L.) Pers. var. *maxima* (Pers.) Baker in Hook. f. Fl. Brit. India 2: 113. 1876.

A prostrate herb, branchlets appressed pubescent. Leaves odd-pinnate, leaflets oblanceolate, 1–1.5 × 0.2–0.5 cm, base cuneate, margin entire, tip obtuse-retuse. Flowers pink in leaf-opposed axillary racemes. Pod sparsely pubescent, 10–12-seeded, strophiolate.

Occasional in open forest. Fl. & Fr.: June–Oct. Central University, *MSM* 15159.

Tephrosia pumila (Lam.) Pers. 2: 330. 1807. *Galega pumila* Lam. Encycl. 2: 599. 1788. *Tephrosia purpurea* (L.) Pers. var. *pumila* (Lam.) Baker in Hook. f. Fl. Brit. India 2: 113. 1876. *T. procumbens* (Buch.-Ham.) Benth. Gen. Index to Trans. of Linn. Soc. 101. 1856; Gamble 320(226).

A perennial prostrate herb, Leaves imparipinnate, leaflets obovate-oblanceolate, 1–2 × 0.2–0.5 cm, base cuneate, margin entire, tip truncate. Flowers white in axillary racemes. Pod pubescent, 10-seeded.

Occasional in open forest. Fl. & Fr.: June–Oct. Central University, *MSM* 15157.

Tephrosia purpurea (L.) Pers. Syn. Pl. 2: 329. 1807; FBI 2: 112, *p.p;* Gamble 320(226). *Cracca purpurea* L. Sp. Pl. 752. 1753.

An erect herb, up to 1 m tall, branchlets sparsely pubescent. Leaves imparipinnate, leaflets 4–9 pairs, obovate-oblanceolate, 1.8–3 × 0.5–1.5 cm, base cuneate, margin entire, tip retuse, glabrescent above, pubescent below. Flowers purple in simple racemes. Pods 4 cm long, downy-pubescent, 4–7-seeded.

Occasional in forests. Fl. & Fr.: All seasons. *Vern.:* Vempali. Mohammadabad beat, *MSM & KH* 10244; Lingampalle, *MSM* 12736; Central University, *MSM* 15137.

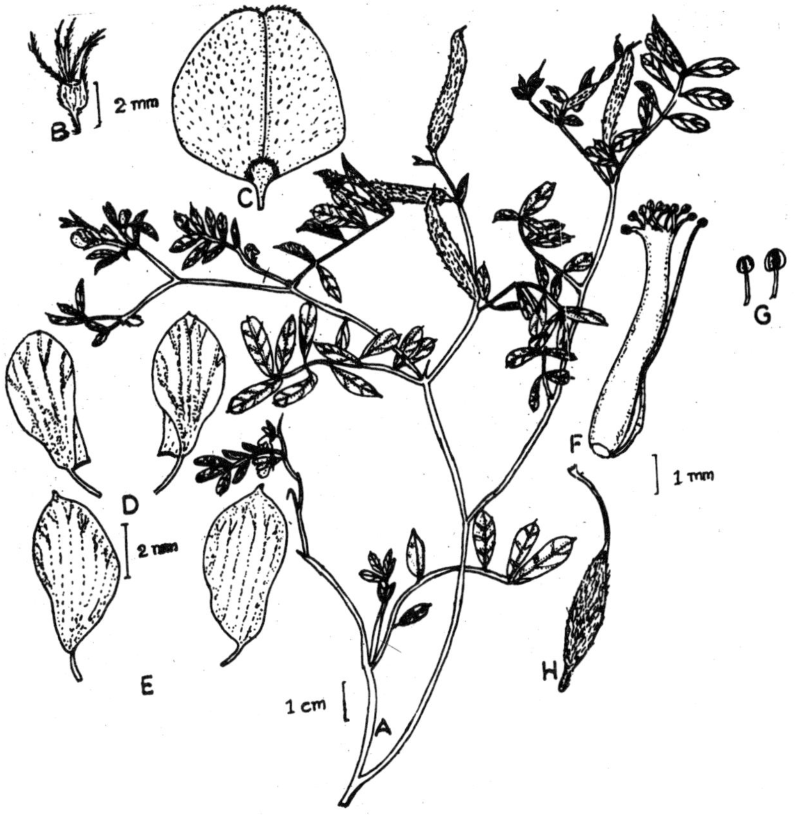

Fig. 10. **Tephrosia pumila** (Lam.) Pers.
A. Twig B. Calyx C. Standard petal D. Wing petals E. Keel petals F. Staminal column G. Anthers H. Pistil.

Tephrosia tinctoria Pers. Syn. 2: 329. 1807; FBI 2: 111; Gamble 319(225).

An annual erect herb, branchlets velvety. Leaves imparipinnate, leaflets oblong, elliptic, 1.5–6 × 0.5–2 cm, base obtuse, margin entire, tip retuse-obtuse. Flowers pink in axillary racemes. Pods linear, velvety. Seeds ellipsoid.

Rare in waste lands (Seshagiri Rao, 1986). Fl. & Fr.: July–Sept.

Tephrosia villosa L.Sp. Pl. 1: 752. 1754. *Tephrosia hirta* (Buch.-Ham.) Benth. Gen. Index to Trans. Linn. Soc. 101. 1866; Gamble 318(225). *Galega hirta* Buch.-Ham. in Trans. Linn. Soc. 13. 546. 1822.

An annual erect herb, up to 50 cm tall, branchlets villous. Leaves imparipinnate, leaflets obovate-oblong, 1–2 × 0.3–0.5 cm, base cuneate, margin entire, tip obtuse, glabrous above, pubescent below. Flowers purple in terminal racemes. Pods 3 cm long, falcate, apex slightly curved, villous, 8-seeded. Seeds sub-quadrate, strophiolate.

Common in forest undergrowth and roadsides. Fl. & Fr.: All seasons. Kusumasamudram beat, *MSM & KH* 11024.

TERAMNUS P. Browne

Teramnus labialis (L.f.) Spreng. in L. Syst. Veg. (ed. 16) 3: 235. 1826; FBI 2: 184; Gamble 352(249. ssp. **labialis** var. **labialis**; Verdc. in Kew Bull. 24. 267. 1970. *Glycine labialis* L.f. Suppl. Pl. 325. 1781.

A climbing herb, branchlets pubescent. Leaves 3-foliolate, terminal leaflet obovate, lateral inequilateral, 3–8 × 1.5–5 cm, base cuneate, margin entire, tip obtuse, glabrous above, appressed pubescent below. Flowers pink in axillary racemes. Pods 7 cm long, linear, pubescent, hooked, 8–12-seeded. Seeds oblong, brown.

Occasional in forests. Fl. & Fr.: Sept.–Jan. Mohammadabad beat, *MSM & KH* 10598.

VIGNA Savi

1. Pods 6–8-seeded .... *V. aconitifolia*
1. Pods 10–16-seeded .... *V. trilobata*

Vigna aconitifolia (Jacq.) Marechal in Bull. Jard. Bot. Nat. Berg. 39. 160. 1969; *Phaseolus aconitifolius* Jacq. Obs. Bot. 3: 2t. 52. 1766; FBI 2: 202; Gamble 363(256).

An annual trailing herb, branchlets hispid. Leaves 3-foliolate, leaflets dissected, 3–6-lobed, elliptic-lanceolate, to 2.5 × 0.5 cm, hairy, acute. Flowers yellow in axillary, hispid, long racemes. Pods cylindrical, pubescent apiculate, 6–8-seeded.

Common weed of waste lands (Seshagiri Rao, 1986). Fl. & Fr.: Nov.–Jan.

Vigna trilobata (L.) Verdc. in Taxon 17: 172. 1968. *Dolichos trilobatus* L. Mant. Pl. 101. 1767. *Phaseolus trilobus* auct. non. L. FBI 2: 201; Gamble 363(256).

An annual trailing herb, branchlets glabrous. Leaves 3-foliolate, leaflets 3-lobed, 2.3–4 × 2.5–3 cm, base rounded, margin entire, tip acute, glabrescent on both sides. Flowers yellow in axillary long racemes. Pods cylindric, pubescent, 10–16-seeded.

Common weed of waste and cultivated land (Seshagiri Rao, 1986). Fl. & Fr.: Sept.–Dec.

ZORNIA J.F. Gmelin

Zornia gibbosa Spanog in Linnaea 15: 192. 1841. *Z. diphylla* auct. non. L., FBI 2: 147; Gamble 325(229).

An annual prostrate herb, branchlets glabrous. Leaves 2-foliolate, leaflets lanceolate, 2–2.6 × 0.5–0.7 cm, base rounded, margin entire, tip acute, stipules to 1 cm. Flowers yellow in long axillary and terminal racemes. Pods moniliform with glochidiate prickles, 3–5-seeded.

Occasional in forests. Fl. & Fr.: July–Oct. Dharur RF, *MSM* & TP 11092; Lingampalle, *MSM* 12741.

Cajanus cajan Millsp. (*Vern.:* Kandulu), *Cicer arietinum* L. (*Vern.:* Senagalu), *Vigna mungo* (L.) Hepp. (*Vern.:* Minumulu) and *V. radiata* (L.) Wilcz. (*Vern.*: Pesalu) are common pulses cultivated in the district. *Arachis hypogaea* L. (*Vern.*: Verusenaga) is cultivated for its oil seeds. *Cyamopsis tetragonoloba* L. (*Vern.* Goruchikkudu), *Lab lab purpureus* (L.) Sweet (*Vern.*: Anapa), *Phaseolus vulgaris* L. (*Vern.:* Beans) are vegetable crops. *Sesbania grandiflora* (L.) Poir (*Vern.:* Avisa) is planted in fields and leaves are used as vegetable. *Trigonella foenum-graecum* L. (*Vern.:* Menthulu) is cultivated for seeds which are used as condiment.

CAESALPINIACEAE

1. Leaves simple or unipinnate:
    2. Leaves simple, deeply emarginate .... ***Bauhinia***
    2. Leaves once pinnate:
        3. Pod with hard rind .... ***Tamarindus***
        3. Pod without hard rind .... ***Cassia***
1. Leaves bipinnate:
    4. Trees:
        5. Branchlets and pods reddish brown .... ***Peltophorum***
        5. Branchlets and pods not as above .... ***Delonix***
    4. Straggling or erect shrubs:
        6. Pods samaroid .... ***Pterocarpus***
        6. Pods not samaroid:
            7. Rachis modified into phyllode .... ***Parkinsonia***
            7. Rachis not modified into phyllode .... ***Caesalpinia***

BAUHINIA Linnaeus

1. Trees or shrubs:
    2. Flowers white. Pod twisted .... ***B. racemosa***
    2. Flowers pink-purple. Pod not twisted .... ***B. purpurea***
1. Lianas .... ***B. vahlii***

Bauhinia purpurea L. Sp. Pl. 375. 1753; FBI 2: 20; Gamble 407(288); Khan in FFH 149.

A deciduous tree, up to 10 m tall, bark grey, branchlets pubescent. Leaves oblong, 8–14 × 8–12 cm, base cordate, margin entire, tip emarginate to half of the length, glabrous on both sides. Flowers purple in axillary and terminal racemes. Pods oblong, glabrous, horned, 6–10-seeded.

Occasional in villages and forests, also planted as an avenue tree. Fl. & Fr.: July–Dec. Lingampalle, *MSM* 12758.

Bauhinia racemosa Lam. Encycl. 1: 390. 1785; FBI 2: 276; Gamble 406(288); Khan in FFH 147. *Piliostigma racemosa* (Lam.) Benth. in Miq. Pl. Jungh. 262. 1852.

A deciduous tree, up to 5 m tall, bark dark brown. Leaves obcordate with two leaflets connate for two third of their length, 2.5–3 × 3.5–5 cm, base cordate, margin entire, tip emarginate, glabrous above, more or less tomentose below. Flowers white in axillary racemes. Pods 15–22 cm long, twisted, glabrous, 10–18-seeded, black.

Common in all forests. Fl. & Fr.: Oct.–June. *Vern.:* Aari. Anantasagar beat, *MSM & NV* 10415.

Bauhinia vahlii Wight & Arn. Prodr. 297. 1834; FBI 2: 279; Gamble 406(288); Khan in FFH 149.

A large leaved straggling shrub. Leaves auriculate, 10–20 × 10–25 cm, bilobed, lobes obtuse, rusty pubescent. Flowers pale yellow in corymbose racemes. Pods flat, rusty tomentose.

Occasional in forest. Fl. & Fr.: Dec.–Mar. *Vern.:* Istaripala. Anantagiri, RF, *MSM & PRSR* 12786.

CAESALPINIA Linnaeus

1. Armed stragglers:
   2. Branches prickled. Pods exceeding 1.5 cm across:
      3. Pinnae 6–7 pairs. Pods prickled .... ***C. bonduc***
      3. Pinnae 10–15 pairs. Pods smooth .... ***C. decapetala***
   2. Branches not prickled, pods not exceeding 1.5 cm across .... ***C. pulcherrima***
1. Trees .... ***C. coriaria***

Caesalpinia bonduc (L.) Roxb. Fl. Ind. 2: 362. 1832, emend. Dandy & Exell in J. Bot. 76. 175. 1938. *Guilandinia bonduc* L. Sp. Pl. 381. 1753. *Caesalpinia crista* L. Sp. Pl. 380. 1753; Gamble 393(278). *C. bonducella* Flem. in Asiat. Res. 11: 159. 1810. *nom. illegit.* FBI: 2: 254; Khan in FFH 139.

An armed straggling shrub, up to 4 m tall, branchlets pubescent, prickles recurved. Leaves bipinnate, leaflets 6–8 pairs, elliptic, oblong, 2–3.3 × 1.5–2.2 cm, base truncate, margin entire, tip mucronate, glabrous on both sides. Flowers yellow in axillary long racemes. Pods oblong, prickled, 1–2-seeded.

Occasional in field hedges. Fl. & Fr.: Aug.–Feb. Mekavarpalli, *MSM & TP* 11050.

Caesalpinia coriaria (Jacq.) Willd. Sp.Pl. 2: 532. 1799; Gamble 394 (277).

An unarmed tree, up to 10 m tall, leaflets oblong, 0.4–0.7 × 0.1–0.3 cm base oblique, margin entire, tip emarginate. Flowers greenish-yellow in terminal panicles. Pods twisted, glabrous.

Occasional in forest. Fl. & Fr. June–Nov. *Vern.:* Divi-divi. Vikarabad, *MSM & PRSR* 12781.

Caesalpinia decapetala (Roth) Alston in Trimen. Handb. Fl. Ceylon 6 (Suppl.): 89. 1931. *Reichardia decapetala* Roth. Nov. Pl. Spec. 212. 1821. *Caesalpinia sepiaria* Roxb. Fl. Ind. 2: 360. 1832; FBI 2: 256; Gamble 394(279); Khan in FFH 140.

An armed straggling shrub, up to 4 m tall, branchlets pubescent, prickles recurved. Leaves bipinnate, leaflets 8–12 pairs, elliptic-oblong, 0.6–1.5 × 0.3–0.5 cm, base rounded, margin entire, tip rounded, sparsely pubescent on both sides. Flowers yellow in axillary and terminal long racemes. Pods oblong, pubescent, apex horned, 6–8-seeded.

Occasional in hedges and along roadsides. Fl. & Fr.: Oct.–Mar. Pargi, *MSM & PRSR* 12799.

Caesalpinia pulcherrima (L.) Sw. Obs. Bot. 165. 1791; FBI 2: 255; Gamble 394(279). *Poinciana pulcherrima* L. Sp. Pl. 380. 1753.

An armed straggler, up to 5 m tall, Leaves bipinnate, leaflets oblong, 1–1.5 × 0.5–0.8 cm, base acute, margin entire, tip retuse, glabrescent above, glaucous below. Flowers red in terminal corymbs. Pods oblong, 8 cm long, compressed, 2–8-seeded. Seeds ovoid.

Occasional on roadsides. Fl. & Fr.: July–Sept. Anantasagar beat, *MSM & NV* 10384.

CASSIA Linnaeus

1. Herbs:
    2. Prostrate herbs:
        3. Branchlets and pods downy villous. Leaflets 10–20 pairs .... ***C. pumila***
        3. Branchlets and pods pubescent. Leaflets 25–40 pairs .... ***C. mimosoides***
    2. Erect herbs:
        4. Leaves with 2 pairs of leaflets. Flowers red. Pod 6-seeded .... ***C. absus***
        4. Leaves with 3 pairs of leaflets. Flowers yellow. Pod many-seeded ..... ***C. tora***
1. Shrubs or trees:
    5. Trees:
        6. Flowers in erect corymbose panicles. Pods 20–25 cm long .... ***C. siamea***
        6. Flowers in drooping panicles. Pods 40–60 cm long .... ***C. fistula***
    5. Shrubs:
        7. Stipules foliaceous .... ***C. auriculata***
        7. Stipules not foliaceous:
            8. Rachis with glands up to third pair .... ***C. suffruticosa***
            8. Rachis not as above:
                9. Pinnae with 5–6 pairs of leaflets. Flowers yellow................ ***C. occidentalis***
                9. Pinnae with 8–10 pairs of leaflets. Flowers red .... ***C. sophera***

Cassia absus L. Sp. Pl. 376. 1753; FBI 2: 265; Gamble 403(285); Khan in FFH 145.

An erect subshrub, up to 60 cm tall, branchlets viscid pubescent. Leaves unipinnate-paripinnate, leaflets ovate-obovate, inequilateral, 1.5–2.5 × 1–1.5 cm, base cuneate, margin entire, tip obtuse, pubescent on both sides. Flowers red in terminal and leaf opposed racemes. Pods short strigose, 6-seeded. Seeds ovoid.

Occasional in forest undergrowth. Fl. & Fr.: Aug.–Oct. Lingampalle, *MSM* 12737; Anantagiri, *MSM & TP* 11066.

Cassia auriculata L. Sp. Pl. 379. 1753; FBI 2: 263; Gamble 402(284). *Senna auriculata* Roxb. Fl. Ind. 2: 349. 1832; Khan in FFH 143.

A perennial erect shrub, up to 2 m tall, branchlets densely tomentose. Leaves paripinnate, leaflets 7–12 pairs, oblong, 2–3 × 0.8–1.5 cm, base cuneate, margin entire, tip mucronate, pubescent on both sides. Flowers yellow in axillary and terminal corymbs. Pods up to 10 cm long, flat, thin, many-seeded. Seeds ovoid, brown.

Common in forest undergrowth. Fl. & Fr.: All seasons. *Vern.*: Thangedu. *Uses*: Bark is used for tanning. Kusumasamudram beat, *MSM & KH* 11021.

Cassia fistula L. Sp. Pl. 377. 1753; FBI 2: 261; Gamble 400(283); Khan in FFH 143. *Cassia rhombifolia* Roxb. Fl. Ind. 2: 334. 1832.

A deciduous tree, up to 8 m tall, bark smooth, pale grey when young, dark brown when old. Leaves paripinnate, leaflets 5 pairs, oblong-broadly ovate, inequilateral, 4–12 × 2–7 cm, base cuneate, margin entire, tip acute, glabrous above, pubescent below. Flowers yellow in drooping panicles. Pods 40–60 cm long, glabrescent, many-seeded.

Common in forests. Fl. & Fr.: May–Oct. *Vern.*: Rela. Mohammadabad beat, *MSM & KH* 10230.

Cassia mimosoides L. Sp. Pl. 379. 1753; FBI 2: 266; Gamble 403(285).

An annual diffuse herb, branchlets sparsely pubescent. Leaves paripinnate, leaflets 25–40 pairs, oblong, inequilateral, 0.2–0.6 × 0.1 cm, obtuse, margin entire, tip obtuse, glabrous on both sides. Flowers yellow, axillary, solitary. Pods 3 cm long, pubescent, 15-seeded. Seeds ellipsoid.

Occasional in forest undergrowth. Fl. & Fr.: All seasons. Rangammagudem beat, *MSM* 10472; Lingampalli, *MSM* 12746.

Cassia occidentalis L. Sp. Pl. 377. 1753; FBI 2: 262; Gamble 401(284); Khan in FFH 145. *Senna occidentalis* Roxb. Fl. Ind. 2: 343. 1832.

An annual erect subshrub, up to 1 m tall, branchlets glabrescent. Leaves unipinnate, paripinnate, leaflets 4–5 pairs, ovate-lanceolate, glands on the base of petiole, 2–6.5 × 1.1–2.5 cm, base oblique, margin entire, tip acuminate, glabrous on both sides. Flowers yellow in terminal racemes. Pods 10 cm long, glabrous, compressed, many-seeded. Seeds brown.

Common in waste places, road sides. Fl. & Fr.: May–Dec. *Vern.*: Cashanda. Rangammagudem, *MSM & KH* 10323; Mohammadabad beat, *MSM & KH* 10596.

Cassia pumila Lam. Encycl. 1: 651. 1785; FBI 2: 266; Gamble 403(285). *Senna prostrata* Roxb. Fl. Ind. 2: 352. 1832.

A prostrate herb, up to 60 cm, branchlets downy villous. Leaves unipinnate, paripinnate, leaflets 10–20 pairs, oblong-lanceolate, 0.5 × 0.2 cm, base cuneate, margin ciliate, tip mucronate, petiole with a stalked gland, glabrous above, sparsely villous below. Flowers yellow, axillary, solitary. Pods 3.5 cm long, compressed, downy villous, horned, 10–15-seeded.

Rare in forests. Fl. & Fr.: All seasons. Kanmahkalva beat, *MSM* 10454.

Cassia siamea Lam. Encycl. 1: 648. 1785; FBI 2: 264; Gamble 402(285); Khan in FFH 144. *Cassia florida* Vahl, Symb. Bot. 3: 57. 1794.

A deciduous tree, up to 10 m tall, branchlets glabrescent. Leaves paripinnate, leaflets elliptic-oblong, 7–10 pairs, 2.5–6.5 × 1.5–2.5 cm, base obtuse, margin entire, tip mucronate, glabrous on both sides. Flowers yellow in terminal corymbose panicles. Pods 20–25 cm long, compressed, many-seeded. Seeds dark brown.

Common in forests. Fl. & Fr.: May–Oct. *Vern.*: Nela thangedu. Mohammadabad beat, *MSM & KH* 10209.

Cassia sophera L. Sp. Pl. 379. 1753; FBI 2: 262; Gamble 401(284); Khan in FFH 145.

A shrub, up to 2 m tall, branches glabrescent. Leaves paripinnate, leaflets 8–10 pairs, ovate-lanceolate or oblong, 2–4 × 0.5–1 cm, base cuneate-obtuse, margin entire, tip acute, glabrous on both sides. Flowers orange-red in axillary and terminal corymbose racemes. Pods oblong, apically horned, 6–10-seeded.

Occasional on roadsides. Fl. & Fr.: Aug.–Dec. Manchanpalle, *MSM & TP* 11100.

Cassia suffruticosa Koenig ex Roth Nov. Pl. Sp. 213. 1821; Singh in Bull. Bot. Surv. India 18: 19. 1976; *C. glauca* Lam. *var. suffruticosa* (Koenig ex Roth.) Baker in Hook. f. Fl. Brit. India 2: 265. 1878; Gamble 402(285).

A shrub, up to 4 m tall. Leaves pinnate, leaflets oblong, elliptic, 1.5–4 × 0.5–1.5 cm, glabrescent below, rachis with glands up to third pair of leaflets from base. Flowers yellow in axillary racemes. Pods flat, compressed.

Rare in deciduous forest. Fl. & Fr.: Oct.–Jan. Anantagiri RF, *MSM & PRSR* 12788.

Cassia tora L. Sp. Pl. 376. 1753; FBI 2: 263; Gamble 40(284); Khan in FFH 145.

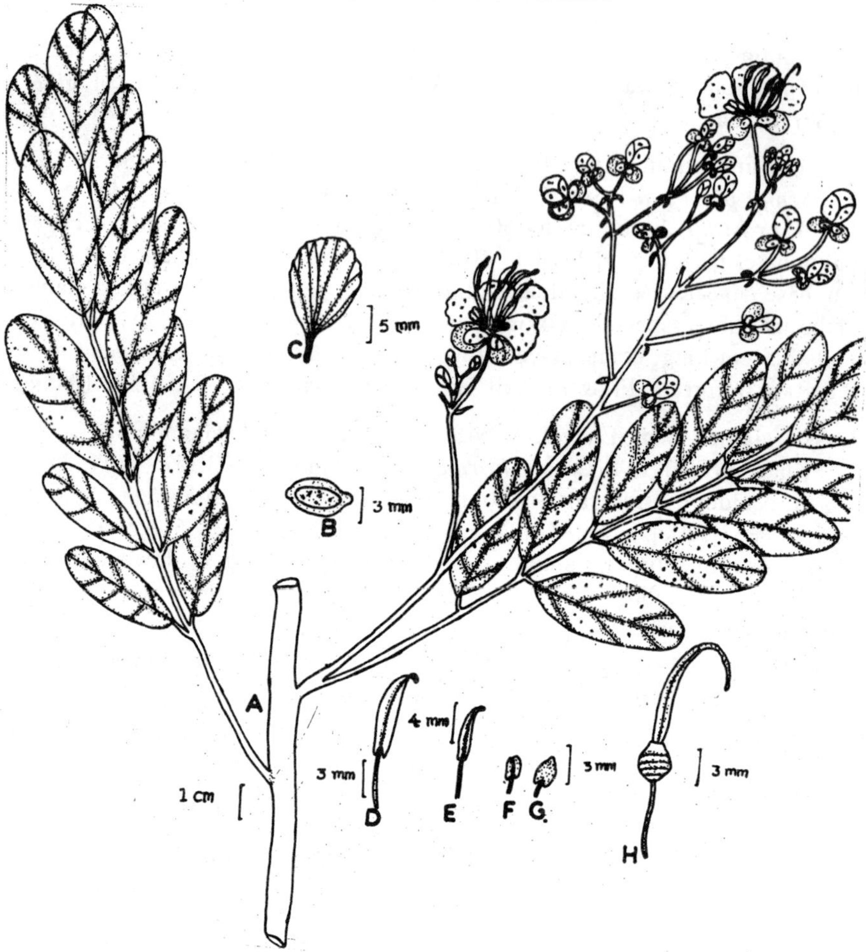

Fig. 11. Cassia siamea L.
A. Twig B. Sepal C. Petal D–F. Stamens G. Staminode H. Pistil.

A subshrub, up to 50 cm tall, branchlets glabrous. Leaves unipinnate-paripinnate, leaflets obovate, 1–3 × 1–1.5 cm, base oblique, margin entire, tip obtuse, glabrous above and pubescent below. Flowers yellow in terminal corymbose racemes. Pods compressed, glabrous, septate between seeds, many-seeded. Seeds oblong.

Common on roadsides. Fl. Fr.: All seasons. Dharur, *MSM* 11093; Gacchibowli, *MSM* 12770.

DELONIX Rafinesque

1. Flowers yellow, petals with short claw. Pod <15 cm long .... *D. elata*
1. Flowers red, petals with long claw. Pod > 20 cm long .... *D. regia*

Delonix elata (L.) Gamble Fl. Pres. Madras 396. 1919 & 280. 1957 (repr. ed.) *Poinciana elata* L. Cent. Pl. 16. 1756; FBI 2: 260.

A tree, up to 10 m tall. Leaves bipinnate up to 15 cm long, pinnae 7–9 pairs, leaflets 20–25 pairs, oblong-elliptic, 0.5–1.3 × 0.2–0.4 cm, base cuneate, margin entire, tip obtuse. Flowers yellow in terminal racemes. Pods 15 cm long, horned, sutures grooved, many-seeded.

An avenue tree occasionally found as wild. Fl. & Fr.: July–Oct.

Delonix regia (Boj. ex Hook.) Rafin Fl. Tellur. 2: 92. 1837; Gamble 396(280). *Poinciana regia* Boj. ex Hook. in Bot. Mag. t. 2884. 1829; FBI 2: 260.

A deciduous tree. Leaves bipinnate up to 30 cm long, leaflets 12–30 pairs, oblong-elliptic, 0.5–1 × 0.1–0.3 cm, base attenuate, margin entire, tip obtuse. Flowers red in corymbs. Pods 20 cm long, flat, many-seeded.

An avenue tree planted in gardens. Fl. & Fr.: Jan.–May.

PARKINSONIA Linnaeus

Parkinsonia aculeata L. Sp. Pl. 375. 1753; FBI 2: 260; Gamble 397(281); Khan in FFH 142.

An armed shrub, up to 4 m tall, branchlets glabrous. Leaves bipinnate, leaflets 30–35 pairs, oblanceolate, 0.2–0.5 × 0.3 cm, base cuneate, margin entire, tip obtuse. Flowers yellow in 10–20 cm long, axillary racemes. Pods torulose, glabrous, 4–6-seeded.

Common in hedges on roadsides. Fl. & Fr. Oct.–Jan. Gacchibowli, *MSM* 12767.

PELTOPHORUM (Vogel) Bentham *nom. cons.*

Peltophorum pterocarpum (DC.) Baker ex K. Heyne Nutt. Pl. Ned. Ind (ed. 2) 2: 755. 1927. *Inga pterocarpa* DC., Prodr. 2: 441. 1825.

An evergreen tree up to 15 m tall, branchlets reddish brown tomentose. Leaves bipinnate, leaflets 11–30 pairs, obliquely oblong, 0.6–1.5 × 0.4–0.6 cm, sessile, base unequal, subcoriaceous. Flowers yellow in terminal long panicled racemes. Pods oblong, narrowed at both ends, 1–4-seeded.

Planted along roadsides (Seshagiri Rao, 1986). Fl. & Fr.: All seasons.

PTEROLOBIUM Robert Brown ex Wight & Arnott

Pterolobium hexapetalum (Roth) Sant. & Wagh in Bull. Bot. Surv. India 5: 108. 1963. *Reichardia hexapetala* Roth, Nov. Pl. Sp. 210. 1821. *Pterolobium lacerans* Wall. ex Wight & Arn. Prodr. 283. 1834. *P. indicum* A. Rich, Tent. Fl. Abyss. 1: 247. 1847; FBI 2: 259; Gamble 395(280).

An armed straggler, up to 10 m, branchlets tomentose, prickles recurved. Leaves bipinnate, leaflets 8–10 pairs, oblong, 1–1.5 × 0.5–0.8 cm, base cuneate, margin entire, tip obtuse, glabrous on both sides. Flowers white in axillary racemes. Pods samaroid, 4 cm long, apically winged, glabrous, 1-seeded. Seeds obovoid.

Common in forest. Fl. & Fr.: May–Sept. Saikunta, *MSM* 10492.

TAMARINDUS Linnaeus

Tamarindus indica L. Sp. Pl. 34. 1753; FBI 2: 273; Gamble 409(290); Khan in FFH 146.

A deciduous tree up to 10 m tall. Leaves bipinnate, leaflets oblong, 1.5–1.8 × 0.4–0.5 cm, base oblique, margin entire, tip obtuse, glabrous on both sides. Flowers yellow in terminal racemes. Pod oblong, compressed.

Common along roadsides. Fl. & Fr.: June–Dec. Central University, *MSM* 19213.

MIMOSACEAE

1. Plants armed:
    2. Flowers 4-merous, stamens 4 or 8 .... *Mimosa*
    2. Flowers 5-merous, stamens 10 or numerous:
        3. Flowers in globose heads:
            4. Pinnae 2 pairs. Pods twisted .... *Pithecellobium*
            4. Pinnae more than 2 pairs. Pods not twisted .... *Acacia*
        3. Flowers in spikes:
            5. Pods turgid, straight. Flowers single coloured .... *Prosopis*
            5. Pods thin, twisted. Flowers bicoloured .... *Dichrostachys*
1. Plants unarmed:
    6. Stamens numerous:
        7. Flowers in globose heads. Pods dehiscent, straight, shiny .... *Albizia*
        7. Flowers in corymbose panicles. Pods indehiscent, curved, not shiny .... *Samanea*
    6. Stamens 10 .... *Leucaena*

ACACIA P. Miller

1. Straggling shrubs:
    2. Branchlets rusty pubescent .... *A. torta*
    2. Branchlets glabrous .... *A. caesia*
1. Trees or erect shrubs:
    3. Flowers in spikes .... *A. chundra*
    3. Flowers in globose heads:
        4. Pods moniliform .... *A. nilotica*
        4. Pods not moniliform:
            5. Flower heads in panicles .... *A. leucophloea*
            5. Flower heads on axillary peduncles .... *A. farnesiana*

Acacia caesia (L.) Willd. Sp. Pl. 4: 1090. 1806; Gamble 428(308). *Mimosa caesia* L. Sp. Pl. 522. 1753. *Acacia intisia* L. var. *caesia* Baker in FBI 2: 297. *A. intsia.* auct. non Willd; FBI 2: 297; Khan in FFH 160.

An armed straggling shrub, branchlets glabrous, prickles straight. Leaves bipinnate, leaflets 15–25 pairs, oblong, 0.5–0.7 × 0.2 cm, base oblique, margin entire, tip obtuse. Flowers yellow, heads in long terminal panicles. Pods oblong, glabrous, horned, 10–12-seeded.

Occasional in deciduous forests and Scrub jungle (Seshagiri Rao, 1986). Fl. & Fr.: July–Dec. *Vern.*: Korinta.

Acacia chundra (Roxb. ex Rottl.) Willd. Sp. Pl. 4: 1978. 1806. *Mimosa chundra* Roxb. ex Rottl. in Ges. Naturf. Freunde Berlin Neue Schriften 4: 207. 1803. *Acacia sundra* DC. Prodr. 2: 458. 1825; FBI 2: 295; Gamble 428(303). *A. catechu* Willd. var. *sundra* Prain in J. Asiat. Soc. Beng. Pl. 2. 66: 508–510. 1898; Khan in FFH 159 (excl. var).

A deciduous armed tree, up to 8 m tall; bark-greyish brown, rough, branchlets glabrous, stipular thorns hooked. Leaves bipinnate, leaflets 30–40 pairs, oblong, 0.3–0.7 × 0.1–0.2 cm, base oblique, margin entire, tip acute. Flowers white in axillary long spikes. Pods oblong, compressed, glabrous, shortly stipitate, 6–8-seeded.

Common in forest. Fl. & Fr.: Aug.–Dec. *Vern.:* Sundra. Anantagiri RF, *MSM & TP* 11074.

Acacia farnesiana (L.) Willd. Sp. Pl. 4: 1083. 1806; FBI 2: 293; Gamble 425(301); Khan in FFH 156. *Mimosa farnesiana* L. Sp. Pl. 521. 1753. *Vachellia farnesiana* (L.) Wight & Arn. Prodr. Fl. Ind. Orient. 272. 1834.

An armed deciduous tree, up to 6 m tall, branchlets warty, bark brown. Leaves bipinnate, paripinnate, leaflets linear-oblong, 10–15 pairs, 0.5 × 0.1 cm, base truncate, margin entire, tip obtuse, stipular thorns unequal, straight, glabrous on both sides. Flowers yellow in axillary globose heads. Pods 5 cm long, glabrous, 5–6-seeded.

Occasional in foot hills and roadsides. Fl. & Fr.: May–Oct. *Vern.*: Murkitumma. Mokarlabad, *MSM & KH* 10262.

Acacia leucophloea (Roxb.) Willd. Sp. Pl. 4: 1083. 1806; FBI 2: 294; Gamble 427(302); Khan in FFH 158. *Mimosa leucophloea* Roxb. Pl. Cor. t. 150. 1800.

An armed tree, up to 10 m tall, branchlets tomentose. Leaves bipinnate, leaflets 20–39 pairs, oblong, 0.3–0.5 × 0.1 cm, base oblique, margin entire, tip acute. Flowers creamish-white in globose heads. Pods linear, pubescent, 8–10-seeded.

Common in forest and waste lands. Fl. & Fr.: June–Oct. *Vern.:* Tella tumma. Lingampalle, *MSM* 12747.

Acacia nilotica (L.) Del. Descr. Egypte. Hist. Nat. 111. 79. 1813. ssp. ***indica*** (Benth.) Brenan in Kew Bull. 12: 84. 1951. *Mimosa nilotica* L. Sp. Pl. 521. 1753. *Acacia arabica* auct. non. (Lam.) Willd. FBI 2: 293; Gamble 425(301); Khan in FFH 156. var. *indica* Benth. in Hook. J. Bot. Kew Gard. Misc. 1: 500. 1842. *Mimosa arabica* Lam. Encycl. 1: 19. 1783.

An armed tree, up to 6 m tall, branchlets grey-pubescent, bark rough, black. Leaves bipinnate, paripinnate, leaflets linear-oblong, 15–20 pairs, 0.5 × 0.1 cm, base oblique, margin entire, tip rotund. Stipular thorns straight, glabrous above, glaucous below. Flowers yellow in axillary globose heads. Pods 18 cm long, glaucous, 10–16-seeded. Seeds brownish black.

Very common in dry arid plains and roadsides. Fl. & Fr.: May–Aug. *Vern.:* Nalla tumma. Gadirayal beat, *MSM & KH* 10251.

Acacia torta (Roxb.) Craib in Bull. Misc. Inform. 1915. 410. 1915: Gamble 428(304). *Mimosa torta* Roxb. Fl. Ind. 2: 566. 1832.

A prickly straggling shrub, up to 6 m, branchlets yellowish brown pubescent. Leaves bipinnate, paripinnate, leaflets linear-oblong, unequal sided, 25–40 pairs, 0.4–0.7 × 0.1–0.4 cm, base truncate, margin entire, tip subacute, pubescent on both sides, hooked prickles present on rachis and peduncle. Flowers yellow, heads in terminal panicles. Pods 8 cm long, glabrous, 5–10-seeded.

Occasional in forests and hedges. Fl. & Fr.: May–Aug. Gadirayal beat, *MSM & KH* 10258.

ALBIZIA Durazzini

1. Pods rusty tomentose .... ***A. amara***
1. Pods glabrous:
   2. Pods smooth shiny .... ***A. lebbeck***
   2. Pods not as above:
      3. Pods 10–12-seeded .... ***A. odoratissima***
      3. Pods 5-seeded .... ***A. chinensis***

Albizia amara (Roxb.) Boivin in Encycl. 19: 34. 1838; FBI 2: 301. 1878; Gamble 432(306); Khan in FFH 163. ssp. **amara**; Brenan in Kew Bull. 10: 189. 1955. *Mimosa amara* Roxb. Pl. Cor. t. 122. 1799.

A deciduous tree, up to 8 m tall, branchlets pubescent, bark dark grey. Leaves bipinnate, paripinnate, leaflets linear-oblong, 20–25 pairs, 0.5 × 0.1 cm, base subacute, margin sparsely ciliate, tip obtuse, glabrous above, pubescent below. Flowers yellow in axillary heads. Pods 15 cm long, rusty tomentose, 8–10-seeded. Seeds ovoid.

Common in forests. Fl. & Fr.: May–June. *Vern.:* Narling. Mohammedabad beat, *MSM & KH* 10210, 10223.

Albizia chinensis (Osbeck) Merr. in Amer. J. Bot. 3: 475. 1916. *Mimosa chinensis* Osbeck Dagb. Ostind resa 233. 1757. *M. emarginata* Lam. Encycl. 1: 12. 1783; Gamble 433(307).

A deciduous tree, up to 20 m tall, branchlets pubescent. Leaves paripinnate, leaflets 25–30 pairs, elliptic, 5–6.5 × 1.5–2 cm,base oblique, margin ciliate, tip acute. Flowers in axillary and terminal panicles. Pods flat, compressed, 5-seeded.

Occasional in forest. Fl. & Fr.: Mar.–May. Anantagiri RF, *MSM & SSR* 18510.

Albizia lebbeck (L.) Benth. in Hook. London J. Bot. 3: 87. 1844; FBI 2: 298; Gamble 432(306); Khan in FFH 161. *Mimosa lebbeck* L. Sp. Pl. 516. 1753.

A deciduous tree, up to 15 m tall, bark grey, branchlets glabrous. Leaves bipinnate, leaflets, 4–8 pairs, oblong-obovate, 1.5–3.5 × 1–2 cm, base obliquely obtuse, margin entire, tip obtuse-retuse. Flowers greenish-yellow in axillary heads. Pods oblong, glabrous, shiny, 6–10-seeded.

Common in villages and roadsides. Fl. & Fr.: June-Dec. *Vern.:* Dirishanam. Central University, *MSM* 19224.

Albizia odoratissima (L.f.) Benth. in Hook. London J. Bot. 3: 88. 1844; FBI 2: 299; Gamble 431(306); Khan in FFH 162. *Mimosa odoratissima* L.f. Suppl. Pl. 437. 1781. *Acacia odoratissima* Willd. Sp. Pl. 4. 1063. 1806.

An unarmed deciduous tree, up to 15 m tall, branchlets appressed-pubescent. Leaves bipinnate, paripinnate, leaflets 8–15 pairs, obliquely oblong-elliptic, 0.7–2.5 × 0.4–0.8 cm, base oblique, margin entire, tip obtuse, pubescent above, glaucous and pubescent below. Flowers pale yellow, heads in terminal panicles. Pods 15 cm long, compressed, glabrous, 10–12-seeded. Seeds yellow.

Common in forests. Fl. & Fr.: May–July. *Vern.:* Chindugu. Chinta elagu. Kanmankalva beat, *MSM & KH* 10304; *MSM* 11069. Anantagiri RF, *MSM & PRSR* 12787.

DICHROSTACHYS (A.P. de Candolle) R. Wight et Arnott

Dichrostachys cinerea (L.) Wight & Arn. Prodr. 371. 1834; FBI 2: 288; Gamble 419(297); Khan in FFH 154. *Mimosa cinerea* L. Sp. Pl. 520. 1753.

A much branched armed tree, up to 3 m tall, branchlets pubescent, spine-tipped, bark light coloured, furrowed. Leaves bipinnate, paripinnate, leaflets linear-oblong, 12–16 pairs, 0.2 × 0.1 cm, base obtuse, margin entire, tip acute, pubescent on both sides. Flowers pinkish-yellow, dimorphous, in dense axillary spikes. Pods 4 cm long, twisted, glabrous, 6–10-seeded. Seeds compressed.

Common in scrub jungles. Fl. & Fr.: May–Oct. Gadirayal beat, *MSM & KH* 10252; Central University, *MSM & TP* 15141.

LEUCAENA Bentham

Leucaena leucocephala (Lam.) de Wit. in Taxon 10: 54. 1961. *Mimosa leucocephala* Lam. Encycl. 1: 12. 1783. *Leucaena glauca* auct. non. Benth. in J. Bot. (Hooker) 4: 416. 1842; FBI 2: 290. 1878; Gamble 2: 419(297).

A deciduous small tree, up to 10 m tall. Leaves bipinnate, leaflets 10–20 pairs, linear-oblong, 0.5–1.5 × 0.2–0.4 cm, base rounded, margin entire, tip acute, glabrous above, hairy below. Flowers white in axillary, globose heads. Pods flat, glabrous, dark brown.

Frequently found in hedges and near villages. Fl. & Fr.: All seasons. Central University, *MSM* 15192.

MIMOSA Linnaeus

1. Herbs, sensitive to touch, pinnae 1–2 .... ***M. pudica***
1. Shrubs or small trees; not sensitive to touch, pinnae more than 2:
   2. Flowers yellow .... ***M. hamata***
   2. Flowers pink or pinkish white:
      3. Branchlets glabrous. Pods horned .... ***M. intsia***
      3. Branchlets pubescent. Pod not horned .... ***M. prainiana***

Mimosa hamata Willd. Sp. Pl. 4. 1033, 1806; FBI 2: 291; Gamble 421 (298).

An armed shrub, up to 2 m tall. Leaves bipinnate, leaflets 6–10 pairs, elliptic-oblong, 2–4 × 1–2 mm. Flowers yellow in axillary globose heads. Pods flat, curved, 4–8-jointed, prickly, seeds brown.

Occasional in scrub jungles (Seshagiri Rao, 1986). Fl. & Fr.: June–Dec.

Mimosa intsia L. Sp. Pl. 522. 1753. *Mimosa rubicaulis* Lam. Encycl. 1: 20. 1784; FBI 2: 291; Gamble 421(299); Khan in FFH 155. *M. octandra* Roxb. Pl. Cor. t. 200. 1805.

An armed small tree, up to 2 m tall, branchlets glabrous, prickles recurved. Leaves bipinnate, paripinnate, leaflets oblong, 0.5 × 0.2 cm, truncate, margin entire, tip acute, petiole and rachis prickled, glabrous on both sides. Flowers pinkish-white in axillary solitary heads. Pods stalked, slightly curved, 8 cm long, prickled, thin along margins, horned, flat, 5–7-seeded.

Common in forests. Fl. & Fr.: Sept.–Jan. Anantasagar beat, *MSM* 10514.

Mimosa prainiana Gamble Fl. Pres. Madr. 21. 1919 & in Kew Bull. 1920: 5. 1920.

A small tree, up to 3 m tall, branchlets pubescent. Leaves bipinnate, leaflets oblong, 0.2–0.3 × 0.1–0.2 cm, base oblique, margin entire, tip obtuse-mucronate, pubescent on both sides. Flowers pinkish white in axillary globose heads. Pods 8 cm long, prickly along sutures, 6–8-seeded.

Common in scrub jungles. Fl. & Fr.: June–Dec. Kusumasamudram beat, *MSM & NV* 10358.

Mimosa pudica L. Sp. Pl. 518. 1753; FBI 2: 291; Gamble 421(298); Khan in FFH 155.

A perennial ascending prickled herb, up to 75 cm tall, branchlets glabrescent. Leaves bipinnate, leaflets 10–20 pairs, oblong, 0.2–0.5 × 0.2 cm, base obtuse, margin entire, tip acute. Flowers pink in axillary globose heads. Pods oblong, compressed, horned, spiny along margin, 2–3-seeded.

Common in moist places. Fl. & Fr.: All seasons. *Vern.*: Attipatti. Central University, *MSM* 19206.

PITHECELLOBIUM C.F.P. Martius *nom. cons.*

Pithecellobium dulce (Roxb.) Benth. in Hook. London J. Bot. 3: 199. 1844; FBI 2: 302; Gamble 434(308); Khan in FFH 163 (*Pithecolobium*). *Mimosa dulcis* Roxb. Pl. Cor. t. 99. 1798.

An armed tree, up to 5 m tall. Leaves bipinnate, leaflets oblong, 1.5–2.5 × 0.5–1 cm, base obtuse, margin entire, tip obtuse. Flowers white in axillary and terminal heads. Pods twisted.

Common in hedges. Fl. & Fr.: All seasons. *Vern.*: Seema chinta. Central University, *MSM* 19205.

PROSOPIS Linnaeus

1. Pods oblong-turgid, pale yellow .... ***P. chilensis***
1. Pods linear-falcate, white .... ***P cineraria***

Prosopis chilensis (Molina) Stunz in U.S.D.A Bur. Pl. Industr. Invent. Seeds 31. 85. 1714. *Ceratonia chilensis* Molina, Sagg. Chil. 172. 1782. *Prosopis juliflora* Sw., Prodr. 85. 1788; Khan in FFH 153.

A deciduous armed tree, up to 5 m tall, stipular thorns conical. Leaves bipinnate, leaflets 15–20 pairs, oblong, 1–1.5 × 0.1–0.2 cm, base cuneate, margin entire, tip obtuse. Flowers yellow in axillary long spikes. Pods linear, oblong-glabrous, 8–15-seeded.

Common in waste lands. Fl. & Fr.: All seasons. *Vern.:* Tumma, Sarkar tumma. Mohammadabad, *MSM* 19236.

Prosopis cineraria (L.) Druce in Bot. Exch. Club Soc. Brit. Isles. 3: 422. 1914; *Mimosa cineraria* L. Sp. Pl. 517. 1753 (*cinerea*). *Prosopis spicigera* L. Mant. Pl. 68. 1767; FBI 2: 288; Gamble 421(297).

Armed tree, up to 8 m tall, prickles short, straight. Leaves pinnate, leaflets 8–10 pairs, oblong, 0.5–1 × 0.2–0.4 cm. Flowers yellow in slender spikes. Pods flat, thick.

Occasional in waste lands. Fl. & Fr.: June–Oct. *Vern.*: Jammi. Central University, *MSM* 19212.

SAMANEA (A.P. de Candolle) Merrill

Samanea saman (Jacq.) Merr. in J. Wash Acad. Sci. 6: 47. 1916. *Mimosa saman* Jacq. Fragm. 15. t. 9. 1800. *Enterolobium saman* (Jacq.) Prain J. Asiat. Soc. Bengal Pt. 2. Nat. Hist. 66: 252. 1897; Gamble 435(308).

A large tree, up to 20 m tall, branchlets yellowish tomentose. Leaves bipinnate, leaflets 1–7 pairs, basal ones oblong, elliptic, upper ones obovate, 1–4 × 0.5–2.5 cm, glabrous above and pubescent below. Flowers pinkish brown in axillary and terminal corymbose panicles. Pods linear, oblong, septate between seeds, indehiscent, brownish black. Seeds ovoid.

Grown as an avenue tree. Fl. & Fr. All seasons. Central University, *MSM* 19202.

DROSERACEAE

DROSERA Linnaeus

1. Leaves rosulate, stipulate. Flowers in terminal racemes .... *D. burmannii*
1. Leaves cauline, exstipulate. Flowers in lateral racemes .... *D. indica*

Drosera burmannii Vahl, Symb. Bot. 3: 50. 1794; FBI 2: 424; Gamble 452(320).

An annual erect herb, up to 5 cm tall. Leaves rosulate, obovate, 0.5–1 × 0.3–0.5 cm, reddish, glandular tentacles elongate, base attenuate, margin entire, tip obtuse, pubescent on both sides. Flowers pink in terminal racemes.

Rare in forest. Fl. & Fr.: Dec.–Jan. Shaikpalle, *MSM & KH* 11019.

Drosera indica L. Sp. Pl. 282. 1753; FBI 2: 424; Gamble 452(320).

An annual erect herb, up to 10 cm tall, branchlets glandular pubescent, branched. Lower leaves deflexed, upper ones alternate, narrow, linear, 1–2.5 cm, glandular tentacled above, glabrous below. Flowers pink in lateral racemes.

Rare in forest. Fl. & Fr.: Sept.–Dec. Dharmapur, *MSM* 10488; Shaikpalle, *MSM & KH* 11020.

COMBRETACEAE

1. Stragglers or lianes, branchlets glabrous. Petals 4 .... *Combretum*
1. Trees, branchlets pubescent or tomentose. Petals 0
    2. Flowers in heads, all bisexual. Fruits small, 2-winged .... *Anogeissus*
    2. Flowers in spikes, basally bisexual, apically male. Fruit medium sized, angular or 5-winged .... *Terminalia*

ANOGEISSUS (A.P. de Candolle) Guillemin & Perrottet

Anogeissus latiofolia (DC.) Wall. ex Beddome, Fl. Sylv. S. India t. 15. 1860; FBI 2: 450; Gamble 466(330); Khan in FFH 171. *Conocarpus latifolia* Roxb. ex DC. Prodr. 3: 17. 1828.

A tree, up to 12 m tall, branchlets grey-pubescent. Leaves oblong-obovate, 5–10 × 2.5–5.5 cm, base cuneate, margin entire, tip retuse or rounded, glabrous above, nerves prominent below. Flowers light yellow or cream in axillary and terminal aggregated cymose heads. Achenes 5 mm long, 2-winged, ending in a persistent calyx tube, 1-seeded.

Common in dry deciduous forests. Fl. & Fr.: Sept.–Dec. Kothapalli beat, *MSM* 10447.

COMBRETUM Loefling *nom. cons.*

Combretum albidum G. Don in Trans. Linn. Soc. London 15: 429. 1827. *C. ovalifolium* Roxb. Fl. Ind. 2: 226. 1832; FBI 2: 458; Gamble 469(332); Khan in FFH 173.

A climbing shrub, branchlets glabrous. Leaves ovate-oblong, 6–10 × 3.5–7 cm, base cuneate, margin entire, tip acuminate, glabrous above, rusty pubescent along nerves below. Flowers greenish-yellow in long panicled spikes. Drupes nearly globose, 4-winged, wings papery, reddish brown.

Rare in deciduous forest (Seshagiri Rao, 1986). Fl. & Fr.: Feb.–May. *Vern.:* Aretiga, Addatiga.

TERMINALIA Linnaeus

1. Fruits winged:
   2. Young leaves tomentose below .... ***T. alata***
   2. Young leaves glabrous below.
      3. Glands close to the base of leaf blade. Fruit wings thick .... ***T. arjuna***
      3. Gland at the base of leaf blade. Fruit wings thin .... ***T. coriacea***
1. Fruits not winged:
   4. Leaves apically clustered. Drupes ridged, tomentose .... ***T. bellirica***
   4. Leaves sub-opposite. Drupes not ridged, glabrous .... ***T. chebula***

Terminalia alata Heyne ex Roth Nov. Pl. Spec. 379. 221. *T. tomentosa* (Roxb. ex DC.) Wight & Arn. Prodr. 314. 1834; FBI 2: 447; Gamble 465(329); Khan in FFH 169. *Pentaptera tomentosa* Roxb. ex DC. Prodr. 3: 14. 1828.

A deciduous tree, up to 5 m tall, bark fissured. Leaves oblong-elliptic, 5–12 × 1.5–5.5 cm, glabrous above and tomentose below, base rounded, margin serrate-crenate, tip obtuse. Flowers yellow in axillary and terminal racemes. Drupes oblong, glabrous, 5-winged.

Occaisonal in deciduous forest. Fl. & Fr.: June–Oct. Central University, *MSM* 15162.

Terminalia arjuna (DC.) Wight & Arn. Prodr. 314. 1834; FBI 2: 447. Gamble 465(329); Khan in FFH 168. *Pentaptera arjuna* Roxb. ex DC. Prodr. 3: 14. 1828.

A deciduous tree, up to 30 m tall, branchlets pubescent, bark thick, smooth, greenish white, flaking off large flat pieces. Leaves elliptic-oblong, 7–12 × 3–5 cm, base obtuse, margin crenate-serrate, tip obtuse, 1–2 glands on petiole close to leaf base, glabrous on both sides. Flowers cream in axillary spikes. Drupes 5 cm long, 5-angled, 5-winged, equal, lines on wings oblique and curling upwards, glabrous.

Common in forests on the banks of rivers and streams. Fl. & Fr.: May–Sept. *Vern.:* Tellamaddi, Erramaddi. Lothuvagu, *MSM & KH* 10294.

Terminalia bellirica (Gaertn.) Roxb. Pl. Cor. t. 198. 1805 (*bellerica*); FBI 2: 445; Gamble 463(328); Khan in FFH 166. *Myrobalanus bellirica* Gaertn. Fruct. Sem. Pl. 2: 90. t. 97. 1791.

A deciduous tree, up to 30 m tall, branchlets warty with persistent leaf scars, rusty tomentose when young. Leaves clustered at the ends of branchlets, red, elliptic-obovate, 8–20 × 3.9 cm, base attenuate, margin entire, tip rotund. Flowers cream in axillary spikes. Drupes 3.5 cm long, brown tomentose.

Common in forests areas. Fl. & Fr.: Nov.–June. *Vern.:* Thani. Anantasagar beat, *MSM & NV* 10391.

Terminalia chebula Retz. Obs. Bot. 5: 31. 1788; FBI 2: 446; Gamble 464(328); Khan in FFH 167. *Myrobalanus chebula* (Retz.) Gaertn. Fruct. Sem. Pl. 2: 91 t. 97 f. 2. 1790.

A deciduous tree, up to 20 m tall, branchlets rusty tomentose when young. Leaves elliptic-obovate, 10–20 × 6–8.5 cm, base rounded, margin entire, tip obtuse, glabrescent on both sides. Flowers cream in axillary spikes. Drupes 4 cm long, glabrous.

Occasional in forests. Fl. & Fr.: Nov.–June. *Vern.*: Karakkai. Anantasagar beat, *MSM & NV* 10410.

Terminalia coriacea (Roxb.) Wight & Arn. Prodr. 315. 1834; Gamble 465(329). *Pentaptera coriacea* Roxb. Fl. Ind. 2: 438. 1824. *Terminalia tomentosa* (DC.) Wight & Arn. var. *coriacea* (Roxb.) Clarke in Hook. f. Fl. Brit. India 2: 448. 1878.

A deciduous tree, up to 20 m tall, branchlets pubescent, bark rough, cracked. Leaves elliptic-oblong, 12–20 × 5–8 cm, base obtuse, cordate, oblique, margin entire, tip obtuse, glabrescent above, velvety pubescent below. Flowers cream in axillary spikes. Drupes 4.5 cm long, 5-angled, glabrous, coriaceous, winged.

Common in forests. Fl. & Fr.: May–Sept. *Vern.*: Nallamaddi. Kothapalli beat, *MSM* 10449.

Terminalia catappa L. (*Vern.:* Badam) is cultivated for its fruits. *Quisqualis indica* L. is planted for ornamental purpose.

MYRTACEAE

SYZYGIUM Gaertner *nom. cons.*

Syzygium cumini (L.) Skeels in U.S.D.A. Bur. Pl. Industr. Bull. 248. 2. 1912. *Myrtus cumini* L. Sp. Pl. 471. 1753. *Eugenia jambolana* Lam. Encycl. 3: 198. 1789; FBI 2: 499; Khan in FFH 176. *Syzygium jambolanum* (Lam.) DC. Prodr. 3: 259. 1828; Gamble 481(340).

An evergreen tree, up to 10 m tall, bark smooth grey. Leaves elliptic-oblong, 7–13 × 4–6 cm, coriaceous, base cuneate, margin entire, tip acuminate. Flowers white in axillary, long cymes. Berries oblong, black when ripe. 1-seeded.

Common along roadsides and in villages. Fl. & Fr.: *Vern.:* Neredu, Gacchibowli, *MSM* 15194.

Callistemon lanceolatus DC., *Psidium guajava* L. (*Vern.:* Jama) are planted for ornamental purpose and edible fruits respectively. *Eucalyptus* spp. are planted for ornamental purpose.

BARRINGTONIACEAE

CAREYA Roxburgh *nom. cons.*

Careya arborea Roxb. Pl. Cor. t. 218. 1811; FBI 2: 511; Gamble 488(345); Khan in FFH 178.

A deciduous tree, up to 15 m tall, branchlets with persistent leaf scars, bark rough, dark grey. Leaves apically clustered, oblong-obovate, 10–15 × 4–9.5 cm, base tapering, margin crenate-denticulate, tip rounded, glabrous on both sides. Flowers white in terminal spikes. Berries 5.5 cm long, glabrous.

Occasional in forest in moist places, in ravines and valleys. Fl. & Fr.: Feb.–May. *Vern.:* Budda Dharmi. Lothuvagu, *MSM & KH* 10297.

LYTHRACEAE

1. Herbs:
   2. Flowers in axillary cymes .... ***Ammannia***
   2. Flowers in axillary panicles .... ***Rotala***
1. Trees and shrubs:
   3. Capsules 0.8 cm long, seeds not winged .... ***Lawsonia***
   3. Capsules 3 cm long, seeds winged .... ***Lagerstroemia***

AMMANNIA Linnaeus

Ammannia baccifera L. Sp. Pl. 120. 1753; FBI 2: 569; Gamble 510(360).

An annual erect herb, up to 60 cm tall, branchlets glabrous, oblong-lanceolate, 1–2.5 × 0.5–1 cm, base cuneate, margin entire, tip acute, glabrous on both sides. Flowers red in axillary dichasial cymes. Capsules 0.1 cm long, exceeding calyx tube, 20-seeded. Seeds turgid.

Common in streams and ponds. Fl. & Fr.: All seasons. Anantasagar beat, *MSM & KH* 11007.

LAGERSTROEMIA Linnaeus

Lagerstroemia parviflora Roxb. Pl. Cor. t. 66. 1796; FBI 2: 575; Gamble 512(362); Khan in FFH 183.

A deciduous tree, up to 8 m tall, branchlets glabrescent, bark ash-coloured. Leaves oblong-ovate, 3–8 × 1–3.5 cm, base cuneate, margin entire, tip obtuse, glabrous above, glaucous below. Flowers white in axillary panicles. Capsules loculicidal, 3 cm long, glabrous, surrounded half by persistent calyx, many-seeded. Seeds winged.

Common in forests. Fl. & Fr.: Mar.–May. *Vern.*: Chenangi. Mohammedabad beat, *MSM & KH* 10214, 10235.

LAWSONIA Linnaeus

Lawsonia inermis L. Sp. Pl. 349. 1753; Gamble 514(363). *L. alba* Lam. Encycl. 3: 106. 1789; FBI 2: 573; Khan in FFH 182.

An armed shrub, up to 4 m tall; branchlets spine tipped. Leaves obovate-elliptic, 1–3.5 × 0.8–1.5 cm, base attenuate, margin entire, tip acute, glabrous on both sides. Flowers yellow, fragrant, in terminal panicles. Capsules 0.8 cm long, glabrous. Seeds pyramidal, not winged, brown.

Rare in forests. Fl. & Fr.: Mar.–June. *Vern.*: Maida-aku, Gorintaku. Anantasagar beat, *MSM & NV* 10386.

ROTALA Linnaeus

1. Capsule dehiscing by 4-valves .... ***R. rotundifolia***
1. Capsule dehiscing by 3-valves .... ***R. densiflora***

Rotala densiflora (Roem. & Schult.) Koehne, Bot. Jahrb. Syst. 1: 164. 1880; Gamble 508(359). *Ammannia densiflora* Roth ex Roem. & Schult. in L. Syst. Veg. 3. 304. 1818. *Rotala roxburghiana* Wight Icon. Pl. Ind. Orient. t. 260. B. 1840 *nom. illegit.*

An annual diffuse herb, up to 15 cm, branchlets glabrous. Leaves linear-lanceolate, 0.5–0.8 × 0.1 cm, base subcordate, margin reflexed, tip acute,

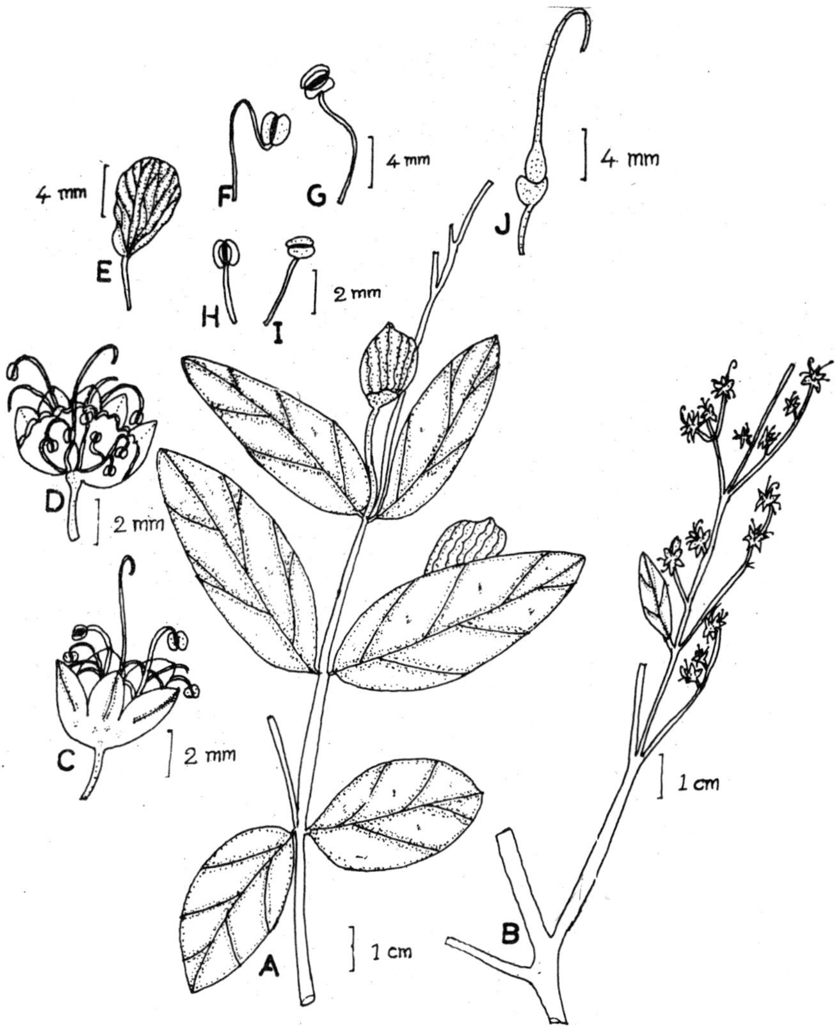

Fig. 12. Lagerstroemia parviflora Roxb.
A. Twig B. Habit with inflorescence C. Flower entire D. Flower split open E. Petal F–I Anthers dorsal and ventral J. Pistil

glabrous on both sides. Flowers pink, axillary, solitary. Capsules globose, 0.2 cm long, glabrous, 3-valved, 10–12-seeded. Seeds granular.

Common in ponds and streams. Fl. & Fr.: All seasons. Kothapalli beat, *MSM & KH* 10562.

Rotala rotundifolia (Roxb.) Koehne, Bot. Jahrb. Syst. 1. 175. 1880; Gamble 509(359). *Ammannia rotundifolia* Buch.-Ham. ex Roxb. Fl. Ind. 1: 446. 1820; FBI 2: 566.

A decumbent herb with thick branched stem at base. Leaves orbicular-ovate, 5–8.5 × 3.5–7 cm, base subcordate, margin entire, tip rotund. Flowers pink in terminal racemes. Capsules dehiscing by 4-valves. 15–20-seeded.

Common near moist places and on the margins of lakes (Seshagiri Rao, 1986). Fl. & Fr.: Apr.–Jan.

Punica granatum L. (*Vern.*: Danimma) (PUNICACEAE) is cultivated for its edible fruits.

ONAGRACEAE

LUDWIGIA Linnaeus

1. Capsules 6 cm long, pubescent .... ***L. octovalvis***
1. Capsules 1.5 cm long, glabrous .... ***L. perennis***

Ludwigia octovalvis (Jacq.) Raven in Kew Bull. 15: 476. 1962. *Oenothera octovalvis* Jacq. Enum. Syst. Pl. 19. 1760. *Jussiaea suffruticosa* L. Sp. Pl. 388. 1753; FBI 2. 587; Gamble 516(365). *J. villosa* Lam. Encycl. 3: 331. 1789. *J. exaltata* Roxb. Fl. Ind. 2: 401. 1832.

An erect subshrub, up to 1 m tall, branchlets pubescent. Leaves elliptic-lanceolate, 3–7 × 0.5–1 cm, base attenuate, margin entire, tip acute, pubescent on both sides. Flowers yellow, axillary, solitary. Capsules loculicidal, 6 cm long, pubescent. Seeds brown.

Occasional along streams. Fl. & Fr.: May–Sept. Malkal Cheruvu, *MSM & KH* 10226; Mohammadabad beat, *MSM* 10440.

Ludwigia perennis L. Sp. Pl. 119. 1753. *L. parviflora* Roxb. Fl. Ind. 1: 419. 1832; FBI 2: 588; Gamble 517(365).

An erect herb, up to 50 cm tall, branchlets glabrous. Leaves narrowly elliptic-lanceolate, 1–9 × 0.2–1 cm, base attenuate, margin entire, tip acuminate, glabrous on both sides. Flowers yellow, axillary, solitary. Capsule 1.5 cm long, 4-ribbed, glabrous, seeds pluriseriate.

Common in open forest. Fl. & Fr.: May–Sept. Mohammadabad, *MSM* 10442.

TURNERACEAE

TURNERA Linnaeus

Turnera ulmifolia L. Sp. Pl. 771. 1753; Gamble 523(369). var. **angustifolia** Willd. ex Urban, Monogr. Turn. 141. 1883.

An annual erect herb, up to 1 m tall, branchlets densely pubescent. Leaves lanceolate-elliptic, 3–6 × 0.5–3.5, base cuneate, margin serrate, tip acuminate, pubescent above, densely pubescent below. Flowers yellow, axillary, solitary. Capsules 8 cm long.

Occasional weed in waste lands. Fl. & Fr.: Mar.–May. Venkat Reddipalli, *MSM & KH* 10309.

CUCURBITACEAE

1. Petals fimbriate .... ***Trichosanthes***
1. Petals entire:
   2. Tendrils entire:
      3. Calyx tube closed at by scales. Fruits tubercled .... ***Momordica***
      3. Calyx tube without scales. Fruits smooth:
         4. Male flowers solitary .... ***Coccinia***
         4. Male flowers in clusters:
            5. Connectives apically crested .... ***Cucumis***
            5. Connectives apically uncrested .... ***Mukia***
   2. Tendrils branched
      6. Fruit globose, 1.5 cm long with vertical greenish and white patches .... ***Diplocyclos***
      6. Fruit oblong, 1.6 cm long, longitudinally white-striped .... ***Luffa***

COCCINIA Wight & Arnott

Coccinia grandis (L.) Voigt. Hort. Suburb. Calc. 59. 1845; Chakravarthy in Fasc. Fl. Ind. 11. 24. 1982. *Bryonia grandis* L. Man. Pl. 1: 126. 1767. *Coccinia indica* Wight & Arn. Prodr. 1: 347. 1834. *nom. illegit*; Gamble 537(379). *Cephalandra indica* Naud. in Ann. Sci. Nat. Bot. (Ser. 5) 5: 16. 1859; FBI 2: 621. *Coccinia cordifolia* auct. non (L.); Cogn. in A. & C. DC. Monogr. Pan 3: 529. 1881.

A climbing herb. Leaves lobed, 4–5 × 5–7 cm, base cordate, margin dentate, tip obtuse. Male and female flowers white, solitary. Berries ovate-oblong, 6–7-seeded.

Wild variety common in hedges. Fl. & Fr.: All seasons. *Vern.*: Donda. Naskal, *MSM & PRSR* 15112.

CUCUMIS Linnaeus

1. Berries 5 cm long .... ***C. trigonus***
1. Berries 60 cm long .... ***C. sativus***

Cucumis sativus L. Sp. Pl. 1: 1012. 1753; FBI 2: 630. 1879; Gamble 1: 535(378). Chakravarthy in Fasc. Fl. Ind. 11: 35. 1982.

An annual monoecious climbing herb. Stems hirsute. Leaves cordate-ovate, palmately 3–5-lobed. Flowers yellow, axillary, solitary. Fruits oblong, dark green or pale yellow. Seeds whitish, oblong.

Cultivated, also found as an escape. Fl. & Fr.: Oct.–Nov. Gacchibowli, *MSM* 15193.

Cucumis trigonus Roxb. Fl. Ind. 3: 722. 1832; FBI 2: 619; Gamble 535(378).

A perennial monoecious prostrate herb, branchlets stiff, hispid, tendrils simple. Leaves palmately 5–7-lobed, lobes ovate-oblong, 2–5 × 1.5–4 cm, base narrow, margin dentate, tip obtuse, scabrid on both sides. Flowers yellow, axillary, solitary. Berries 5 cm long, glabrous. Seeds white.

Occasional in scrub jungles. Fl. & Fr.: Mar.–May. Malkalcheruvu, *MSM & KH* 10277.

DIPLOCYCLOS (Endlicher) Post et Kuntze

Diplocyclos palmatus C. Jeffrey in Kew. Bull. 15. 352. 1962. *Bryonia palmata* L. Sp. Pl. 1012. 1753. excl. Syn., *Bryonopsis laciniosa* auct. non. (L.) Nandin; Gamble 534(377).

A climbing herb. Leaves orbicular, 5-lobed. 7–10 × 3–5 cm, lobes elliptic-oblong, base cordate, margin denticulate, tip acute. Male flowers yellow in axillary clusters. Female flowers green, solitary. Berries globose, green.

Common in hedges. Fl. & Fr.: All seasons. Pargi, *MSM & PRSR* 12795.

LUFFA P. Miller

Luffa acutangula (L.) Roxb. Fl. Ind. 3: 713. 1832; FBI 2: 615; Gamble 533(377). *var. acutangula;* Chakrav. 69. 1982. *Cucumis acutangulus* L.Sp. Pl. 1011. 1753.

A trailing herb. Leaves palmately 5–7-lobed, 10–20 × 10–15 cm, base cordate, margin dentate, tip apiculate. Flowers yellow, unisexual, axillary. Male in fascicles, female solitary. Fruit angled, oblong, many-seeded.

Wild variety on hedges. Fl. & Fr.: Aug.–Dec.

MOMORDICA Linnaeus

1. Plants dioecious, branchlets glabrescent. Fruits obtuse at the ends, not ribbed but softly spinous .... ***M. dioica***
1. Plant monoecious, branchlets villous. Fruit narrowed at the ends, ribbed and tubercled .... ***M. charantia***

Momordica charantia L. Sp. Pl. 1009. 1753; FBI 2: 616; Gamble 532(380). Chakravarthy in Fasc. Fl. Ind. 11: 84. 1982.

An annual climbing herb. Stem hairy. Leaves 5–7-lobed. 4–5 × 5–7 cm, lobes ovate-denticulate. Male and female flowers yellow on long peduncles. Berries oblong-tapering at both ends, many-seeded.

Cultivated, rarely as an escape. Fl. & Fr.: Oct.–May. *Vern.:* Kakara. Mohammadabad, *MSM* 19237.

Momordica dioica Roxb. ex Willd. Sp. Pl. 4: 605. 1805; FBI 2: 617; Gamble 532(375).

An annual dioecious climbing herb, branchlets glabrescent, tendrils simple. Leaves broadly ovate, 3–5-lobed, 4–10 × 3.5–6 cm, base cordate, margin denticulate, tip obtuse, pubescent above, glabrous below. Flowers yellow, axillary, solitary. Capsules 4.5 cm long, clothed with blunt spines, many-seeded. Seeds white.

Occasional in forest on hill tops. Fl. & Fr.: Apr.–June. *Vern.:* Adavikakara. Kusumasamudram beat, *MSM & NV* 10347.

MUKIA Arnott

Mukia maderaspatana (L.) M. Roem. Fam. 2. Peponif 47. 1846. *Cucumis maderaspatana* L. Sp. Pl. 1012. 1753. *Bryonia cordifolia* L. loc.cit. *B. scabrella* L.f. Suppl. Pl. 424. 1781. *Mukia scabrella* (L.f.) Arn. in Hook. J. Bot. 3: 276. 1841; FBI 2: 623. 1879. *Melothria maderaspatana* (L.) Cogn. in A. & C. DC. Monogr. Phan. 3: 623. 1881; Gamble 539(381).

A climbing shrub, branchlets hispid. Leaves ovate-deltoid, 3–5-lobed, 2–5.5 × 1.5–4.5 cm, base cordate, margin denticulate, tip acuminate, scabrid above and hispid below. Male flowers in clusters, female flowers yellow, axillary, solitary. Berries 1 cm long, glabrous, 1-seeded. Seeds rugose.

Common along hedges in forests. Fl. & Fr.: Apr.–Sept. Mohammadabad beat, *MSM* 10498.

TRICHOSANTHES Linnaeus

Trichosanthes tricuspidata Lour. Fl. Cochinch. 589. 1790. *T. bracteata* (Lam.) J. Voigt. Hort. Subterb. Calc. 58. 1845. *Modecca bracteata* Lam. Encycl. 4: 210. 1797. *Trichosanthes palmata* Roxb. Fl. Ind. 3: 704. 1832. *nom. illegit. non* L.; FBI 2: 606; Gamble 529(374).

A climbing herb, branchlets glabrescent. Leaves ovate, deeply 3-lobed, 8–10 × 5–7 cm, base cordate, margin denticulate, tip acuminate, scabrid on both sides. Flowers white, male in leaf opposed racemes, female solitary. Berries 9 cm long, smooth, glabrous, red when ripe, pulp black, many-seeded. Seeds oblong, brown.

Occasional in forest. Fl. & Fr.: June–Sept. Mohammadabad beat, *MSM* 10497.

Benincasa hispida (Thunb.) Cogn. (*Vern.:* Budida gummadi), *Citrullus lanatus* (Thunb.) Mats. & Nakai, *Cucurbita maxima* Duch. ex Lam., *C. pepo* L. (Gummadi), *Lagenaria siceraria* (Molina) Standl (*Vern.:* Sorakai), *Luffa acutangula* (L.) Roxb. (*Vern.:* Beera), *L. cylindrica* (L.) Roem. (*Vern.:* Nethibeera) and *Trichosanthes anguina* L. (*Vern.:* Potlakai) are cultivated for fruit vegetables.

CACTACEAE

OPUNTIA Linnaeus

1. Fruits purple. Spines some curved, the largest stout .... *O. stricta*
1. Fruits red. Spines all straight, slender .... *O. elatior*

Opuntia elatior Mill. Gard. Dict. ed. 8. n. 4. 1768; Gamble 1: 548(387).

An erect shrub, up to 1.5 m tall. Stem jointed. Joints flat, oblong-obovate, 5–20 × 3–12 cm, succulent, leafless, glaucous green with pale-yellowish spines. Flowers yellow at first, later change to red or purple, axillary, solitary. Berries obovoid, dark red, prickly, fleshy, edible.

Occasional in scrub jungle as an escape (Seshagiri Rao, 1986). Fl. & Fr.: Jan.–May.

Opuntia stricta (Haw.) Haw., Syn. Pl. Succ. 191. 1812. var. *dillenii* Ker-Gawl, L. Benson in Cact & Succ. J. Amer. 41: 126. 1969. *Cactus strictus* Haw. Misc. 188. 1803. *C. dillenii* Ker.-Gawl., Edward's Bot. Reg. 3. 255. 1818. *Opuntia dillenii* (Ker.-Gawl.) Haw., Suppl. Syn. Pl. Succ. 79. 1819; FBI 2: 657; Gamble 548(387), Khan in FFH 189.

A thorny bush, up to 1 m tall. Stems jointed, modified into fleshy cladode like plates, 10–20 × 6–12 cm. Leaves modified into spines, up to 3 cm long, hard, 3–7 per areole. Flowers yellow, solitary. Berries obovoid, fleshy, purple, areolar, many-seeded.

Common in waste lands. Fl. & Fr.: All seasons. *Vern.*: Naga jamudu. Naskal, *MSM & PRSR* 15120.

AIZOACEAE

1. Styles 2–5 .... *Zaleya*
1. Style 1 .... *Trianthera*

TRIANTHEMA Linnaeus

Trianthema portulacastrum L. Sp. Pl. 233. 1753; Gamble 550(389). *T. monogyna* L. Mant. Pl. 69. 1767; FBI 2: 680.

An annual prostrate somewhat succulent herb, up to 1 m, branchlets angular, glabrous. Leaves unequal, larger one obovate, smaller ones oblong, 1–3.5 × 0.8–2.5 cm, base cuneate, margin entire, tip obtuse, glabrous on both sides. Flowers pinkish-white, axillary, solitary. Capsules upper part come off as a lid, 3–5-seeded. Seeds dull black.

Occasional in waste places and forests. Fl. & Fr.: May–Aug. Dharmapur, *MSM & KH* 10286.

ZALEYA N. Burman

Zaleya decandra (L.) Burm. f. Fl. Ind. 110. t. 31. f. 3. 1768. *Trianthema decandra* L. Mant. Pl. 7. 1767; FBI 2: 661; Gamble 551 (389).

Prostrate herb. Leaves exstipulate, elliptic-oblong, 2–3 × 1–1.5 cm, subsucculent, papillose, base acute, margin entire, apex obtuse. Flowers pinkish in dense axillary or terminal subumbellate cymes or fascicles. Capsule truncate separating into 2 parts. Seeds orbicular, rugulose.

Rare in waste lands and roadsides. Fl. & Fr.: July–Nov. Chevella, *TP & MSM* 19255.

MOLLUGINACEAE

1. Leaves all cauline, flowers in clusters. .... *Glinus*
1. Leaves radical and cauline, flowers in dichasial cymes .... *Mollugo*

GLINUS Linnaeus

1. Leaves stellate tomentose on both sides. Pedicels usually very short, sepals densely stellate hairy .... *G. lotoides*
1. Leaves glabrous on both sides. Pedicels long, filiform, sepals glabrous .... *G. oppositifolius*

Glinus lotoides L. Sp. Pl. 463. 1753. *Mollugo lotoides* (L.) O. Kuntze, Rev. in Gen. Pl. 214. 1891; Gamble 552(390). *M. hirta* Thunb. Prodr. in Pl. Cap. 24. 1794; FBI 2: 662.

An annual prostrate herb, up to 60 cm, branchlets stellate-tomentose. Leaves obovate, spathulate, 1–2.5 × 1–2 cm, base cuneate, margin entire, tip obtuse, stellate-tomentose above, grey stellate tomentose below. Flowers yellow in axillary fascicles. Capsules loculicidal, many-seeded. Seeds black.

Occasional in dried tanks, ponds, streams. Fl. & Fr.: May–Aug. Mohammadabad, *MSM & KH* 10227.

Glinus oppositifolius (L.) A. DC. Bull. Herb. Boissier (Ser. 2) 1: 552. 1901. *Mollugo oppositifolia* L. Sp. Pl. 89. 1753; Gamble 552(390). *M. spergula* L. Syst. Nat. (ed. 10). 881. 1759; FBI 2: 662.

An annual prostrate herb, up to 25 cm, branchlets pubescent. Leaves in whorls of 4–5, unequal, obovate-spathulate, base attenuate, margin entire, tip apiculate, glabrous on both sides. Flowers pink in axillary clusters. Capsules many-seeded. Seeds brown.

Occasional in rivers, tanks and ponds. Fl. & Fr.: May–Aug. Mohammadabad, *MSM & KH* 228.

MOLLUGO Linnaeus

1. Leaves radical .... ***M. nudicaulis***
1. Leaves cauline:
   2. Flowers pink .... ***M. pentaphylla***
   2. Flowers white:
      3. Stems terete. Flowers in terminal umbellate cyme, pedcels long, filiform .... ***M. cerviana***
      3. Stems angular. Flowers in leaf opposed cymes, pedicels short. ..... ***M. disticha***

Mollugo cerviana (L.) Ser. in DC Prodr. 1: 392; FBI 2: 663; Gamble 553(390). *Pharnaceum cervianum* L. Sp. Pl. 272. 1753.

An erect herb, up to 10 cm tall, branchlets glaucous. Leaves radical and cauline, linear or acicular, 1–1.5 cm long, base attenuate, margin entire, tip obtuse, glaucous below. Flowers white in terminal polychasial umbellate cymes. Capsule ellipsoid, many-seeded.

Occasional in waste lands (Seshagiri Rao, 1986). Fl. & Fr.: Jan.–Mar.

Mollugo disticha (L.) Ser. in DC. Prodr. 1.392; FBI 2: 663; Gamble 553(390); *Pharnaceum distichum* L. Mant. Fl. 221. 1771.

An annual erect herb, up to 10 cm tall, branchlets pubescent. Leaves radical and cauline. Cauline leaves obovate-oblanceolate, 0.5–2 × 0.2–0.4 cm, base cuneate, margin entire, tip acute. Flowers white in leaf opposed or terminal cymes. Capsule ovoid, many-seeded, compressed, pale pink.

Common in waste lands. Fl. & Fr. Jan.–Mar. Mekavarpalli, *MSM & TP* 11045.

Mollugo nudicaulis Lam. Encycl. 4: 234. 1797; FBI 2: 664; Gamble 553(390).

An annual erect herb, up to 20 cm tall, branchlets angular, glabrous. Leaves radical, rosette at the base, spathulate-obovate, 3–4 × 1.5–2.5 cm, base attenuate, margin entire, tip obtuse, glabrous on both sides. Flowers white in many peduncled, trichotmously branched cymes. Capsules 2 cm long, glabrous, many-seeded. Seeds reddish black.

Common in sandy areas, roadsides. Fl. & Fr.: Mar.–May. Dharampuram, *MSM & KH* 10284.

Mollugo pentaphylla L. Sp. Pl. 89. 1753; Gamble 553(390). *M. stricta* L. Sp. Pl. (ed. 2) 131. 1762; FBI 2: 663.

An annual erect herb, up to 20 cm tall, branchlets dichotomous, angular, glabrous. Leaves in whorl of 2–9, linear-lanceolate, 0.5–2.5 × 0.3–1 cm, base attenuate, margin entire, tip apiculate, glabrous on both sides. Flowers pink in terminal corymbs. Capsules 0.2 cm long, glabrous, many-seeded. Seeds dark brown.

Common in waste lands. Occasional in forests. Fl. & Fr.: Apr.–June. Kusumasamudram beat, *MSM & NV* 10330, 10331.

APIACEAE (*nom. alter.* Umbelliferae)

CENTELLA Linnaeus

Centella asiatica (L.) Urban in C. Martius. Fl. Braz. 11(1): 287. t. 78. f. I. 1879; Gamble 556(392). *Hydrocotyle asiatica* L. Sp. Pl. 234. 1753; FBI 2: 669.

A prostrate herb. Stem with long stolon and roots at nodes. Leaves reniform, 1.5–2 × 3–4.5 cm, base cordate, margin crenate, dentate, tip rotund. Flowers purple in axillary umbels. Fruits ovate, mericarps compressed.

Occasional along streams. Fl. & Fr.: Aug.–Nov. Rudraram RF, *TP & MSM* 12723.

Coriandrum sativum L. (*Vern.:* Daniyalu) is cultivated for its leaves and shoots which are used as flavouring agents in curries and dried fruits as aromatic stimulant. *Daucus carota* L. (*Vern.:* Carrot) and *Trachyspermum ammi* (L.) Spr. are cultivated for edible roots and seeds respectively.

ALANGIACEAE

ALANGIUM Lamarck *nom. cons.*

Alangium salvifolium (L.f.) Wang. in Engl. Pflanzenf. 4(220b). 41: 9. 1910; Gamble 572(404). *Grewia salvifolia* L.f. Suppl. Pl. 409. 1781. *Alangium lamarckii* Thw. Enum. Pl. Zeyl. 133. 1859; FBI 2: 741; Khan in FFH 191.

A deciduous tree, up to 4 m tall, branchlets glabrous, spinescent, bark grey. Leaves ovate-lanceolate, 2–11 × 1.5–3.5 cm, base oblique, margin entire, tip acute, glabrous above, pubescent below. Flowers white in axillary fascicles. Drupes 2.5 cm long, tomentose with persistent calyx lobes, 1-seeded. Seeds ovoid.

Common in forests and roadsides. Fl. & Fr.: Feb.–May. *Vern.:* Uduga. Mohammadabad beat, *MSM* & KH 10211.

RUBIACEAE

1. Herbs:
    2. Flowers in dense sessile heads .... ***Borreria***
    2. Flowers not as above:
        3. Leaves not exceeding 0.5 cm across; Flowers white .... ***Hedyotis***
        3. Leaves exceeding 0.5 cm across; Flowers purple .... ***Neanotis***
1. Shrubs or trees:
    4. Armed:
        5. Leaves pubescent .... ***Catunaregum***
        5. Leaves glabrous:
            6. Fruit a berry .... ***Benkara***
            6. Fruit a drupe .... ***Canthium***

4. Unarmed:
    7. Flowers 4-merous:
        8. Style as long as corolla tube; stigma 2-fid. Cymes exclusively terminal .... ***Ixora***
        8. Style 1 1/2 to 2 times longer than corolla-tube; stigma fusiform. Cymes both terminal and axillary........ ***Pavetta***
    7. Flowers 5-merous:
        9. Flowers in paniculate racemes; seeds winged .... ***Hymenodictyon***
        9. Flowers solitary or globose heads, seeds not winged:
            10. Flowers solitary .... ***Gardenia***
            10. Flowers in heads:
                11. Leaves glabrous .... ***Mitragyna***
                11. Leaves pubescent beneath:
                    12. Leaves cordate at base .... ***Haldinia***
                    12. Leaves cuneate-obtuse at base .... ***Morinda***

BENKERA Adanson

Benkera malabarica Tirveng. in Taxon 32: 440. 1983. *Randia malabarica* Lam. Encycl. Meth. 3: 25. 1789; FBI 3: 111; Gamble 616(435); Khan in FFH 198.

An armed shrub, up to 3 m tall, branchlets glabrous, reduced to short sharp spines. Leaves coriaceous, obovate-oblanceolate, 1.5–3.5 × 1–1.5 cm; base cuneate, margin entire, tip obtusely apiculate, glabrous on both sides. Flowers white in axillary corymbs. Berries 0.8 cm long, glabrous, 6–8-seeded. Seeds compressed.

Common in open forests and scrub jungles. Fl. & Fr.: May–July. *Vern.:* Pedda Manga. Rangammagudem beat, *MSM & KH* 10274.

BORRERIA G.F.W. Meyer *nom. cons.*

1. Prostrate herb, leaves elliptic-ovate .... ***B. articularis***
1. Erect herb, leaves linear-lanceolate .... ***B. pusilla***

Borreria articularis (L.f.) F.N. Will. in Bull. Herb. Boise. Ser. 2, 5: 956. 1905. *Spermacoce articularis* L.f. Suppl. Pl. 119. 1781. *Borreria hispida* (L.) K. Schum. in Engl. Pflanzenf. 4(4): 144. 1891. non Spruce & Schum. 1888; Gamble 654(461). *Spermacoce hispida* L. Sp. Pl. 102. 1753; FBI 3: 200.

An annual procumbent herb, up to 50 cm, branchlets quadrangular, hispid. Leaves elliptic-ovate, 1–3 × 0.5–2 cm, base cuneate, margin ciliate, tip acute, scabrid on both sides. Flowers white in axillary fascicles. Capsules 0.5 cm long, pubescent, 2-seeded. Seeds brown.

Occasional in forests. Fl. & Fr.: June–Oct. Kusumasamudram beat, *MSM & NV* 10340.

Borreria pusilla (Willd.) DC. Prodr. 4: 543. 1830. *Spermacoce pusilla* Wall. in Roxb. Fl. Ind. 1: 379; 1820. *Borreria stricta* auct. non K. Schum. Gamble 654(461). *Spermacoce stricta* auct. non L.f.: FBI 3: 200.

An annual erect herb, up to 70 cm tall, branchlets rugose. Leaves whorled, linear-lanceolate, 2–5 × 0.4–0.6 cm, base cuneate, margin revolute, tip acute, rugose on both sides. Flowers white in axillary whorls. Capsules oblong, 0.3 cm long, rugose.

Occasional in forests. Fl. & Fr.: May–Sept. Kanmankalva beat, *MSM* 10465.

CANTHIUM Lamarck

Canthium parviflorum Lam. Encycl. 1: 602. 1785; FBI 3: 136. *Plectronia parviflora* (Lam.) Beddome, Fl. Sylv. S. India 124. 1872; Gamble 625(441); Khan in FFH 202.

An armed woody shrub, up to 5 m tall, branchlets glabrous with supra axillary straight sharp spines. Leaves elliptic-obovate, 1.5–3.5 × 0.8–2.6 cm, base attenuate, margin entire, tip acute, glabrous on both sides. Flowers greenish-white in axillary cymes. Drupes green, brown when ripe, 1 cm long, glabrous.

Common in scrub jungles. Fl. & Fr.; Mar.–May. *Vern.:* Balusu. Anantasagar beat, *MSM & NV* 10400; Central University, *MSM* 15153.

CATUNAREGAM Wolf

Catunaregam spinosa (Thunb.) Tirveng. in Bull. Mus. Nat. Hist. Paris. 3: Ser. 521. Bot. 35: 13. 1978. *Gardenia spinosa* (Thunb.) Keay in Bull. Jard. Bot. Etat. Brux. 28: 37–38. 1958. *Randia dumetorum* (Retz.) Poir. var. *floribunda* (DC); Gamble 616(434). *Randia floribunda* DC. Prodr. 41: 386. 1830. *Randia longispina* Wight & Arn. Prodr. 398. 1834; Gamble 616(434). *R. brandisii* Gamble Fl. Pres. Madras 2: 616(434) 1921; Khan in FFH 198.

An armed woody shrub, up to 3 m tall, branchlets grey, glabrous with strong straight sharp decussate spines. Leaves obovate, 1–3.5 × 0.5–2 cm, base attenuate, margin entire, tip obtuse, pubescent above, densely pubescent below. Flowers whitish-yellow in axillary clusters. Berries crowned with calyx limb, 4.5 cm long, shining, smooth, many-seeded. Seeds dark brown.

Common in scrub-jungles. Fl. & Fr.: Mar.–May. *Vern.:* Manga. Mohammabad beat, *MSM & KH* 10222; Gadirayal, *MSM & KH* 10254.

Note: Trivengadum (1978) re-established the genus *Catunaregum* for the old world *Randia sensu lato*. He said that *Randia* is an exclusively American genus characterised by its one-celled ovary with parietal placentation. He transfered the South Indian *Xeromphis* to *Catunaregum* (Thunb.) Trivengadum.

GARDENIA Ellis *nom. cons.*

1. Flowers sessile. Cayx lobes triangular. Corolla 4.5 cm across .... ***G. gummifera***
1. Flowers pedicellate. Calyx lobes lanceolate. Corolla 7–8 cm across:
   2. Secondary nerves of leaves stout, not parallel and very prominent beneath. Berries globose .... ***G. latifolia***
   2. Secondary nerves of leaves slender, parallel, very prominent beneath. Berries ellipsoid .... ***G. resinifera***

Gardenia gummifera L.f. Suppl. Pl. 164. 1781; FBI 3: 117; Gamble 618(436); Khan in FFH 200.

A deciduous tree up to 5 m tall, bark grey, branchlets glabrous. Leaves obovate-oblong, 2–5 × 0.7–3 cm, base rounded, margin entire, tip acute-obtuse, glabrous on both sides. Flowers yellow, axillary and terminal, solitary. Berries oblong.

Common in deciduous forest. Fl. & Fr.: Mar.–Sept. *Vern.:* Chitmit. Dharur RF, *MSM & TP* 11097.

Gardenia latifolia Aiton Hort. Kew. 1: 296. 1789; FBI 3: 116; Gamble 619(436); Khan in FFH 200.

A deciduous tree up to 8 m tall. Leaves elliptic, coriaceous, 10–20 × 5–15 cm, base attenuate, margin entire, tip acute, glabrous on both sides. Flowers white axillary, solitary. Berries globose.

Occasional in forest. Fl. & Fr.: May–Aug. *Vern.:* Chitmit. Anantagiri RF, *MSM & TP* 11062.

Gardenia resinifera Roth, Nov. Pl. Spec. 150. 1821. *G. lucida* Roxb. Fl. Ind. 2: 553. 1824; FBI 3: 115; Gamble 618(436).

A small tree up to 5 m tall; branchlets yellow. Leaves elliptic, ovate-obovate, 6–12 × 4–8 cm, base obtuse, margin entire, tip acute. Flowers white, axillary, solitary. Berries ellipsoid with a crown of persistent calyx. Seeds rugose.

Common in dry deciduous forest. Fl. & Fr.: Aug.–Dec. Rudraram RF, *MSM & TP* 12708.

HALDINIA Ridsdale

Haldinia cordifolia (Roxb.) Ridsd. in Blumea 24: 361. 1978. *Nauclea cordifolia* Roxb. Pl. Cor. t. 53. 1796. *Adina cordifolia* (Roxb.) Hook. f. ex Brandis, Forest. Fl. N.W. India t. 33. 1784; FBI 3: 34; Gamble 584(412); Khan in FFH 193.

A deciduous tree, up to 20 m, branchlets pubescent, bark brownish-grey. Leaves cordate-orbicular, 9–22 × 9–20 cm, base cordate, margin entire, tip acuminate, hispid above, pubescent below. Flowers cream in axillary globose heads. Capsules 0.4 cm long, densely downy. Seeds elongate, tailed above.

Common in dry deciduous forests. Fl. & Fr.: June–Nov. *Vern.:* Rudraganam, Paspuganapu. Anantasagar, *MSM & NV* 10413.

Note: Earlier floristic workers treated this taxon under *Adina cordifolia* (*loc. cit.*). But Ridsdale (*loc. cit.*) in his revision of tribe *Naucleae* s.s. feels that Indian taxon does not belong to the genus *Adina* Salisbuary and has raised the monotypic new genus *Haldinia* to accommodate it.

HEDYOTIS Linnaeus

1. Flowers in umbellate cyme .... ***H. puberula***
1. Flowers not in umbellate cyme:
   2. Flowers solitary .... ***H. herbacea***
   2. Flowers 2–4:
      3. Leaf base decurrent. Flowers white........ ***H. corymbosa***
      3. Leaf base attenuate. Flowers purple .... ***H. aspera***

Hedyotis aspera Heyne ex Roth, Nov. Pl. Sp. 94. 1821. *Oldenlandia aspera* (Roth) DC. Prodr. 4: 428. 1830; FBI 3: 68. 1880. Gamble 602(424). *Kohautia aspera* (Heyne ex Roth) Bremek in Verh. Kon. Ned. Akad. Wetensch. Afd. Natuurk. Tweede Sect. Ser. 2. 48: 113. 1952.

An annual erect herb up to 40 cm tall. Leaves whorled, linear, 2.5–4.5 × 0.3–0.5 cm, base attenuate, margin enitre, tip acute. Flowers purple in terminal lax racemes. Capsule loculicidal, scabrid. Seeds reticulate.

Occasional in waste lands. Fl. & Fr.: Mar.–July. Mohammadabad, *MSM & NV* 10421.

Hedyotis corymbosa L. Sp. Pl. 119. 1753; FBI 3: 64; Gamble 600(423).

An annual erect herb, up to 1 m tall, branchlets quadrangular. Leaves linear, lanceolate, 1–3 × 0.1–0.3 cm, base decurrent, margin revolute, tip acute, glabrous on both sides. Flowers white in corymbs. Capsules loculicidal, 0.3 cm long. Seeds 3-gonous, reticulate.

Occaisonal in forest undergrowth. Fl. & Fr.: Sept.–Dec. Mohammadabad beat, *MSM & KH* 10496.

Hedyotis herbacea L. Sp. Pl. 102. 1753. *Oldenlandia herbecea* (L.) Roxb. Fl. Ind. 1: 445. 1820; Gamble 601(424).

An erect herb, up to 40 cm tall, branchlets glabrous, quadrangular. Leaves linear-lanceolate, 1–2.5 × 0.1–0.3 cm, base decurrent, margin revolute, tip acute, glabrous on both sides. Flowers white, axillary, solitary. Capsules loculicidal, 0.2 cm long, crown protruding. Seeds reticulate.

Common in forests. Fl. & Fr.: Sept.–Dec. Kothapalli beat, *MSM & KH* 10563; Central University, *MSM* 15150.

Hedyotis puberula (G. Don) Arn. Pugill. Pl. Ind. Orient. 342. 1836. *Oldenlandia puberula* G. Don Gen. Syst. 3: 530. 1834. *O. umbellata* L.

Sp. Pl. 119 (L.) Lam. Tabl. Encycl. 1: 272. 1792. non Wall. 1788. *Oldenlandia wightii* Hook. f. Fl. Brit. India 3: 66. 1880; Gamble 2. 601(424). *Hedyotis wightii* (Hook. f.) K.K.N. Nair in J. Bombay Nat. Hist. Soc. 79. 453. 1982.

An annual erect herb, up to 20 cm tall. Leaves clustered, linear, 1.5 × 0.4 cm, base decurrent, margin entire, tip acute. Flowers white in terminal capitate clusters. Capsules loculicidal.

Common weed of waste lands and open forests. Fl. & Fr.: Sept.–Jan. *MSM* 15147.

HYMENODICTYON Wallich *nom. cons.*

Hymenodictyon orixense (Roxb.) Mabb. in Taxon 31: 66. 1982. *Cinchona orixensis* Roxb. Bot. Descr. Swietania 21. 1793 & Med. Facts. Obs. 6: 152. 1795; *Hymenodictyon excelsum* (Roxb.) Wall. in Roxb. Fl. Ind. 2: 149. 1824; FBI 3: 35; Gamble 589(416). *Cinchona excelsa* Roxb. Pl. Cor. t. 106. 1798.

A deciduous tree, up to 15 m tall. Leaves elliptic-ovate, 7–15 × 7–12 cm, base attenuate, margin entire, tip acute, glabrescent above and pubescent below. Flowers green in axillary and terminal paniculate racemes. Capsules ovoid.

Occasional in deciduous forest. Fl. & Fr.: Aug.–Dec. Anantagiri RF, *MSM & TP* 11077.

IXORA Linnaeus

Ixora arborea Roxb. ex Sm. in Rees, Cyclop. 19: 5. 1811. *I. parviflora* Vahl, Symb. Bot. 3: 11. t. 52. 1794 (non Lam.); FBI 3: 142; Gamble 631(445), Khan in FBI 203. *I. pavetta* Andrews Bot. Repos t. 78. 1799.

An evergreen shrub, up to 5 m tall, branchlets glabrous, bark dark-brown, exfoliating in irregular patches. Leaves elliptic-oblanceolate, 9–15 × 2–6 cm, base cuneate, margin entire, tip obtuse, glabrous on both sides. Flowers white in axillary cymes, fragrant. Drupes subglobose, 6 cm long, glabrous. Seed brown, smooth.

Common in forests. Fl. & Fr.: May–Nov. *Vern.:* Ponna. Lothuvagu, *MSM & KH* 10291; Kusumasamudram beat, *MSM & KH* 11023.

MITRAGYNA Korthals *nom. cons.*

Mitragyna parvifolia (Roxb.) Korth. Observ. Naud. Ind. 19. 1839; Gamble 585(413). *Nauclea parvifolia* Roxb. Pl. Cor. t. 52. 1796. *Stephygyne parvifolia* (Roxb.) Korth. Verhn. Nat. Gesch. Ned. Bezilt. Bot. 161. 1842; FBI 3: 25; Khan in FFH 194.

A deciduous tree up to 10 m tall. Leaves oblong-obovate. 5–10 × 3–9 cm, base cuneate, margin entire, tip obtuse. Flowers cream in axillary heads. Capsules aggregated into heads, many-seeded.

Common in dry deciduous forest. Fl. & Fr.: Oct.–Dec. *Vern.*. Battaganupu. Anantagiri, *MSM & PRSR* 12790.

MORINDA Linnaeus

Morinda pubescens Sm. in Rees. Cyclop. 24n. 3. 1813. *M. tomentosa* Roxb. Nov. Pl. Sp. 147. 1821. *M. tinctoria* Roxb. Fl. Ind. 1: 543. 1832; FBI 3: 156; Gamble 65(459).

A deciduous tree, up to 12 m tall, branchlets tomentose, bark dark-brown. Leaves elliptic-ovate, 4.5–20 × 4–11 cm, base cuneate, margin entire, tip acuminate, pubescent above, grey tomentose below. Flowers white in globose umbellate heads. Syncarpia of pyrenes 2 cm long.

Common in all forests. Fl. & Fr.: Mar.–May. *Vern.:* Mogilli, togaru. Mohammadabad beat, *MSM & KH* 10240.

NEANOTIS W.H. Lewis

Neanotis lancifolia (Hook. f.) Lewis in Ann. Missouri Bot. Gard. 53. 59. 1966. *Anotis lancifolia* Hook. f. Fl. Brit. India 3: 73. 1880; Gamble 605(427). *Hedyotis lancifolia* Dalz. in Hook. J. Bot. 2: 135. 1850 non. Schumm & Thonn. 1827. *Neanotis concanensis* Daniel & Vajravelu in J. Econ. Tax. Bot. 3: 675. 1982.

An annual erect herb, up to 25 cm tall, branchlets sparsely hairy. Leaves ovate-lanceolate, 3–5 × 0.5–1.5 cm, base acute, margin entire, tip acuminate, sparsely hairy above, hairy on nerves below. Flowers purple in axillary and terminal cymes. Capsules compressed, seeds pitted, black.

Rare in open forest. Fl. & Fr.: Aug.–Dec. Anantagiri RF, *MSM & TP* 11082.

PAVETTA Linnaeus

Pavetta tomentosa Roxb. ex Smith in Rees Cyclop. 26. n. 2. 1819; Gamble 1879(1301). *P. indica L. var. tomentosa* (Smith) Hook. f. Fl. Brit. India 3: 150. 1880; Gamble 683(446).

A shrub, up to 3 m tall. Leaves oblong, 8–10 × 4–5 cm, base rounded, margin entire, tip acute, pubescent above. Flowers white in terminal branched corymbs. Berry globose.

Common in forest. Fl. & Fr.: Aug.–Dec. Anantagiri RF, *MSM & TP* 12791.

ASTERACEAE

1. Achenes paired, capitula unisexual .... ***Xanthium***
1. Achenes single, capitula homo/heterogamous:
   2. Involucral bracts 1-seriate:
      3. Capitula homogamous:
         4. Florets pink .... ***Emilia***
         4. Florets other than pink .... ***Launaea***
      3. Capitula heterogamous... ***Senecio***
   2. Involural bracts more than 1-seriate:
      5. Heads homogamous, not rayed, pappus present:
         6. Leaves opposite .... ***Ageratum***
         6. Leaves alternate:
            7. Involucral bracts not spinescent, style branches elongated ...... ***Vernonia***
            7. Involucral bracts spinescent, style branches not elongated:
               8. Involucral bracts spinescent:
                  9. Heads simple. Leaves not spinescent, dentate or lyrately lobed .... ***Oligochaeta***
                  9. Heads compound. Leaves spinescent, deeply lobed ...... ***Echinops***
               8. Involucral bracts aristate-acuminate or acuminate .... ***Tricholepis***
      5. Heads heterogamous, rayed or not (except *Caesulia* and *Lagascea*), pappus present or absent:
         10. Receptacle not paleaceous:
            11. Leaves opposite.. ***Flaveria***
            11. Leaves usually alternate:
               12. Anthers base obtuse .... ***Grangea***
               12. Anthers base sagittate or tailed:
                  13. Heads aggregated into globose or ovoid clusters .... ***Sphaeranthus***
                  13. Heads not aggregated into clusers:
                     14. Plants marshy herbs, heads solitary .... ***Caesulia***
                     14. Plants other than marshy, heads in racemes:
                        15. Heads rayed .... ***Pulicaria***
                        15. Heads not rayed:
                           16. Involucral bracts broad, scarious .... ***Gnaphalium***
                           16. Involucral bracts other than above:
                              17. Stem woody, involucral bracts broad, rigid .... ***Pluchea***
                              17. Stem herbaceous, involucral bracts narrow, slender .... ***Blumea***
         10. Receptacle paleaceous:
            18. Leaves opposite, all cauline
               19. Inner involucral bracts partly or completely enclosing the achenes of ray florets, pappus of ray and disc florets absent, bisexual florets sterile, ray florets fertile .... ***Acanthospermum***
               19. Inner involucral bracts do not enclose the achenes of ray florets, almost free from achenes, pappus of either ray and disc florets present or absent, bisexual florets fertile, ray florets usually sterile:
                  20. Heads one-flowered, clustered together or aggregated in compound fashion, subtended by leafy bracts, ligule absent .... ***Lagascea***
                  20. Heads separate, not aggregated in compound fashion, not subtended by leafy bracts, ligule present:
                     21. Pappus of awns:
                        22. Pappus awns plumose, achenes densely silky .... ***Tridax***

22. Pappus awns retrorsely barbellate, achenes hispid or glabrous .... ***Bidens***
21. Pappus of scales, bristles or absent:
23. Stems erect, heads on conspicuous strong peduncles, ligules yellow ... ***Wedelia***
23. Stems often decumbent, heads subsessile to sessile or with weak peduncles, white:
24. Bisexual florets white, palea narrow, acute to gland tipped .... ***Eclipta***
24. Bisexual florets yellow, palea broad, dentate at apex .... ***Blainvillea***
18. Leaves alternate, radical and cauline or cauline only:
25. Prostrate herbs:
26. Pappus of awns .... ***Glossocordia***
26. Pappus absent .... ***Chrysanthellum***
25. Erect herbs:
27. Leaves simple. Florets yellow .... ***Sonchus***
27. Leaves pinnatifid. Florets white .... ***Parthenium***

ACANTHOSPERMUM Schrank

Acanthospermum hispidum DC. Prodr. 5: 522. 1836; Gamble 704(495).

An annual erect herb, up to 50 cm tall, branchlets hispid. Leaves obovate-oblanceolate, 0.5–2 × 0.5–1.5 cm, base cuneate, margin serrate, tip acute, scabrous on both sides. Flowers yellow, in axillary, solitary heterogamous heads. Achenes compressed, echinate, two horn like spines at the tip. Pappus absent.

Common on roadsides. Fl. & Fr. All seasons. Kusumasamudram beat, *MSM & KH* 11025.

AGERATUM Linnaeus

Ageratum conyzoides L. Sp. Pl. 839. 1753; FBI 3: 243; Gamble 677(476).

An annual erect herb, up to 70 cm tall, branchlets downy pubescent. Leaves ovate, 4–10 × 2–5 cm, base cuneate, margin bluntly serrate, tip gradually acute, puberulous on both sides. Flowers white in axillary and terminal homogamous heads. Achenes slightly curved, sparsely scabrous.

Common in forests and banks of streams and rivers. Fl. & Fr.: Sept.–Dec. Kanmankalva beat, *MSM* 10467.

BIDENS Linnaeus

Bidens biternata (Lour.) Merr. & Shreff ex Shreff in Bot. Gaz. 88: 293. 1929. *Coreopsis biternata* Lour. Fl. Cochinch. 508. 1790. *Bidens pilosa* auct. non L. 1753; FBI 3: 309; Gamble 709(499).

An erect herb, up to 70 cm tall, branchlets glabrous, striate. Leaves 3-foliolate, terminal leaflets larger than lateral leaflets, obovate, 3–7 × 2–4 cm, base acute, margin crenate-serrate, tip•acuminate, pubescent on both sides. Flowers yellow in terminal solitary, heterogamous head, involucre campanulate. Achenes linear, 1 cm long, glabrous, 2–4-barbellate.

Common in forests and hedges. Fl. & Fr.: May–Sept. Mohammadabad beat, *MSM* 10444.

BLAINVILLEA Cassini

Blainvillea acmella (L.) Phillipson in Blumea 6: 350. 1950. *Verbesina acmella* L. Sp. Pl. 901. 1753. *Eclipta latifolia* L.f. Suppl. Pl. 378. 1781. *Blainvillea latifolia* (L.f.) DC. in Wight. Contrib. Bot. India 17: 1834; FBI 3: 305. *B. rhomboidea* Cass. in F. Cuvier. Sci. Nat. 29: 494. 1823; Gamble 706(496).

An annual erect herb, up to 1 m tall, branchlets scabrid. Leaves ovate-lanceolate, 7–10 × 3–5 cm, base obtuse, margin serrate, tip gradually acuminate, hirsute on both sides. Flowers yellow in axillary, heterogamous heads. Achenes 0.4 cm long.

Occasional weed in waste places and open forests. Mohammadabad beat, *MSM & NV* 10428.

BLUMEA A.P. de Candolle

1. Heads solitary:
   2. Leaves oblong-spathulate .... ***B. obliqua***
   2. Leaves obovate-oblanceolate .... ***B. mollis***
1. Heads in clusters, not solitary .... ***B. hieracifolia***

Blumea hieracifolia (D. Don) DC. in Wight. Contrib. Bot. Ind. 15. 1834. var. *macrostachya* (DC.) Hook. f. Fl. Brit. Ind. 3. 263. 1881; Gamble 687(483).

A scapigerous herb. Basal leaves in rosette, oblanceolate-spathulate, shortly dentate, cauline leaves smaller, oblong, white woolly. Heads in terminal clusters. Achenes ribbed, pubescent first and glabrous later on.

Rare along streams (Seshagiri Rao, 1986). Fl. & Fr.: Aug.–Jan.

Blumea mollis (D. Don) Merr. in Phillipp. J. Sci. 5: 395. 1910. *Erigeron molle* D. Don, Prodr. Fl. Nepal 172. 1825. *Blumea wightiana* DC. in Wight, Contrib. Bot. Ind. 14. 1834; FBI 3: 261; Gamble 686(483). *B. neilgherrensis* Hook. f. Fl. Brit. India 3: 261. 1881; Gamble 686(483).

An annual erect herb, up to 60 cm tall; branchlets ribbed, pubescent. Leaves obovate, 3.5–9.5 × 1–3.5 cm, base attenuate, margin irregularly serrate, tip acute, pubescent on both sides. Heads in axillary and terminal corymbs. Achenes angled, sparsely hairy, not ribbed.

Common in ponds. Fl. & Fr.: Sept.–Jan. Dharmapur *MSM & KH* 10547.

Blumea obliqua (L.) Druce in Bot. Exch. Club. Soc. Brit. Isles 4: 609. 1917; Randeria in Blumea 10. 286. 1960. *Erigeron obliquam* L. Mant.

Pl. 573. 1771. *Blumea amplectans* DC. in Wight. Contrib. Bot. Ind. 13. 1834; FBI 3: 260; Gamble 686(482).

An ascending herb up to 60 cm, stem striate, pubescent. Leaves oblong-spathulate, 1.5–3 × 0.5–1.5 cm, base semi-amplexicaul, margin dentate. Flowers purple and yellow in axillary and terminal heterogamous heads. Achenes angular, hairy, pappus absent.

Occasional in fields and waste lands (Seshagiri Rao, 1986). Fl. & Fr.: Sept.–Mar.

CAESULIA Roxburgh

Caesulia axillaris Roxb. Pl. Cor. 1: 64. t. 93. 1795; FBI 3: 291; Gamble 704(494).

An annual prostrate herb, up to 60 cm, stem glabrous. Leaves lanceolate, 3.5–8 × 0.5–1.5 cm, base attenuate, margin serrate, tip acute, glabrous on both sides. Flowers white in axillary, solitary, homogamous heads. Achenes obovoid, 2-winged, pappus with 2 ovate scales.

Common weed of moist places and streams in forests. Fl. & Fr.: Sept.–Jan. Mohammadabad beat, *MSM* 10493, Kondapur beat, *MSM & KH* 10587; Naskal, *MSM & PRSR* 15114.

CHRYSANTHELLUM L.C. Richard

Chrysanthellum americanum (L.) Vatke. Abh. Nat. Vern. Brem. 9: 122. 1887; *Artemisia americana* L. Sp. Pl. 895. 1753. *Chrysanthellum indicum* DC. Prodr. 5: 631. 1836; FBI 3: 310.

A prostrate herb. Stem glabrous. Leaves pinnatifid, 1–2.5 × 0.5–1.5 cm, lobes sparsely hairy, tip acute. Flowers yellow in axillary and terminal solitary heterogamous heads. Achenes compressed, stipitate, pappus auriculate.

Rare in dry deciduous forest. Fl. & Fr.: Sept.–Jan. Anantagiri RF, *MSM & TP* 11083.

ECHINOPS Liannaeus

Echinops echinatus Roxb. Fl. Ind. 3: 447. 1832; FBI 3: 358; Gamble 274(509).

An annual armed erect herb, up to 50 cm tall, branchlets white cottony pubescent. Leaves lanceolate, pinnatifid, 5–9 × 1–2.5 cm, base amplexicaul, lobes spine-tipped, tip acute, scabrid above, cottony below. Heads greenish-white, 1-flowered, aggregated in terminal clusters, homogamous. Achenes elongate, villous, pappus barbellate.

Common weed of waste lands and roadsides. Fl. & Fr.: Jan.–May. Lankalakunta, *MSM & KH* 10269.

ECLIPTA Linnaeus

Eclipta prostrata (L.) L. Mant. Pl. 286. 1771. *Verbesina prostrata* L. Sp. Pl. 902. 1753. *E. alba* (L.) Hassk. Pl. Jav. Rar. 528. 1848; FBI 3: 304; Gamble 705(496).

An annual prostrate herb with ascending branches, stem hirsute. Leaves ovate-lanceolate, 3–13 × 1–2 cm, base cuneate, margin serrate-dentate, tip acute, hirsute on both sides. Heads white, axillary, solitary, heterogamous. Achenes of ray florets triquetrous, of disc florets compressed, pappus present.

Abundant in moist places. Fl. & Fr.: Mar.–June. Kusumasamudram beat, *MSM & NV* 10350.

EMILIA Cassini

Emilia sonchifolia (L.) DC. in Wight Contrib. Bot. Ind. 24. 1834; FBI 3: 336; Gamble 716(503). *Cacalia sonchifolia* L. Sp. Pl. 835. 1753.

An annual sub-erect herb, up to 60 cm tall, branchlets glabrescent. Leaves both radical and cauline, upper one lanceolate, 5–10 x 1–2.5 cm, base amplexicaul, margin crenate, tip acute, lower one ovate, 4–6 × 2–3.5 cm, base attenuate, margin crenate, tip acute, glabrescent on both sides. Heads purple, in terminal lax corymbs, homogamous. Achenes angled, 5-ribbed, scabrid, pappus present.

Common in all forests and waste lands. Fl. & Fr.: Mar.–May. Anantasagar beat, *MSM & KH* 11005.00

FLAVERIA A.L. Jussieu

Flaveria trinervia (Spreng.) C. Mohr. U.S.D.A. Contrib. Nat. Herb. 6: 810. 1901. *Oedera trinervia* Spreng. Bot. Gart. Halle. 63. 1800. *Flaveria australasica* Hook. in T.L. Mitch. J. Exped. Trop. Austr. 118. 1848; FBI 3: 546; Gamble 711(501).

An annual erect herb, up to 50 cm tall, branchlets glabrous. Leaves oblong-spathulate, 2–4 × 0.5–1 cm, base cuneate, margin dentate-serrate, tip acute, glabrescent above, glaucous below. Heads yellow in terminal corymbs, heterogamous. Achenes laterally compressed, ribbed. Pappus of scarious fimbriate scales.

Occasional weed in waste lands and roadsides. Fl. & Fr.: July–Jan. Mohammadabad beat, *MSM & NV* 10431.

GLOSSOCORDIA Cassini

Glossocordia bosvallea (L.f) DC. in Wight Contrib. Bot. 19. 1934; Gamble 609(499). *Verbesina bosvallea* L.f. Suppl. Pl. 379. 1781. *Glossocordia linearifolia* Cass. in Dict. Sci. Nat. 19: 62. 1821; FBI 3: 308.

An annual prostrate herb. Stem glabrous. Leaves pinnatisect, pinnules 1–2 cm long, filiform, tip acuminate. Flowers yellow in axillary, solitary, heterogamous heads. Achenes compressed, margin ciliate. Pappus of 2 awns.

Common weed in open deciduous forest. Fl. & Fr.: Aug.–Nov. Anantagiri RF, *MSM & TP* 11073.

GNAPHALIUM Linnaeus

Gnaphalium polycaulon Pers. Syn. Pl. 2: 421. 1807. *G. indicum* auct. non L.: FBI 3: 289; Gamble 698(491).

An annual erect herb, up to 10 cm tall, branchlets woolly. Leaves oblong-spathulate, 1–3.5 × 0.3–1 cm, base attenuate, margin entire, tip obtuse, white pilose on both sides. Flowers white in terminal heads, heterogamous. Achenes oblong, beaked, pappus hairy.

Occasional in dried ponds. Fl. & Fr.: Sept.–Jan. Rangammagudem beat, *MSM & KH* 10578; Anantasagar beat, *MSM & KH* 11008.

GRANGEA Adanson

Grangea maderaspatana (L.) Poir. in Lam. Encycl. (Suppl. 2) 825. 1812; FBI 3: 247; Gamble 680(478). *Artemisia maderaspatana* L. Sp. Pl. (ed. 2) 1190. 1763.

An annual prostrate herb, up to 20 cm, branchlets pubescent. Leaves oblong-obovate, pinnatifid, 2–4 × 1–3 cm, base decurrent, margin entire, tip apiculate, pubescent on both sides. Heads yellow, terminal, solitary, heterogamous. Achenes compressed, truncate, pappus margin fimbriate.

Common in dry tanks and ponds. Fl. & Fr.: Feb.–May. Venkatammacheruvu, *MSM & KH* 10268.

LAGASCEA Cavanilles

Lagascea mollis Cav. Anal. Cienc nat. 6: 332. t. 44. 1803; FBI 3: 302; Gamble 703(494).

An annual erect herb, up to 60 cm, branchlets pubescent. Leaves ovate-lanceolate, 3–6 × 1–2.5 cm, base acute, margin serrate, tip acuminate, softly pubescent on both sides. Heads white, terminal, aggregated in dense globose heads, homogamous. Achenes 3-angled, pappus fimbriate.

Common in waste lands. Fl. & Fr.: May–Aug. Rangammagudem beat, *MSM & KH* 10313.

LAUNAEA Cassini

Launaea procumbens (Roxb.) Ramayya & Raja. in Kew Bull. 23: 465. 1969. *Prenanthes procumbens* Roxb. Fl. Ind. 3: 404. 1832. *Launaea nudicaulis* sensu Hook. f. FBI 3. 416. 1881 non *Chondrilla nudicaulis* L., Gamble 733 (515).

An annual suberect herb, up to 60 cm tall. Leaves radical ones rosette, base auricled, margin runcinate, tip apiculate, cauline ones a few, pinnatifid. Flowers cream, homogamous head in terminal racemes. Achenes 4-gonous.

Occasional in hills. Fl. & Fr.: Oct.–Dec. Naskal, *MSM & PRSR* 15116.

OLIGOCHAETA Koch

Oligochaeta ramosa (Roxb.) Wagenitz, Veroff. Geobot. Inst. ETH Stiftung Rubel Zurich 37: 323. 1962. *Carduus ramosus* Roxb. Pl. Ind. 3: 407. 1832. *Volutarella ramosa* (Roxb.) Kitamura, Pl. W. Pakistan & Afghan. 159. 1964. *V. divaricata* Benth. in Benth. & Hook. f. Gen. Pl. 2: 476. 1873; FBI 3: 383; Gamble 727 (571).

An annual erect herb, up to 50 cm tall, stem grooved, pubescent. Leaves lanceolate, pinnatifid, 2–5 × 0.3–1 cm, base amplexicaul, lobes spine tipped, tip acute, pubescent on both sides. Heads purple, terminal, solitary, heterogamous. Achenes ribbed, pappus present.

Occasional in open forests. Fl. & Fr.: June–Sept. Kusumasamudram beat, *MSM & NV* 10352.

PARTHENIUM Linnaeus

Parthenium hysterophorus L. Sp. Pl. 988. 1753; Roll S. Rao in J. Bombay Nat. Hist Soc. 54: 219. 1956.

An annual erect sub-shrub, up tol m tall, branchlets angled, pubescent. Leaves oblong-lanceolate, 3–7 × 0.5–2.5 cm, base decurrent, margin entire, tip acute, pubescent on both sides. Heads white in terminal panicles, heterogamous. Achenes compressed, 3-quetrous, puberulous along sides. Pappus of 2 awns.

Common weed of roadsides, waste lands and open forests. Fl. & Fr.: May–July. Mohammadabad beat *MSM & KH* 10236.

PLUCHEA Cassini

Pluchea tomentosa DC. in Wight Contrib. Bot. Ind. 16. 1834; FBI 3: 272; Gamble 689(485).

A woody under shrub up to 50 cm tall, branchlets tomentose. Leaves obovate-elliptic, 4–6 × 2–5 cm, base obtuse, margin serrate-dentate, tip

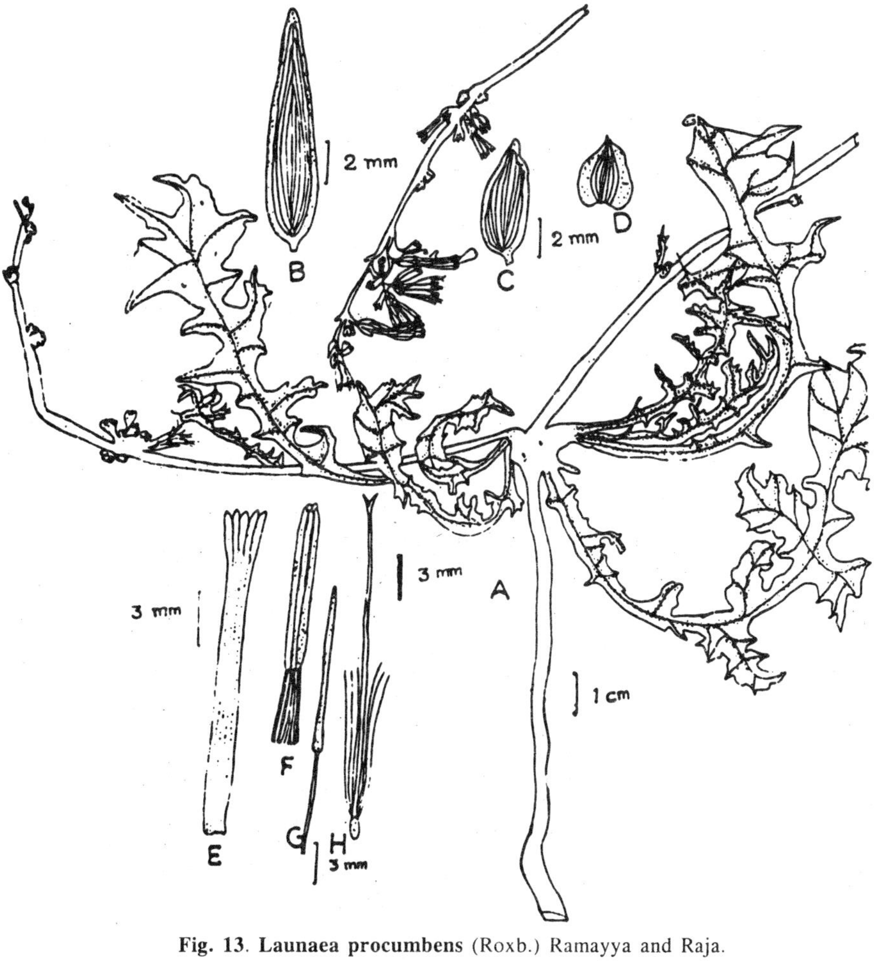

Fig. 13. **Launaea procumbens** (Roxb.) Ramayya and Raja.
A. Twig B.Outer bract C. Middle bract D. Inner bract E. Corolla split open F. Androecium G. Stamen H. Pistil.

obtuse. Flowers yellow or pinkish white in terminal heads. Achenes obscurely angled.

Occasional in forest. Fl. & Fr.: Oct.–Dec. Lingampalle, *MSM* 12743.

PULICARIA Gaertner

Pulicaria wightiana (DC.) Clarke Comp. Ind. 118. 1876; Gamble 702(493). *Poloa wightiana* DC. in Arch. Bot. 2: 514. 1833. *Callistephus wightianus* (DC.) DC Prodr. 5: 275. 1836.

An annual erect herb, up to 60 cm tall, branchlets pubescent. Leaves lanceolate, 2–4 × 0.5–0.8 cm, base amplexicaul, margin serrate, tip acute, pubescent on both sides. Flowers yellow in terminal solitary, heterogamous heads. Achenes ribbed, pappus of filiform hairs.

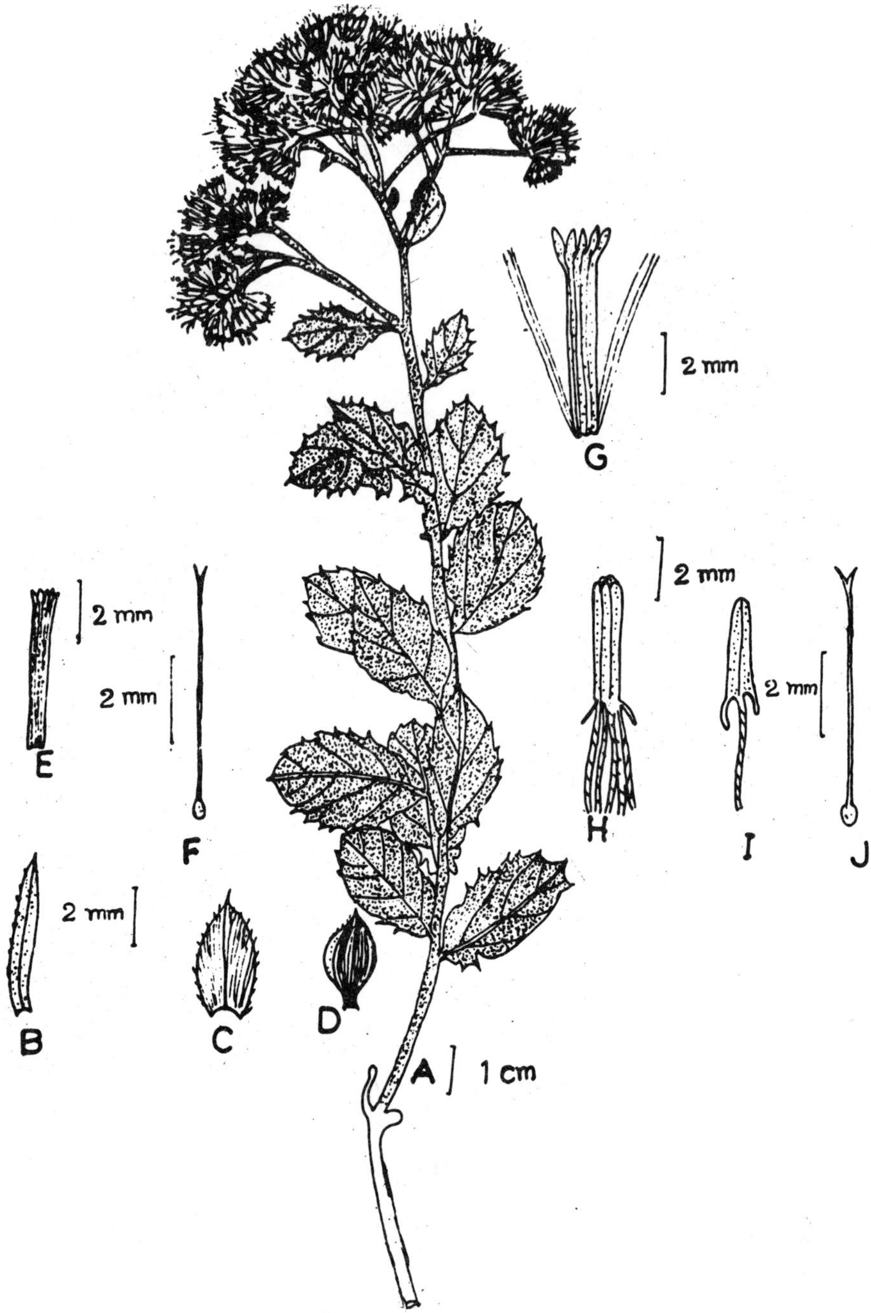

Fig. 14. Pluchea tomentosa DC.
A. Twig B–D. Bracts E.Corolla split open F. Pistil G. Corolla of bisexual flower split open H. Androecium I. Stamen J. Pistil.

Occasional in forests. Fl. & Fr.: Sept.–Jan. Mohammadabad, *MSM* 10437.

SENECIO Linnaeus

Senecio tenuifolius Burm. f. Fl. India 181. t. 60. f. 4. 1768; FBI 3: 345; Gamble 722(508). *S. multidus* Willd. Sp. Pl. 3: 1986. 1803. *Doronicum tenuifolium* Wight, Icon. Pl. Ind. Orient. t. 1129. 1846.

An annual erect herb, up to 50 cm tall, branchlets pubescent. Leaves pinnatifid, 4–6 × 1–2.5 cm, base truncate, margin reflexed, tip acute, glabrescent on both sides. Heads yellow in terminal panicles. Achenes 0.5 cm long, 10-striate, glabrous, pappus yellowish, copious, as long as the achene.

Occasional in forests. Fl. & Fr.: June–Sept. Kusumasamudram beat, *MSM & NV* 10365.

SONCHUS Linnaeus

Sonchus oleraceus L. Sp. Pl. 774. 1753; FBI 3: 414; Gamble 732(514).

An annual erect herb, up to 80 cm tall. Stem grooved. Leaves oblong-obovate, base amplexicaul, margin entire-runcinate-pinnatifid, tip acute. Flowers yellow in irregular umbellate cymes. Achenes compressed, oblong, light brown.

Occasional in forest and waste lands. Fl. & Fr.: Oct.–Dec. Rangammagudem, *MSM & KH* 10314.

SPHAERANTHUS Linnaeus

Sphaeranthus indicus L. Sp. Pl. 927. 1753; FBI 3: 275, *p.p.*; Gamble 692(487).

An annual ascending herb, up to 40 cm, branchlets pubescent, winged, dentate. Leaves ovate-spathulate, 1.5–3.5 × 0.5–2.5 cm, base decurrent, margin coarsely serrate-dentate, tip obtuse, hirsute above, pubescent below. Heads violet, terminal, solitary, heterogamous. Achenes glabrous. Pappus absent.

Common weed in waste lands, open forests, dry tanks, ponds. Fl. & Fr.: May–Aug. Mohammadabad beat, *MSM & KH* 10205.

TRICHOLEPIS A.P. de Candolle

Tricholepis radicans DC. Prodr. 6: 564. 1838; FBI 3: 381; Gamble 726(511).

An annual prostrate herb up to 50 cm. Leaves basal obovate, cauline oblong, 2–5 × 0.5–2.5 cm, base obtuse, margin minutely toothed, tip subacute. Flowers blue in woolly heads. Achenes oblong, smooth, pappus scales many.

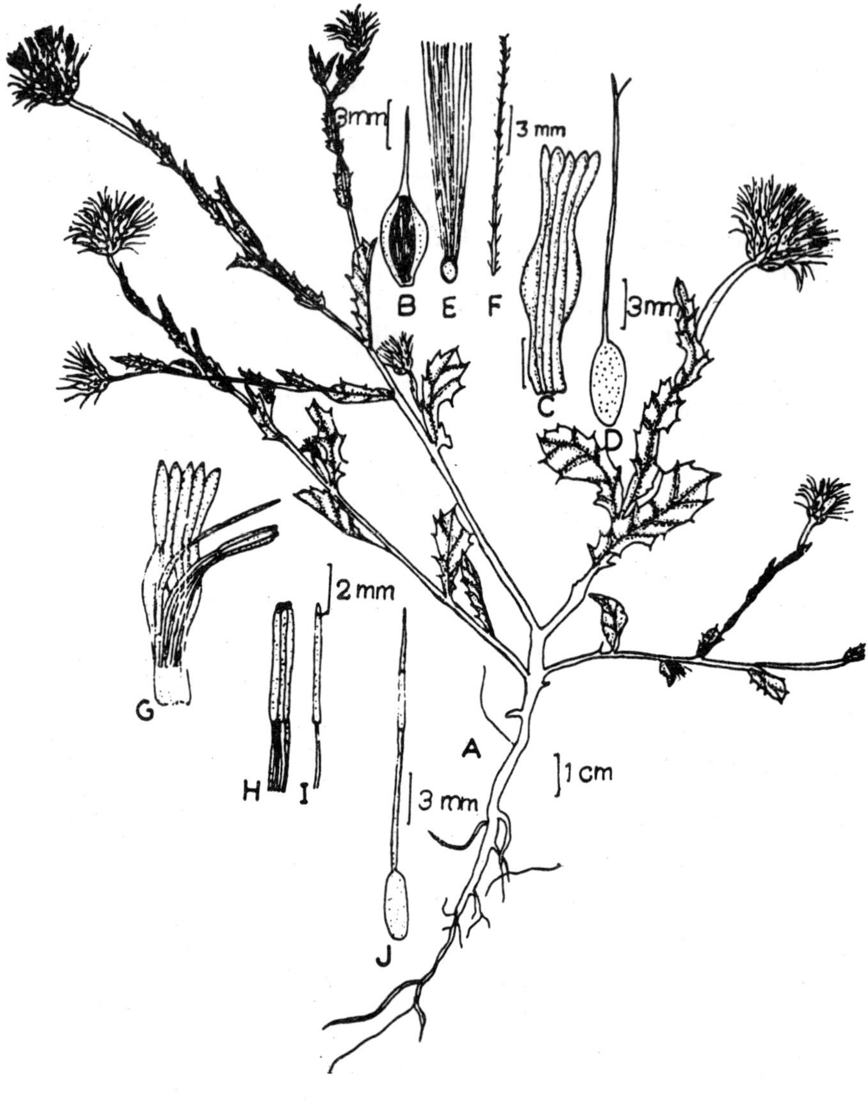

Fig. 15. Tricholepis radicans DC.
A. Twig B. Bract C. Corolla split open D. Pistil E–F. Pappus G. Corolla of bisexual flower split open H. Androecium I. Stamen J. Pistil.

Occasional in waste lands. Fl. & Fr.: Dec.–Mar. Naskal, *MSM & PRSR* 15115.

TRIDAX Linnaeus

Tridax procumbens L. Sp. Pl. 900. 1753; FBI 3: 311; Gamble 711(500).

A perennial decumbent herb up to 50 cm, stem hirsute. Leaves ovate, 2–6 × 1–2.5 cm, base cuneate, margin dentate, tip acute. Flowers yellow in terminal solitary heterogamous head. Achenes oblong, pappus of feathery hair.

A common weed of waste lands. Fl. & Fr.: All seasons. *Vern.*: Gaddi chamanti. Vikarabad, *MSM* 19238.

VERNONIA Schreber *nom. cons.*

1. Leaves white, tomentose below .... *V. albicans*
1. Leaves pubescent below:
   2. Achenes ribbed .... *V. anthelmintica*
   2. Achenes not ribbed .... *V. cinerea*

Vernonia albicans DC. in Wight Contrib. Bot. Ind. 6. 1834; Gamble 676(475). *V. cinerea* sensu FBI 3: 233. *p.p.*

An annual erect herb, up to 60 cm tall. Leaves ovate-rhomboid, 3–5 × 0.5–2.5 cm, base truncate, margin serrate, tip obtuse, pubescent above tomentose below. Flowers yellow in axillary and terminal panicled heads. Achenes villous, angled.

Occasional in waste lands. Sept.–Jan. Central University, *MSM* 15144; Gacchibowli, *MSM* 15199.

Vernonia anthelmintica (L.) Willd. Sp. Pl. 3. 1634. 1803; FBI 3: 236. *Conyza anthelmintica* L. Sp. Pl. (ed. 2) 1207. 1763. *Centratherum anthelminticum* (L.) O. Kuntze Rev. Gen. Pl. 1: 320. 1891; Gamble 667(469).

An annual erect herb, up to 1 m tall, stem grooved, hairy. Leaves elliptic-oblong, 8–10 × 3–4 cm, base cuneate, margin serrate, tip acute-obtuse, glabrescent above and thinly pubescent below. Flowers pink or violet in terminal peduncled solitary heads. Achenes oblong, ribbed, pappus woolly, white.

Occasional in open forest. Fl. & Fr.: Dec.–Mar. Central University, *MSM & TP* 15144.

Vernonia cinerea (L.) Less. in Linnaea 4: 291. 1829; FBI 3: 233; Gamble 676(475). *Conyza cinerea* L. Sp. Pl. 862. 1753.

An annual erect herb, up to 60 cm tall, branchlets ribbed, pubescent. Leaves ovate, 2–6 × 1–2.5 cm, base decurrent into petiole, margin serrate-dentate, tip acute, sparsely hairy above, pubescent below. Heads violet in terminal corymbose panicles, homogamous. Achenes villous, pappus present.

Common in forests and waste lands. Fl. & Fr.: July–Feb. Mohammadabad beat, *MSM & NV* 10417.

WEDELIA N.J. Jacquin

Wedelia urticaefolia DC. in Wight. Contrib. Bot. Ind. 18. 1834; FBI 3. 306. Gamble 707(497).

An erect herb up to 50 cm tall. Leaves ovate, 3–5 × 2–4 cm, base truncate, margin serrate, tip acute. Flowers yellow in axillary and terminal heads. Achenes angled, pubescent.

Occasional in moist places. Fl. & Fr.: Aug.–Nov. Anantagiri RF, *MSM & TP* 11055.

XANTHIUM Linnaeus

Xanthium strumarium L. Sp. Pl. 987. 1753; FBI 3: 303; Gamble 702(494).

An under shrub up to 1 m tall. Stem grooved. Leaves cordate, 3-lobed, 10–12 × 10–15 cm, rough, irregularly toothed. Flowers yellowish green in axillary and terminal heads. Achenes enlarged in prickly glandular envelope.

Common in moist places. Fl. & Fr.: All seasons. Gacchibowli *MSM* 12774.

Helianthus annuus L. (*Vern.* Suryakanthi) is commonly cultivated for its seed oil. *Chrysanthemum indicum* L. (*Vern.:* Chamanti) species of *Cosmos, Dahlia, Solidago canadenis* L., *Tagetus erecta* L., *T. patula* L. (Banti), *Tithonia diversifolia* (Hems.) A. Gray and *Vernonia elaeagnifolia* DC. are planted for ornamental purpose.

LOBELIACEAE

LOBELIA Linnaeus

Lobelia alsinoides Lam. Encycl. 3: 588. 1792. *L. trigona* Roxb. Fl. Ind. 2: 111. 1824; FBI 3: 423; Gamble 736(518).

An annual ascending herb, up to 15 cm, branchlets glabrous. Leaves ovate, 0.5–1.5 × 0.5–1 cm, base truncate, margin crenate-dentate, tip acute, glabrous on both sides. Flowers blue, axillary, solitary. Capsules 0.4 cm long, glabrous, many-seeded.

Occasional in marshy places of forests. Fl. & Fr.: Dec.–Feb. Mohammadabad beat, *MSM & KH* 10600; Anantasagar beat, *MSM & KH* 11011.

PLUMBAGINACEAE

PLUMBAGO Linnaeus

Plumbago zeyianica L. Sp. Pl. 151. 1753; FBI 3: 480; Gamble 744(524).

A perennial climbing shrub, up to 1 m, branchlets striate, glabrous. Leaves elliptic-ovate, 2–6.5 × 0.5–3 cm, base attenuate, margin entire, tip acute, glabrous on both sides. Flowers white in terminal spikes. Calyx tube 1.5 cm long, conspicuously glandular. Capsules 1 cm long, glabrous, 1-seeded.

Occasional in hedges and forest outskirts. Fl. & Fr.: June–Dec. Kommi Reddipally, *MSM & NV* 10326.

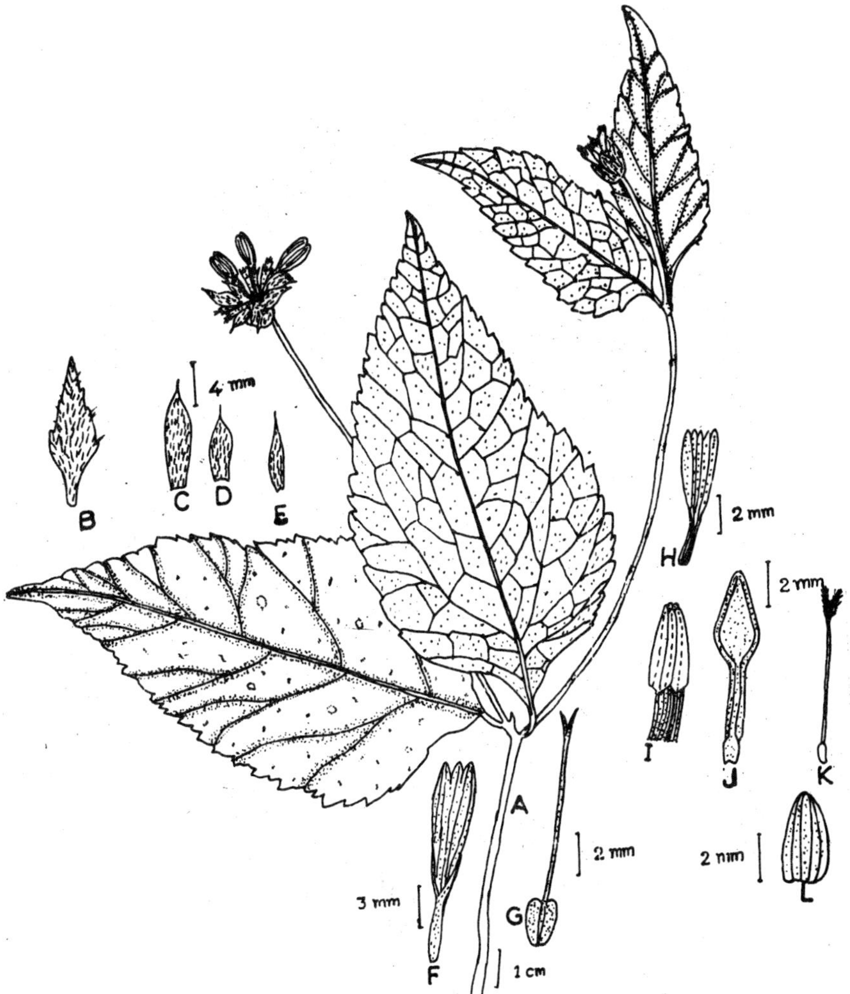

Fig. 16. Wedelia urticaefolia DC.
A. Twig B–E. Bracts F. Corolla split open G. Pistil H. Corolla of bisexual flower split open I. Androecium J. Stamen K. Pistil L. Seed.

SAPOTACEAE

MADHUCA Hamilton ex J.P. Gmelin

Madhuca indica Gmel. in L. Syst. Nat. 2(1): 799. 1791. *Madhuca longifolia* (Koenig) Macbr. var. *latifolia* A. Chev. Rev. Int. Bot. Appl. Agric. Trop. 23: 149. 1943. *Bassia latifolia* Roxb. Pl. Cor. t. 19. 1975; FBI 3: 544; Gamble 763(536); Khan in FFH 211.

A deciduous tree, up to 20 m tall, branchlets tomentose, bark thick, dark coloured, cracked, inner bark red. Leaves elliptic-ovate, 10–23 × 7–12 cm, base cuneate, margin entire, tip apiculate, glabrous on both sides. Flowers creamish-white in axillary dense fascicles. Berries tawny-tomentose, 1–4-seeded.

Abundant in all forests. Ocassional in hills and villages. Fl. & Fr.: Dec.–May. *Vern.:* Ippa. *Uses:* Flowers are eaten either raw or cooked and wine is extracted from the flowers. Mohammadabad beat, *MSM & KH* 10224.

Manilkara zapota (L.) P. Royen (*Vern.:* Sapota) is planted for its delicious fruits.

EBENACEAE

DIOSPYROS Linnaeus

1. Armed trees. Female flowers sessile, calyx lobes hardly enlarged in fruit .... ***D. chloroxylon***
1. Unarmed trees. Female flowers pedicellate, calyx lobes enlarged in fruit:
   2. Flowers yellow, 2–2.5 cm across. Stamens 2–4 .... ***D. peregrina***
   2. Flowers whitish, 0.5–1.5 cm across, stamens 16:
      3. Leaves tomentose, drying green. Style 1. Fruiting calyx slightly enlarged .... ***D. melanoxylon***
      3. Leaves glabrous, drying black. Styles 4. Fruiting calyx spreading or reflexed .... ***D. montana***

Diospyros chloroxylon Roxb. Pl. Cor. t. 49. 1795; FBI 3: 560; Gamble 776(546); Khan in FFH 217.

An armed tree, up to 5 m tall, branchlets rusty-tomentose. Leaves elliptic-ovate or obovate, 1.5–7 × 1–3.5 cm, base truncate-rounded, margin entire, tip obtuse, appressed hairy on both sides. Flowers white in axillary cymes. Male in clusters of 4–6, subsessile. Female solitary, sessile. Berries 1–2 cm long, glabrous, 2–3-seeded.

Common in all forests. Fl. & Fr.: May–Dec. *Vern.*: Illinta. Mohammadabad beat, *MSM & KH* 10247, 10248.

Diospyros melanoxylon Roxb. Pl. Cor. t. 46. 1795; FBI 3: 564; Gamble 772(543); Khan in FFH 218. *D. tupru* Buch.-Ham. in Trans. Linn. Soc. London 15: 111. 1827; FBI 3: 563.

A deciduous tree, 2 m tall, branchlets rusty tomentose, bark grey. Leaves elliptic-ovate, 5–20 × 3–10 cm, base acute-rounded, margin entire, tip obtuse, tomentose above and densely tomentose below. Flowers white, male in clusters and female solitary. Berries globose, 3 cm long, glabrous, 3–8-seeded.

Common in dry deciduous forests. Fl. & Fr.: Feb.–July. *Vern.:* Tunki, Abnoos. *Uses.*: Leaves used in making local cigarettes called beedies. Kusumasamudram beat, *MSM & KH* 11035.

Diospyros montana Roxb. Pl. Cor. t. 48. 1795; FBI 3: 555; Gamble 776(545); Khan in FFH 216.

A deciduous tree, up to 10 m tall, branchlets glabrous, bark bluish-grey. Leaves elliptic-ovate, 3.5–13 × 2–6 cm, base truncate, margin entire, tip obtusely acuminate, glabrescent above, thinly tomentose below. Flowers white in axillary cymes, male 2–8 in long cymes. Female solitary on drooping peduncle. Berries 2 cm long, glabrous with reflexed calyx, 2–8-seeded.

Rare in dry deciduous forests. Fl. & Fr.: Jan.–May. *Vern.:* Muchi-tunki. Mohammadabad beat, *MSM & KH* 10225.

Diospyros peregrina (Gaertn.) Guerke in Engl. Nat. Pflanzenfam. 1: 164. 1891; Gamble 797(546). *Embryopteris peregrina* Gaertn. Fruct. 1: 145. t. 29. f. 2. 1788. *Diospyros embryopteris* Pers. Syn. Pl. 2: 624. 1807; FBI 3: 556. 1882.

An evergreen tree, up to 10 m tall, bark dark grey smooth. Leaves oblong, thick, base rounded, margin entire, tip acute. Flowers yellow, dioecious. Male flowers few in axillary cymes. Female solitary. Berries globose.

Occasional in forests along streams (Seshagiri Rao, 1986). Fl. & Fr.: Mar.–May.

OLEACEAE

JASMINUM Linnaeus

1. Erect shrubs. Leaf base narrowed .... *J. cuspidatum*
1. Climbing shrubs. Leaf base rounded or truncate .... *J. auriculatum*

Jasminum auriculatum Vahl, Symb. Bot. 3: 1. 1794; FBI 3: 600; Gamble 795(555).

A climbing shrub, branchlets pubescent. Leaves unipinnate, paripinnate, leaflets ovate, 3–4 pairs, 1.5–3 × 1–2 cm, base rounded or truncate, margin entire, tip acute, sparsely pubescent on both sides. Flowers white in terminal cymes. Berries 0.7 cm long, glabrous, black when ripe, 1-seeded.

Occasional in forests. Fl. & Fr.: June–Dec. *Vern.:* Adavi malli. Kusumasamudram beat, *MSM & NV* 10356.

Jasminum cuspidatum Rottl. Ges. Naturf. Freunde. Berlin Neue Schriften 4: 192. 1803. *J. rigidum* Zenker Pl. Ind. 5. t. 6. 1835; FBI 3: 598; Gamble 790(555).

A woody shrub, up to 4 m tall. Leaves elliptic, 2–8 × 1–3 cm, base narrowed, margin entire, tip acute. Flowers white in axillary and terminal cymes. Berries globose, black when ripe.

Occasional in forests (Seshagiri Rao, 1986). Fl. & Fr.: Aug.–Dec. *Vern.:* Adavi malli. Central University, *KSR* 68.

Jasminum azoricum L. (*Vern.:* Sannajaji), *J. grandiflorum* L. (*Vern.:* Jaji) and *J. sambac* (L.) Ait. (*Vern.* Malli) are commonly planted for their fragrant flowers.

NYCTANTHACEAE

NYCTANTHES Linnaeus

Nyctanthes arbor-tristis L. Sp. Pl. 6. 1753; FBI 3: 603; Gamble 792(556); Khan in FFH 222.

A deciduous tree, up to 6 m tall, branchlets quadrangular, pubescent. Leaves ovate-lanceolate, 3–11 × 1.5–6.5 cm, base rounded, margin entire, tip acuminate, scabrid above, pubescent below. Flowers white in axillary and terminal fascicles. Capsules 1.5 cm long, glabrous, 2-seeded.

Common in forests. Fl. & Fr.: Sept.–Jan. *Vern.:* Parijatham. Kothapalli beat, *MSM & KH* 10446.

APOCYNACEAE

1. Trees or shrubs:
   2. Fruits long follicles:
      3. Flowers cream .... ***Plumeria***
      3. Flowers white:
         4. Corolla with coronary scales and exerted stamens. Follicles stout .... ***Wrightia***
         4. Corolla without coronary scales and inserted stamens. Follicles slender .... ***Holarrhena***
   2. Fruits bloated drupes .... ***Cascabela***
1. Shrubs or herbs:
   5. Armed shrubs. Berries black .... ***Carissa***
   5. Unarmed shrubs:
      6. Erect shrubs .... ***Rauvolfia***
      6. Climbing shrubs:
         7. Liane. Flowers white .... ***Vallaris***
         7. Climber. Flowers cream .... ***Ichnocarpus***

CARISSA Linnaeus *nom. cons.*

Carissa spinarum L. Mant. 559. 1771; FBI 3: 631; Gamble 805(565).

A shrub, up to 1 m tall, branchlets zig-zag, thorns divaricate, straight. Leaves ovate-elliptic, 3–3.5 × 1–2 cm, base cuneate, margin entire, tip acute, young leaves pubescent. Flowers creamish-white in terminal and axillary cymes. Berries globose, black when ripe.

Occasional in scrub jungle. Fl. & Fr.: Dec.–Mar. Central University, *MSM* 15154.

CASCABELA Rafinesque

Cascabela thevitia (L.) Lippold in Feddes, Repert. 91: 52. 1980. *Cerbera thevetia* L. Sp. Pl. 209. 1753. *Thevetia peruviana* (Pers.) Merr. in Phillip. J. Sci. 9: 130. 1914. *Cerbera peruviana* Pers., Syn. Pl. 1: 267. 1809. *Thevetia nerifolia* Juss. ex Steud. Nom. Bot. (ed. 2) 2: 680. 1841 (*nereifolia*), Gamble 821(577).

A small tree, up to 5 m tall. Leaves linear lanceolate, 4–10 × 0.2–0.5 cm, base cuneate, margin entire, tip acute. Flowers yellow in axillary cymes. Drupes elliposid-oblong, bloated, laterally compressed, glabrous.

Occasional in villages. Fl. & Fr.: Sept.–Dec. *Vern.*: Devaganneru, Mohammadabad, *MSM & PRSR* 15125.

CATHARANTHUS G. Don

1. Leaves elliptic lanceolate. Flowers small .... ***C. pusillus***
1. Leaves ovate-obovate. Flowers large .... ***C. roseus***

Catharanthus pusillus (Murr.) G. Don Gen. Syst. 4: 95. 1837. *Vinca pusilla* Murr. Nov. Comment. Soc. Regiae Sci. Gott. 3: 66. t. 2. f. 1. 1773; FBI 3: 640. *Lochnera pusilla* (Murr.) Schum. in Engl. Nat. Pflanzenf. IV 2. 145. 1895; Gamble 809(568).

An annual erect herb, up to 40 cm tall, branchlets terete below, quadrangular above, glabrous. Leaves elliptic-lanceolate, 3.5–6 × 1–2 cm, base acute, margin entire, tip acuminate, glabrous above, glaucous below. Flowers white in axillary corymbs. Follicles 4 cm long, glabrous, many-seeded.

Common weed of marshy places. Fl. & Fr.: Feb.–June. Kusumasamudram beat, *MSM & NV* 10349.

Catharanthus roseus (L.) G. Don Gen. Syst. Hist. 4: 1837. *Vinca rosea* L. Syst. Nat. ed. 10. 944. 1759; FBI 3: 640. *Lochnera rosea* (L.) Reichb. Consp. Regni. Veg. 134. 1828; Gamble 809(569).

A perennial erect herb, up to 50 cm tall, branchlets glabrous. Leaves elliptic-obovate, 3–7.5 × 2–4 cm, base cuneate, margin entire, tip mucronate. Flowers white or pink, axillary, solitary. Follicles linear, puberulous, many-seeded.

Common in grave yards and tombs, also cultivated as an ornamental. Fl. & Fr.: All seasons. *Vern.:* Billa ganneru. Central University, *MSM* 19210.

HOLARRHENA Robert Brown

Holarrhena pubescens (Buch.-Ham.) Wall. ex G. Don Gen. Syst. 4: 78. 1837. *Echites pubescens* Buch.-Ham. in Trans. Linn. Soc. London 13: 521. 1821. *Holarrhena antidysenterica* (Roth) Wall. ex DC. Prodr. 8:

413. 1844; FBI 3: 644; Gamble 811(570); Khan in FFH 230. *Echites antidysenterica* Roth Nov. Pl. Sp. 138. 1821.

A deciduous tree, up to 5 m tall, branchlets pubescent. Leaves elliptic-oblong, 9–20 × 3.5–10 cm, base rounded, margin entire, tip acuminate, sparsely pubescent above, densely pubescent below. Flowers white in axillary and terminal cymes, fragrant. Follicles paired, 14 cm long, glabrous, curved, many-seeded.

Common in dry deciduous forests. Fl. & Fr.: May–Dec. Mohammadabad beat, *MSM & KV* 10202; Nayanigutta, *MSM & NV* 10328; Anantasagar beat, *MSM & NV* 10405.

ICHNOCARPUS Robert Brown *nom. cons.*

Ichnocarpus frutescens (L.) R. Br. in Ait. Hort. Kew. ed. 2. 2: 69. 1811; FBI 3: 669. 1882; Gamble 820(577). *Apocynum frutescens* L. Sp. Pl. 213. 1753.

A climbing shrub, branchlets rusty tomentose. Leaves elliptic-oblong, 4–9 × 1.5–4.5 cm, base rounded, margin entire, tip acuminate, glabrous on both sides, flowers cream in axillary or terminal cymes. Follicles paired, 10 cm long, rusty pubescent, many-seeded.

Common in dry deciduous forests. Fl. & Fr.: Sept.–Jan. Anantasagar beat, *MSM & KH* 10510.

PLUMERIA Linnaeus

Plumeria rubra L. Sp. Pl. 209. 1753. *P. acuminata* R. Br. in Ait. Hort. Kew. 2. 2. 1789. *P. acutifolia* Poir. in Lam. Encycl. Suppl. 2: 667. 1812; FBI 3: 641; Gamble 2: 821(577).

A small tree, up to 5 m tall, branchlets fleshy with milky sap. Leaves oblong-lanceolate, 20–25 × 7–8 cm, base acute, margin entire, tip acuminate. Flowers cream in terminal many-flowered thyrsi. Follicles paired, fistulate. Seeds flat, winged.

Cultivated mostly in temples (Seshagiri Rao, 1986). Fl. & Fr.: Jan.–Aug. *Vern.:* Deva ganneru.

RAUVOLFIA Linnaeus

Rauvolfia tetraphylla L. Sp. Pl. 208. 1753. *R. canescens* L. Sp. Pl. (ed. 2) 303. 1762; Gamble 803(568).

A shrub, up to 2.5 m tall, branchlets young pubescent. Leaves 3 or 4 whorled, elliptic-ovate, 1–2 × 1–3 cm. Flowers white in terminal, 5–7-flowered corymb. Drupes brown or red when ripe.

As an escape in forest. Fl. & Fr.: July–Dec. Central University, *MSM* 19203.

VALLARIS N.L. Burman

Vallaris solanacea (Roth) O. Kuntze Rev. Gen. Pl. 2: 447. 1891; Gamble 85(573). *Peltanthera solanacea* Roth Nov. Pl. Sp. 132. 1821. *Vallaris heynei* Spreng. in L. Syst. Veg. 1: 635. 1824; FBI 3: 650. *V. dichotoma* (Roxb.) Wall. ex G. Don, Gen. Syst. 4: 79. 1837. *Echites dichotoma* Roxb. Fl. Ind. 2: 19. 1832.

A liane, up to 20 m long, tender parts pubescent. Leaves elliptic-oblong, 3–8 × 2–4 cm, base attenuate, tip acuminate. Flowers white in axillary 3–10-flowered lax cymes.

Occasional in forest (Seshagiri Rao, 1986). Fl. & Fr.: Mar.–Dec.

WRIGHTIA Robert Brown

1. Leaf apex caudate-acuminate. Follicles connnate throughout .... *W. arborea*
1. Leaf apex abruptly acuminate. Follicles connate apically .... *W. tinctoria*

Wrightia arborea (Dennst.) Mabb. in Taxon 26: 533. 1977. *Periploca arborea* Dennst. Schlusseel Hortus Malab. 13: 23 & 25. 1818 & in Fortsetz. Allg. Tent. Sch. Gart. Mag. 3: 32. 41 & 83. 1818. *Wrightia tomentosa* Roem. & Schult. in Linn. Syst. Veg. 4. 144. 1819; FBI 3: 653; Gamble 816(573).

A deciduous tree, up to 5 m tall, branchlets tomentose. Leaves elliptic-oblong, 8–10 × 3.5–5 cm, base acute, margin entire, tip acuminate, silky tomentose on both sides. Flowers yellow in terminal cymes. Follicles connate throughout, 15–20 cm long, pustulate, apically beaked.

Occasional in forest. Fl. & Fr.: June–Dec. Lothuvagu, *MSM & KH* 10292; Anantagiri, *MSM & TP* 11088.

Wrightia tinctoria R. Br. Mem. Wern. Nat. Hist. Soc. 1: 74. 1811; FBI 3: 653; Gamble 815(573); Khan in FFH 232.

A deciduous tree, up to 10 m tall, branchlets pubescent. Leaves elliptic, lancelolate, 5–15 × 2.5–8 cm, base obtuse, margin entire, tip acuminate, grey pubescent on both sides. Flowers greenish-white in terminal cymes. Follicles paired, 25–30 cm long, terete, glabrous, curved, apically coherent, many-seeded.

Frequent in dry deciduous forests. Fl. & Fr.: May–Dec. *Vern.*: Palakodisha. Mohammadabad beat, *MSM & KH* 10203; Kondapuram beat, *MSM & KH* 10320.

Tabernaemontana divaricata (L.) R. Br. (*Vern.*: Nandivardhanam) is planted for ornamental purpose.

ASCLEPIADACEAE

1. Erect herbs or shrubs:
  2. Mature plant succulent, angular, leafless .... ***Caralluma***
  2. Mature plant woody, terete, leafy .... ***Calotropis***
1. Climbers or straggling lianes:
  3. Leafless plants, branchlets jointed .... ***Sarcostemma***
  3. Leafy plants, branchlets not jointed:
    4. Leaves cordiform:
      5. Peduncle to 12 cm long. Corona double. Follicles 6 × 2 cm, softly echinate .... ***Pergularia***
      5. Peduncle to 2.5 cm long. Corona single. Follicles 9 × 2.5 cm, smooth .... ***Wattakaka***
    4. Leaves other than cordiform:
      6. Corolla tube inflated basally .... ***Ceropegia***
      6. Corolla tube not inflated:
        7. Inflorescence an umbel:
          8. Leaf apex apiculate or mucronate. Pedicel to 1.5 cm .... ***Tylophora***
          8. Leaf apex acute. Pedicel 0.7–1 cm long:
            9. Leaves elliptic to obovate, 1.5–5 × 1.4 cm .... ***Gymnema***
            9. Leaves elliptic-oblong or lanceolate, 4–7 × 1.5–3 cm ... ***Leptadenia***
        7. Inflorescence not an umbel:
          10. Corona double:
            11. Leaves longer than broad, base truncate .... ***Oxystelma***
            11. Leaves as broad as long, base attenuate-truncate ... ***Decalepis***
          10. Corona single:
            12. Anthers free, distinct .... ***Cryptostegia***
            12. Anthers adnate to stigma, indistinct .... ***Hemidesmus***

CALOTROPIS Robert Brown

1. Flowers pale white .... ***C. gigentea***
1. Flowers purple .... ***C. procera***

Calotropis gigantea (L.) R. Br. in Ait. f. Hort. Kew. ed. 2. 2: 78. 1811; FBI 4: 17; Gamble 832(585); Khan in FFH 237. *Asclepias gigantea* L. Sp. Pl. 214. 1753.

A shrub, up to 1 m tall. Leaves ovate-obovate, 4–10 x 2–5 cm, base amplexicaul, margin entire, tip acuminate. Flowers white in axillary and terminal umbellate cymes. Follicles oblong, pubescent.

Occasional in waste lands and roadsides. Fl. & Fr.: All seasons. *Vern.:* Tella jilledu. Central University, *MSM* 15182.

Calotropis procera (Ait.) R. Br. in Ait. f. Hort. Kew. ed. 2. 2: 78. 1811; Gamble 832(585); Khan in FFH 238. *Asclepias procera* Ait. Hortus Kew. ed. 1. 1: 305. 1789.

A shrub, up to 2 m tall, branchlets white woolly. Leaves obovate, 3–7 × 2–5 cm, base amplexicaul, margin entire, tip acute. Flowers purple in umbellate cymes, follicles oblong, pubescent.

Common in waste lands and on roadside. Fl. & Fr.: All seasons. *Vern.:* Jilledu, Chevella *MSM* 19239.

CARALLUMA Robert Brown

Caralluma adscendens (Roxb.) Haw. Syn. Pl. Succ. 47. 1812. var. **attenuata** (Wight) Grav. & Mayuranathan in Bull. Madras Govt. Museum. 4. 13. 1931. *C. attenuata* Wight, Ic. t. 1268. 1848; FBI 4: 76; Gamble 861(605).

A perennial erect leafless herb, up to 50 cm tall, stem 4-angled, succulent, leaf-scars spine like. Flowers purple in terminal solitary or paired racemes. Follicles linear, 12 cm long, terete, glabrous. Seeds with silky white coma.

Rare in open forests and scrub jungles. Fl. & Fr.: May–Dec. Kusumasamudram beat, *MSM & NV* 10362; Rangammagudem beat, *MSM & KH* 10317.

CEROPEGIA Linnaeus

Ceropegia candelabrum L. Sp. Pl. 211. 1753; FBI 4: 70; Gamble 857(603).

A climbing herb. Stem glabrous. Leaves ovate-oblong, 2–5 × 0.5–2.5 cm, base acute, margin entire, tip apiculate. Flowers greenish purple in long lateral umbellate cymes. Follicles paired, glabrous.

Rare in forest. Fl. & Fr.: Aug.–Nov. Anantagiri RF, *MSM & TP* 11059.

CRYPTOSTEGIA Robert Brown

Cryptostegia grandiflora R. Br. Bot. Reg. t. 435. 1820; FBI 4: 6; Gamble 862(606); Khan in FFH 237.

A climbing shrub. Leaves elliptic-obovate, 5–10 × 3–5 cm, base cuneate, margin entire, tip acute. Flowers purple in terminal cymes. Follicles in pairs, ovate-oblong, angled.

Occasional in hedges. Fl. & Fr.: All seasons. *Vern.:* Rabbarupulu. Central University, *MSM* 19218.

DECALEPIS Wight & Arnott

Decalepis hamiltonii Wight & Arn. in Wight Contrib. Bot. Ind. 64. 1834; FBI 4: 11: Gamble 828(582).

A deciduous straggler, latex milky. Leaves oblong-orbicular, 3–6 × 1.5–5 cm, base truncate, margin entire, tip acute. Flowers yellow in dichotmous cymes. Follicles oblong, cylindric. Seeds ovate.

Rare in forests (Seshagiri Rao, 1986). Fl. & Fr.: Mar.–Sept.

GYMNEMA Robert Brown

Gymnema sylvestre (Retz.) R. Br. ex Roem. & Schult. in L. Syst. Veg. 6: 57. 1820; FBI 4: 29; Gamble 839(590). *Periploca sylvestris* Retz. Obs. Bot. 2; 15. 1781.

A climbing shrub, branchlets yellowish tomentose. Leaves elliptic-ovate, 2.5–6 × 1.5–3.5 cm, base rounded, margin entire, tip acuminate, pubescent on both sides. Flowers whitish cream in axillary umbels. Follicles 7 cm long, glabrous, one follicle often suppressed. Seeds brown.

Common in dry deciduous forests. Fl. & Fr.: June–Oct. Kusumasamudram beat, *MSM & NV* 10367.

HEMIDESMUS Robert Brown

1. Leaves appressed pubescent on both sides .... ***H. indicus*** var. ***pubescens***
1. Leaves glaucous above, pubescent below .... ***H. indicus*** var. ***indicus***

Hemidesmus indicus (L.) R. Br. in Ait. Hort. Kew. ed. 2. 2: 75. 1811 *var.* **indicus;** FBI 4: 5. 1883; Gamble 825(580). *Periploca indica* L. Sp. Pl. 211. 1753.

A climbing shrub, branchlets glabrous. Leaves elliptic-oblong, 3–8 × 1–4 cm, base round, margin entire, tip obtuse-mucronate, glaucous above, pubescent below. Flowers yellowish-brown crowded in axillary cymes. Follicles slender, tapering to a point at the tip, glabrous.

Occasional in both plains and forest. Fl. & Fr.: Sept.–Feb. *Vern.:* Sugandhapala, Nannari. Central University, *MSM* 15152.

Hemidesmus indicus (L.) R. Br. var. **pubescens** (Wight & Arn.) Hook. f. in FBI 4: 5. 1883; Gamble 825 (580). *H. pubescens* Wight & Arn in Wight Contrib. Bot. 63. 1834.

A climbing shrub, branchlets pubescent. Leaves elliptic-oblong, 1.5–5.5 × 1–3.5 cm, base truncate, margin entire, tip mucronate, appressed pubescent on both sides. Flowers purple in axillary fascicles. Follicles paired, 0.8 cm long, striate, glabrous.

Common in all forests and in hedges. Fl. & Fr.: June–Dec. Kusumasamudrum beat, *MSM & NV* 10333.

LEPTADENIA Robert Brown

Leptadenia reticulata (Retz.) Wight & Arn. in Wight. Contrib. Bot. Ind. 47. 1834; FBI 4: 63; Gamble 850(598); Bullock in Kew Bull. 10. 291. 1955. *Cynanchum reticulatum* Retz. Obs. Bot. 2: 15. 1781.

A diffuse straggling shrub. Leaves ovate, 4–7 × 2–7 cm, base obtuse, margin entire, tip acute. Flowers green in axillary or terminal umbellate cymes. Follicles paired, cylindric with curved beak. Seeds brown.

Common in forests (Seshagiri Rao, 1986). Fl. & Fr.: May–Dec.

OXYSTELMA Robert Brown

Oxystelma esculentum (L.f.) R. Br. ex Schult. in L. Syst. Veg. 6: 89. 1820; FBI 4: 17; Gamble 833 (586). *Periploca esculenta* L.f. Suppl. Pl. 168. 1781.

A twining herb, latex milky. Leaves linear-lanceolate, 5–8 × 0.5–1 cm, base truncate, margin entire, tip mucronate, glabrous on both sides. Flowers purple or pink, axillary, solitary. Follicles ovoid. Seeds ovate with silky coma.

Occasional on hedges. Fl. & Fr.: Mar.–Oct. Central University, *MSM* 15198.

PERGULARIA Linnaeus

Pergularia daemia (Forssk.) Chiov. Result. Sc. Miss. Stefanpaoli. Somal. Hal. 1: 115. 1916. *Asclepias daemia* Forssk. Fl. Aegypt. - Arab. 51. 1775. *Cynanchum extensum* Jacq. Misc. Austr. 2: 353. 1782. *Daemia extensa* (Jacq.) R. Br. in W.T. Aiton, Hort. Kew. ed. 2. 2: 76. 1811; FBI 4: 20. *Pergularia extensa* (Jacq.) N.E. Br. in Harvey & Sonder. Fl. Cap. 7: 758. 1908; Gamble 836(588).

A climbing shrub, branchlets hispid. Leaves ovate, 1.5–5.5 × 2–3 cm, base cordate, margin entire, tip acuminate, pubescent on both sides. Flowers cream in lateral cymes. Follicles paired, 6 cm long, softly echinate.

Common in hedges. Fl. & Fr.: Sept.–Jan. Anantasagar beat, *MSM & KH* 11003.

SARCOSTEMMA Robert Brown

Sarcostemma acidum (Roxb.) Voigt, Hort. Sub. Calc. 542. 1845. *Asclepias acida* Roxb. Fl. Ind. 2: 31. 1832. *Sarcostemma brevistigma* Wight & Arn. in Wight Contrib. Bot. Ind. 59.1834; FBI 4: 26; Gamble 838(589).

A climbing leafless jointed shrub, branchlets glabrous. Flowers white in terminal umbels. Follicles tapering to both ends, straight. Seeds flattened.

Common in open forests and on hedges. Fl. & Fr.: June–Oct. Anantasagar beat, *MSM & NV* 10404.

TYLOPHORA Robert Brown

1. Leaf tip acuminate. Flowers white in axillary cymes .... *T. indica*
1. Leaf tip acute. Flowers bluish-pink in umbellate cymes .... *T. fasciculata*

Tylophora fasciculata Buch.-Ham. ex Wight & Arn. in Wight Contrib. Bot. Ind. 50. 1834; FBI 4: 40; Gamble 842(592).

A sub-erect herb, up to 50 cm tall, stem pubescent. Leaves ovate-lanceolate, 2.5–5 × 0.5–3 cm, base acute-rounded, margin entire, tip acute. Flowers bluish-pink in long umbellate cymes. Follicles fusiform, striate, glabrous.

Rare in deciduous forest. Fl. & Fr.: Aug.–Nov. Rudraram RF, *MSM & TP* 12718.

Tylophora indica (Burm. f.) Merr. in Philipp. J. Sci. 19: 373. 1921. *Cynanchum indicum* Burm. f. Fl. Indica 70. 1768. *Asclepias asthmatica* L.f. Suppl. Pl. 171. 1781. *Tylophora asthmatica* (L.f.) Wight & Arn. in Wight, Contrib. Bot. Ind. 51. 1834; FBI 4: 44; Gamble 843(593).

A climbing shrub, stem tomentose. Leaves ovate-cordate, 1–7.5 × 2–4.5 cm, base cordate, margin entire, tip acuminate, glabrescent above, grey-tomentose below. Flowers white in axillary cymes. Follicles paired, often solitary, 10 cm long, striate, glabrous.

Common in hedges and outskirts of forests. Fl. & Fr.: All seasons. Anantasagar beat, *MSM & NV* 10390; Mohammadabad beat, *MSM & KH* 10216.

WATTAKAKA Hasskarl

Wattakaka volubilis (L.f.) Stapf in Curtis, Bot. Mag. Sub. t. 8976. 1922. *Asclepias volubilis* L.f. Suppl. Pl. 170. 1781. *Hoya viridiflora* R. Br. Mem. Wern. Nat. Hist. Soc. 1: 27. 1811. *Dregea volubilis* (L.f.) Benth. ex Hook. f. FBI 4: 46. *Marsdenia volubilis* (L.f.) T. Cooke, Fl. Bombay 2: 166 (280). 1904; Gamble 846(595); Khan in FFH 240.

A climbing shrub. Stem corky, pubescent. Leaves ovate, 2.5–11 × 1.5–7 cm, base cordate, margin entire, tip acuminate, pubescent on both sides. Flowers whitish-cream in axillary umbels. Follicles paired, 4 cm long, brownish-tomentose when young. Seeds flattened.

Occasional in hedges and forests. Fl. & Fr.: June–Nov. Gadirayl beat, *MSM & KH* 10272; Kusumasamudram beat, *MSM & NV* 10355.

LOGANIACEAE

STRYCHNOS Linnaeus

1. Trees up to 15 m tall, berries orange when ripe, seeds 1–4 .... ***S. nux-vomica***
1. Trees up to 8 m tall, berries blue when ripe, seeds 1 or 2 .... ***S. potatorum***

Strychnos nux-vomica L. Sp. Pl. 189. 1753; FBI 4: 90; Gamble 868(610). Khan in FFH 241.

A deciduous tree, up to 15 m tall, bark yellowish-grey, smooth, branchlets glabrous. Leaves orbicular-ovate, 5–8 × 4–7.5 cm, base attenuate, margin entire, tip obtuse-acute. Flowers green in axillary and terminal short cymes. Berries globose, orange when ripe, 1–4-seeded.

Occasional in deciduous forest (Seshagiri Rao, 1986). Fl. & Fr.: Mar.–Aug. *Vern.*: Musthi.

Strychnos potatorum L.f. Suppl. Pl. 148. 1781; FBI 4: 90; Gamble 868(610).

A deciduous tree, up to 8 m tall. Leaves narrowly ovate, 5–10 × 2–5 cm, base truncate, margin entire, tip acute, glabrescent on both sides. Flowers white in axillary and terminal cymes. Berries globose, 1.5 cm long, deep blue when ripe, 1-2 seeded.

Occasional in forest. Fl. & Fr.: Apr.–Sept. *Vern.*: Chilla. Anantagiri RF, *MSM & TP* 11068.

GENTIANACEAE

1. Flowers sessile:
   2. Flowers white in axillary fascicles .... ***Enicostemma***
   2. Flowers cream in lax cymes .... ***Hoppea***
1. Flowers pedicellate:
   3. Flowers blue .... ***Exacum***
   3. Flowers pink .... ***Canscora***

CANSCORA Lamarck

1. Capsules 1.5 cm long .... ***C. decurrens***
1. Capsules 0.6 cm long .... ***C. diffusa***

Canscora decurrens Dalz. in Hooker's J. Bot. Kew Misc. 2: 136. 1850; FBI 4: 103; Gamble 878(617).

An annual erect herb, up to 50 cm tall, stem 4-angled, glabrous. Lower leaves elliptic-oblong, 2–5 × 0.5–1 cm, upper leaves ovate-lanceolate, 1–2 × 0.5–1 cm, base attenuate, margin entire, tip acute, glabrous on both sides. Flowers pink in terminal panicles. Capsules 1.5 cm long, 2-valved, glabrous, many-seeded. Seeds rugose.

Occasional in moist places of all localities. Fl. & Fr.: All seasons. Kanmankalva beat, *MSM & KH* 10305.

Canscora diffusa (Vahl) R. Br. Prodr. 451. in Obs. 1810. *Gentiana diffusa* Vahl, Symb. Bot. 3: 47. 1794; FBI 4: 103; Gamble 878(617).

An annual erect herb, up to 40 cm tall, branchlets quadrangular, winged. Leaves ovate, 1–4 × 0.5–2 cm, base rounded, margin entire, tip acute, glabrous on both sides. Flowers pink in axillary panicles. Capsules 0.6 cm long, oblong, compressed, many-seeded. Seeds angular, rugose.

Occasional in streams and ponds. Fl. & Fr.: Dec.–Mar. Rangammagudem beat, *MSM & KH* 10577.

ENICOSTEMA Bl. *nom. cons.*

Enicostema axillare (Lam.) A. Raynal in Adansonia 9: 75. 1969. *Gentiana axillaris* Lam. Tabl. Encycl. 1(2): 487. 1793, non Raf. 1828, nec. Reichb. 1828. *Enicostema hyssopifolium* Willd. Sp. Pl. 640. 1798. *E. littorale* auct. non. Bl., FBI 4: 101; Gamble 875(615).

A perennial prostrate herb. Stem glabrous. Leaves lanceolate-oblong, 2–5.5 × 0.2–0.8 cm, base cuneate, margin entire, tip acute. Flowers sessile, white, in axillary fascicles. Capsules oblong, 2-valved, glabrous, many-seeded.

Common in all forests. Fl. & Fr.: Oct.–Dec. Central University, *MSM* 12761.

EXACUM Linnaeus

Exacum pedunculatum L. Sp. Pl. 112. 1753; FBI 4: 97; Gamble 874(614).

An annual erect herb, up to 20 cm tall. Stem 4-angled, glabrous. Leaves lanceolate, 0.5–3 × 0.1–0.5 cm, base attenuate, margin entire, tip acute. Flowers pedicillate, blue, in long dichasial cymes. Capsules globose, 2-valved, glabrous, many-seeded.

Occasional in waste lands. Fl. & Fr.: Sept.–Dec. Naskal, *MSM & PRSR* 15109.

HOPPEA Willdenow

Hoppea dichotoma Willd. in Ges. Naturf. Freunde Berlin Neue Schriften. 3: 435. 1801; FBI 4: 100; Gamble 877(616).

An annual erect herb, up to 10 cm tall, branchlets 4-angled, glabrous, divaricate. Leaves ovate, 0.2–0.5 × 0.1–0.2 cm, base cuneate, margin entire, tip apiculate. Flowers cream in lax cymes. Capsules 0.4 cm long, glabrous, many-seeded.

Common in moist places. Fl. & Fr.: All seasons. Anantasagar beat, *MSM & KH* 11010.

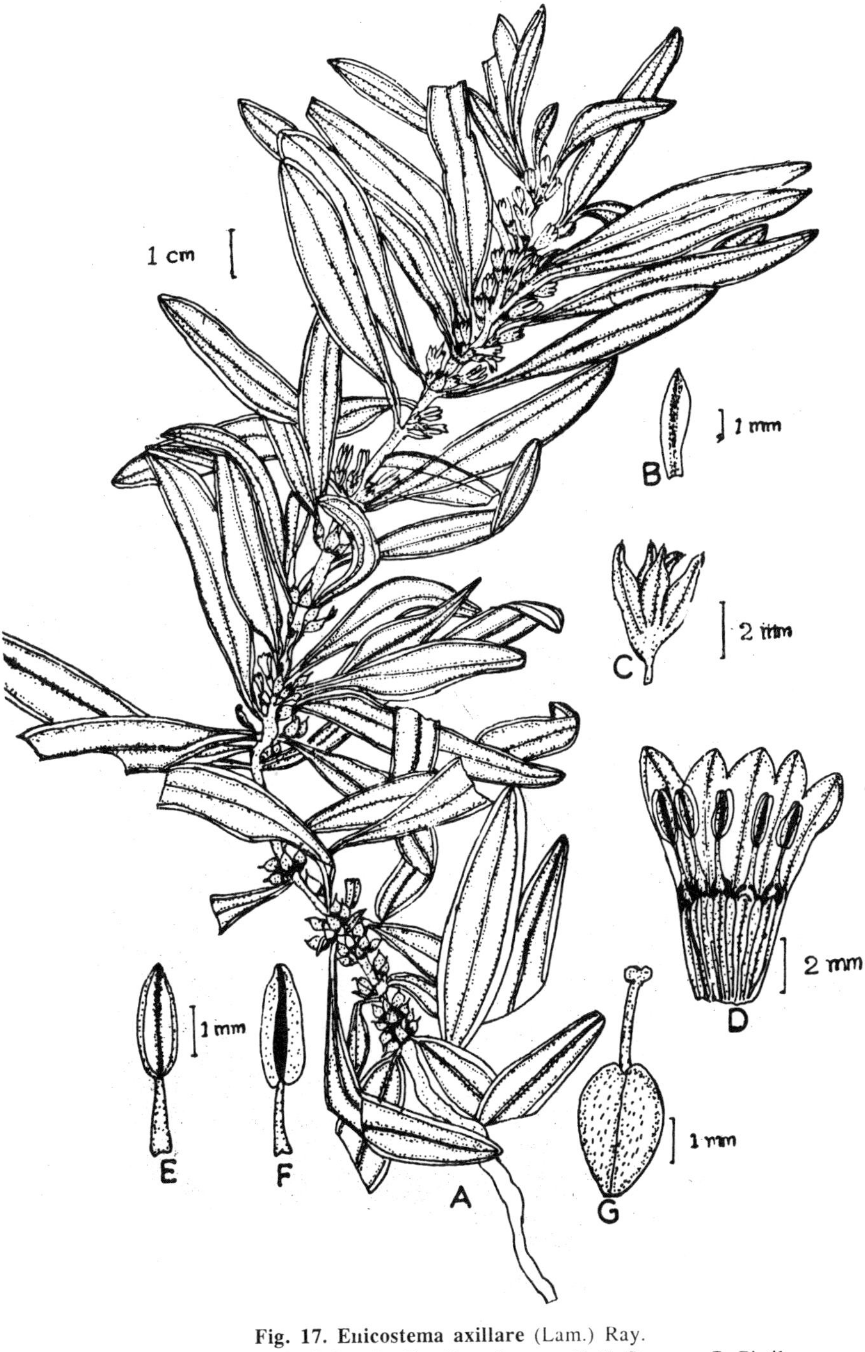

Fig. 17. Enicostema axillare (Lam.) Ray.
A. Twig B. Bract C. Calyx D. Corolla split open E–F. Stamens G. Pistil.

BORAGINACEAE

1. Flowers in spikes .... ***Heliotropium***
1. Flowers solitary or in simple cymes:
   2. Prostrate herb, flowers yellow .... ***Coldenia***
   2. Erect herb, flowers blue .... ***Trichodesma***

COLDENIA Linnaeus

Coldenia procumbens L. Sp. Pl. 125. 1753; FBI 4: 144; Gamble 892(627).

An annual prostrate herb, branchlets densely pubescent. Leaves oblong-ovate-spathulate, 1–3 × 0.5–1.5 cm, base attenuate, margin coarsely serrate, tip rounded, pilose above and densely so below. Flowers yellow, axillary, solitary. Drupes of 4-pyrenes, 0.3 cm long, pubescent.

Common in waste lands and roadsides. Fl. & Fr.: Jan.–June. Kusumasamudram beat, *MSM & NV* 10339.

HELIOTROPIUM Linnaeus

1. Erect herbs:
   2. Leaves linear-lanceolate:
      3. Flowers ebracteate .... ***H. zeylanicum***
      3. Flowers bracteate .... ***H. bracteatum***
   2. Leaves ovate-obovate:
      4. Branchlets hirsute .... ***H. indicum***
      4. Branchlets white sericeous .... ***H. ovalifolium***
1. Prostrate herbs:
   5. Leaf margin entire, drupes 1.3 cm long .... ***H. scabrum***
   5. Leaf margin reflexed, drupes 0.3 cm long .... ***H. strigosum***

Heliotropium bracteatum R. Br. Prodr. 493. 1810; FBI 4: 151. 1883; Gamble 897(630).

An annual erect hairy herb, up to 40 cm tall, branchlets hirsute. Leaves linear-lanceolate, 2–3.5 × 0.3–0.5 cm, base cuneate, margin revolute, tip acute. Flowers white, bracteate in terminal racemes. Drupes rounded, smooth, sparsely hirsute.

Common in waste lands and forest. Fl. & Fr.: July–Dec. Kusumasamudram, *MSM & NV* 10337.

Heliotropium indicum L. Sp. Pl. 130. 1753; FBI 4: 152; Gamble 896(630).

An annual erect herb, up to 1 m tall, branchlets hirsute. Leaves elliptic-ovate, 2–10 × 1.5–5 cm, base cuneate, margin undulate-serrate, tip acute, hirsute on both sides. Flowers violet white in extra axillary or terminal spikes. Drupes conical, 0.3 cm long, combined in pairs, beaked above, glabrous

Common in dry ponds and lakes. Fl. & Fr.: Mar.–May. Malkalcheruvu, *MSM & KH* 10275, Peddakunta, *MSM & NV* 10418.

Heliotropium ovalifolium Forssk. Fl. Aegypt.-Arab. 38. 1775; FBI 4: 150; Gamble 896(629); Kazmi in J. Arnold Arbor. 51: 178. 1970.

An annual erect herb, up to 60 cm tall, branchlets softly white sericeous. Leaves obovate, spathulate, 0.5–2 × 0.2–1.2 cm, base attenuate, margin entire, tip mucronate, white sericeous on both sides. Flowers white in terminal spikes. Drupes conical, 0.1 cm long, nutlets 4, silky.

Common in waste lands, open forests. Fl. & Fr.: All seasons. Kusumasamudram beat, *MSM & NV* 10335, Peddakunta *MSM & NV* 10420.

Heliotropium scabrum Retz. Obs. Bot. 2: 8. 1781; FBI 4: 152; Kazmi in J. Arnold Arbor. 51: 154. 1970.

An annual prostrate herb, up to 15 cm, branchlets strigose-hairy, closely appressed to substratum. Leaves linear-oblong, 1–1.5 × 0.3–0.5 cm, base attenuate, margin entire, tip acute, sericeous on both sides. Flowers white in terminal racemes. Drupes conical, 1.3 cm long, appressed hirsute.

Occasional in forests and common in dried ponds and lakes. Fl. & Fr.: May–Sept. Malkalcheruvu *MSM & KH* 10282.

Heliotropium strigosum Willd. Sp. Pl. 1: 743. 1797; FBI 4: 151. *H. brevifolium* Wall. ex Roxb. Fl. Ind. 2: 2. 1824; Gamble 897(630).

An annual prostrate herb, up to 40 cm, branchlets white, strigose-hispid. Leaves linear-lanceolate, 0.5–1.5 × 0.2–0.5 cm, base acute, margin reflexed, tip acute, sessile, hirsute on both sides. Flowers white in terminal hispid spikes. Drupes conical, 0.3 cm long, nutlets 4, hispid.

Occasional in water courses and open forests. Fl. & Fr.: All seasons. Mohammadabad beat, *MSM & NV* 10423.

Heliotropium zeylanicum (Burm. f.) Lam. Encycl. 3: 94. 1789; Gamble 896(630); ssp. **zeylanicum**; Kazmi in J. Arnold Arbor. 51: 156. 1970. *H. paniculatum* R. Br. Prodr. 494. 1810; FBI 4: 151. *H. curassavicum* L. var. *zeylanicum* Burm. f. Fl. Indica 41. 5. 16. f. 2. 1768.

An annual erect herb, up to 50 cm tall, branchlets sparsely pubescent. Leaves linear, 1–2 × 0.3–0.4 cm, base cuneate, margin revolute, tip acute, glabrous above, appressed strigose below. Flowers white in terminal elongate racemes. Drupes 0.1 cm long, nutlets 4, sparsely hairy.

Occasional in open forests. Fl. & Fr.: All seasons. Kusumasamudram beat, *MSM & NV* 10337.

TRICHODESMA Robert Brown *nom. cons.*

1. Calyx lobes rounded at base, stamens exserted ..... *T. zeylanicum*
1. Calyx lobes cordate at base, stamens inserted:
   2. Lower leaves petiolate, elliptic ..... *T. sedgewickianum*
   2. Lower leaves sessile, oblong ..... *T. indicum*

Trichodesma indicum (L.) R. Br. Prodr. 149. 1810; FBI 4: 153; Gamble 899(631); Kazmi in J. Arnold Arbor. 52: 51. 1971. *Borago indica* L. Sp. Pl. 137. 1753.

An annual erect herb, up to 70 cm tall, branchlets hispid. Leaves oblong-lanceolate, 2–6 × 0.5–2 cm, base cordate, margin entire, tip acute, sessile, appressed pubescent above, hirsute below. Flowers bluish-white, axillary, solitary. Drupes pyramidal, 0.6 cm long, nutlets 4, hirsute.

Common in waste lands on roadsides and dried tanks. Fl. & Fr.: Feb.–July. Mohammadabad beat, *MSM & NV* 10429; Central University, *MSM* 15178.

Trichodesma sedgewickianum S.P. Bererjee in Bull. Bot. Soc. Bengal. 16: 10. 1962. *T. amplexicaule* auct. non Roth; FBI 4: 154. *p.p.* Gamble 899(631).

An annual erect herb, up to 60 cm tall. Stem hispid. Leaves elliptic, 2–8 × 0.5–3.5 cm, base cordate, margin entire, tip acute, scabrid above, hispid below. Flowers violet-blue or white in terminal, simple cymes. Drupes pyramidal, smooth. Nutlets 4.

Occasional in forest and waste lands. Fl. & Fr.: Oct.–Jan. Central University, *TP & MSM* 19249.

Trichodesma zeylanicum (Burm. f.) R.Br. Prodr. 496. 1810; FBI 4: 154. 1883; Gamble 899(632). *Borago zeylanica* Burm. f. Fl. Indica 41. 1768.

An annual erect herb, up to 75 cm tall, branchlets white hirsute, leaves oblong-lanceolate, 2–8 × 1–3 cm, base acute-cordate, margin entire, tip acute, white hirsute above, scabrous along nerves below. Flowers light-blue in terminal panicles. Drupes 0.5 cm long, nutlets 4, smooth on the outer, spotted and rugose along inner face.

Rare in open forest, foot hills and waste lands. Fl. & Fr.: Mar.–July Mohammadabad beat, *MSM & KH* 10324.

CORDIACEAE

1. Tree, fruit a drupe covered to some extent by calyx .... *Cordia*
1. Shrub, fruit a drupe not covered by calyx .... *Carmona*

CARMONA Cavanilles

Carmona retusa (Vahl) Masam. in Trans. Nat. Hist. Soc. Taiwan 30: 61. 1940. *Cordia retusa* Vahl, Symb. Bot. 2: 42. 1791. *Ehretia buxifolia* Roxb. Pl. Cor. t. 57. 1796; FBI 4: 144. *E. microphylla* Lam. Tabl. Encycl. 1: 425. 1792; Gamble 891(626).

A shrub, up to 80 cm tall, branchlets pubescent. Leaves 3–6 fascicled at nodes, obovate 0.5–2 × 0.5–1 cm, base cuneate, margin entire, tip rounded-truncate, scabrid above and glabrous below. Flowers white or pale yellow, axillary, solitary. Drupes globose of 4-pyrenes, glabrous, apiculate.

Common in open forest. Fl. & Fr.: All seasons. *Vern.:* Barranki. Rudraram RF, *MSM & TP* 12721.

CORDIA Linnaeus

Cordia obliqua Willd. Phytogr. 4. t. 4. f. 1. 1794; FBI 4: 137; Gamble 387(623).

A deciduous tree, up to 10 m tall. Leaves elliptic-obovate, 7–10 x 2–5 cm, base rounded, margin entire, tip acute. Flowers white in axillary and terminal cymes. Drupes ovoid, creamish with gummy pulp.

Occasional in forest and villages. Fl. & Fr.: Dec.–June. *Vern.:* Bankiri. Central University, *MSM* 15151.

Note: Merrill in Trans. Ann. Phil. Soc. (*n.s.*) 24: 329–30. 1935, treated *Cordia obliqua* Willd. as a synonym of *Cordia dichotoma.*

CONVOLVULACEAE

1. Prostrate herbs:
   2. Leaves linear-oblong. Flowers blue .... ***Evolvulus***
   2. Leaves 3–5-lobed. Flowers yellow .... ***Merremia***
1. Twinners or stagglers:
   3. Calyx of fruit woody .... ***Rivea***
   3. Calyx of fruit not woody:
      4. Pollen smooth:
         5. Capsule 8-valved .... ***Jacquemontia***
         5. Capsule 4-valved .... ***Hewettia***
      4. Pollen spinulose .... ***Ipomoea***

EVOLVULUS Linnaeus

1. Leaves suborbicular. Flowers cream .... ***E. nummularius***
1. Leaves linear-oblong. Flowers violet-blue .... ***E. alsinoides***

Evolvulus alsinoides (L.) L. Sp. Pl. (ed. 2) 392. 1762; FBI 4: 220; Gamble 923(648); Khan in FFH 249. *Convolvulus alsinoides* L. Sp. Pl. 157. 1753.

A perennial prostrate herb, up to 25 cm, branchlets sericeous. Leaves linear-oblong, 0.6–3 × 0.2–0.5 cm, base acute, margin entire, tip apiculate, sericeous on both sides. Flowers violet-blue, in a few-flowered cymes. Capsules globose, 0.5 cm long, glabrous, 2–4-seeded.

Common in all places. Fl. & Fr.: All seasons. Malkalcheruvu, *MSM & KH* 10278; Kusumasamudram beat, *MSM & NV* 10342, 10343.

Evolvulus nummularius (L.) L. Sp. Pl. ed. 2. 391. 1762. *Convolvulus nummularius* L. Sp. Pl. 157. 1753.

A prostrate herb, branchlets hairy. Leaves sub-orbicular, 0.5–2 × 0.6–1.6 cm, base subcordate-truncate, margin entire, tip obtuse. Flowers cream, axillary, solitary, or 2–3 in clusters. Capsules subglobose, 2–4-seeded.

Occasional in moist places. Fl. & Fr.: All seasons. Central University, *MSM* 15145; *TP & MSM* 15635.

HEWETTIA Wight et Arnott

Hewettia scandens (Milne) Mabb. in Manilal Bot. Hist. Hort. Malab. 84. 1980. *Convolvulus scandens* Milne Reser. cut. 2. 1773. *C. bicolor* Vahl Symb. Bot. 3: 25. 1794. *Hewettia bicolor* (Vahl) Wight & Arn. in Madras J. Lit. Sci. 5: 22. 1837; FBI 4: 216; Gamble 924(649).

A woody straggler. Leaves ovate, 5–10 × 5–10 cm, base cordate, margin entire, tip acute, pubescent above, rough hairy below. Flowers pink in axillary umbellate cymes. Capsules globose, 4-valved, seeds glaucous.

Occasional in forest. Fl. & Fr.: Aug.–Dec. Pargi, *MSM* 12796.

IPOMOEA Linnaeus

1. Plants aquatic ..... ***I. aquatica***
1. Plants terrestrial:
    2. Climbing herbs or shrubs:
        3. Flowers cream or whitish cream:
            4. Leaves 5-foliolate ..... ***I. cairica***
            4. Leaves 1-foliate:
                5. Leaves reniform, tip obtuse ..... ***I. asarifolia***
                5. Leaves obovate, tip acuminate ..... ***I. obscura***
        3. Flowers pink or other than pink:
            6. Leaves lobed:
                7. Leaves 3-lobed ..... ***I. nil***
                7. Leaves 5–7-lobed:
                    8. Capsule 3-seeded ..... ***I. coptica***
                    8. Capsule 4-seeded ..... ***I. pestigridis***
            6. Leaves not lobed ..... ***I. eriocarpa***
    2. Erect shrubs ..... ***I. carnea***

Ipomoea aquatica Forssk. Fl. Aegypt.-Arab. 44. 1775; FBI 4: 210. *I. reptans* Poir. in Lam. Encycl. (Suppl. 3) 469. 1814 non L.; Gamble 916(643).

An annual prostrate herb, branchlets glabrous. Leaves hastate-lanceolate, 5–9 × 4–5 cm, base cordate, margin entire, tip acuminate, glabrous on both sides. Flowers pink, solitary. Capsules 1 cm long, glabrous, 3–4-seeded.

Common in ponds. Fl. & Fr.: All seasons. *Vern.:* Koila teega. Kusumasamudram pond, *MSM & KH* 11027.

Ipomoea asarifolia (Desr.) Roem. & Schult. in L. Syst. Veg. 4. 251. 1819. *Convolvulus asarifolius* Desr. in Lam. Encycl. 3: 562. 1792. *Ipomoea repens* Lam. Tabl. Encycl. 1: 467. 1791. non Roth 1821; Gamble 916(643). *I. beladamboe* Roem. & Schult. in L. Syst. Veg. 4. 233. 1819; FBI 4: 209.

An annual climbing herb, branchlets hirsute. Leaves cordiform to reniform, 2–3 × 1.5–3 cm, base deeply cordate, margin entire, ciliate, tip obtuse, glabrous above, tomentose below. Flowers whitish-cream in axillary lax cymes. Capsules globose, 1–2 cm long, glabrous, 3–4-seeded. Seeds brown, rugose.

Common on hedges. Fl. & Fr.: Sept.–Dec. Mohammadabad beat, *MSM & KH* 10597.

Ipomoea cairica (L.) Sweet Hort. Brit. 287. 1827; Gamble 918(645). *Convolvulus cairicus* L. Syst. Nat. ed. 10. 922. 1759. *Ipomoea palmata* Forssk. Fl. Aegypt.-Arab. 43. 1775; FBI 4: 214.

A perennial climbing herb, branchlets glabrous. Leaves 5-foliolate, oblanceolate, 1.5–4 × 0.5–1 cm, base attenuate, margin entire, tip acute, glabrous on both sides. Flowers whitish cream in solitary or 2–3 in lax cymes. Capsules globose, 1 cm long, glabrous, 4-seeded. Seeds rugose.

Common on hedges. Fl. & Fr.: Sept.–Dec. Anantasagar beat, *MSM & KH* 11004.

Ipomoea carnea Jacq. Enum. Syst. Pl. 13. 1760; Gamble 919(645); ssp. **fistulosa** (Choisy) D. Austin in Taxon 26: 237. 1977. *I. fistulosa* C. Martius ex Choisy in DC. Prodr. 9: 349. 1845.

A perennial erect shrub, up to 2 m tall, branchlets hollow, glabrous. Leaves ovate-hastate, 10–23 × 6–9.5 cm, base cordate, margin entire, tip acuminate, glabrous on both sides. Flowers violet in axillary dichasial cymes. Capsules ovoid, 2 cm long, glabrous, 2–4-seeded.

Common in all places. Fl. & Fr.: July–Jan. Mohammadabad beat, *MSM & NV* 10432.

Ipomoea coptica (L.) Roem. & Schult in L. Syst. Veg. 4: 208. 1819. *Convolvulus copticus* L. Mant. Pl. 559. 1771. *Ipomoea dissecta* Willd. Phytographia 5. t. 2. f. 3. 1794; FBI 4: 200; Gamble 918(645).

An annual prostrate herb, stem glabrous. Leaves pedately 5–7-lobed, lobes linear-lanceolate, 0.5–2 × 0.1–0.3 cm, base attenuate, margin entire, tip apiculate, pubescent on both sides. Flowers yellow, axillary, solitary. Capsules subglobose, glabrous, 3-seeded.

Rare in open forest. Fl. & Fr.: Oct.–Jan. Lingampalle, *MSM* 12740, 10740.

Ipomoea eriocarpa R. Br. Prodr. 484. 1810; FBI 4: 204. *I. hispida* (Vahl) Roem. & Schult. in L. Syst. Veg. 4: 238. 1819. non Zucc. 1809; Gamble 915(643). *Convolvulus hispidus* Vahl, Symb. Bot. 3: 29. 1794.

An annual climbing herb, branchlets hispid. Leaves ovate, 3–5 × 1.5–2.5 cm, base deeply cordate, margin entire, tip acuminate, glabrous on both sides. Flowers pink in clusters. Capsules 1.5 cm long, hairy, 3-seeded.

Common on hedges. Fl. & Fr.: Sept.–Dec. Rangammagudem beat, *MSM & KH* 10575; Naskal, *MSM & PRSR* 15119.

Ipomoea nil (L.) Roth Catal. Bot. 1. 36. 1797. *Convolvulus nil* L. Sp. Pl. ed. 2. 1. 219. 1762. *Ipomoea hederacea* auct. non (L.) Jacq., FBI 4: 199; Gamble 917(644); Khan in FFH 249.

A climbing shrub, stem puberulous. Leaves 3-lobed, ovate, 8–18 × 8–12 cm, base cordate, margin entire, tip acuminate, puberulous on both sides. Flowers whitish-pink, axillary, solitary. Capsules 1.5 cm long, glabrous; 5–7-seeded.

Occasional in hedges and forests. Fl. & Fr.: Sept.–Jan. Gadirayal beat, *MSM* 10517.

Ipomoea obscura (L.) Ker.-Gawl. Bot. Reg. 3: t. 239. 1817; FBI 4: 207; Gamble 916(643). *Convolvulus obscurus* L. Sp. Pl. (ed. 2) 220. 1762.

An annual climbing herb, stem sparsely hairy. Leaves ovate, 2.5–11 × 3–6.5 cm, base deeply cordate, margin entire, tip acuminate, glabrous above, glaucous below. Flowers cream, axillary, solitary. Capsules 0.8 cm long, glabrous, 4–6-seeded.

Common in hedges. Fl. & Fr.: July–Dec. Gidda kunta, *KSM & NV* 10424. Gacchibowli, *MSM* 12771.

Ipomoea pes-tigridis L. Sp. Pl. 162. 1753; FBI 4: 204; Gamble 918(644).

An annual climbing herb. Stem hirsute. Leaves palmately 5-lobed, lobes, ovate, 2–4 × 0.6–1.5 cm, base attenuate, margin entire, tip acuminate. Flowers whitish-pale pink in clusters. Capsules globose, glabrous, 4-seeded.

Occasional in open forest. Fl. & Fr.: Oct.–Jan. Lingampalle, *MSM* 12748.

JACQUEMONTIA J.D. Choisy

Jacquemontia paniculata (Burm. f.) Hallier f. in Bot. Jahrb. Syst. 16: 541. 1893; Gamble 926(650). *Convolvulus parviflorus* Vahl Symb. Bot. 3: 29. 1794; non Desr. 1791; FBI 4: 220.

A straggling shrub. Leaves cordate-ovate, 6–10 × 5–7 cm, base cordate, margin enitre, tip acuminate. Flowers white in axillary cymes. Capsules subglobose, 8-valved. Seeds light winged.

Occasional in forests. Fl. & Fr.: July–Dec. Anantagiri RF, *MSM & PRSR* 12780.

MERREMIA Dennsted ex Endlicher *nom. cons.*

1. Leaves oblong-spathulate, base with toothed auricles .... ***M. tridentata***
1. Leaves ovate-reniform, base not toothed .... ***M. emarginata***

Merremia emarginata (Burm. f.) Hallier f. in Bot. Jahrb. Syst. 16: 552. 1893; Gamble 928(652). *Evolvulus emarginatus* Burm. f. Fl. Ind. 77 t. 30. f. 1. 1768. *Ipomoea reniformis* (Roxb.) Choisy in Mem. Soc. Phys. Geneve 6: 446. 1834; FBI 4: 206. *Convolvulus reniformis* Roxb. Fl.

Ind. 2: 67. 1824. *Merremia gangetica* (L.) Cuf. in Bull. Jard. Bot. Etat. 31 (Suppl.) 743. 1961. *Convolvulus gangeticus* L. Cent. Pl. 2: 9. 1756.

A spreading herb, rooting at nodes, branchlets hispid. Leaves ovate-reniform, 0.8–2.4 × 0.5–2 cm, base cordate, margin crenate, tip obtuse-emarginate. Flowers yellow, axillary, solitary. Capsules subglobse, glabrous, 4-seeded.

Common in wet lands and waste lands. Fl. & Fr.: Oct.–Jan. Anantagiri RF, *MSM & TP* 11084.

Merremia tridentata (L.) Hallier f. Bot. Jahrb. Syst. 16. 552. 1893; ssp. **tridentata**; Gamble 928(652). *Convolvulus tridentatus* L. Sp. Pl. 157. 1753. *Ipomoea tridentata* (L.) Roth in Roem. Arch. Bot. 1. 238. 1798; FBI 4: 205.

A perennial prostrate herb, branchlets many, radiating from root stock, rusty brown. Leaves oblong-spathulate, 0.5–2.8 × 0.5–1.5 cm, base with toothed auricles, margin entire, tip acute, 3-toothed or emarginate, glabrous on both sides. Flowers cream, axillary, solitary. Capsules 0.4 cm long, glabrous, 4-seeded.

Common in waste lands, scrub jungles. Fl. & Fr.: All seasons. Kusumasamudram beat, *MSM & NV* 10346.

RIVEA Choisy

Rivea hypocrateriformis (Desr.) Choisy in Mem. Soc. Phys. Geneva 6: 408. 1834; FBI 4: 184; Gamble 903(635); Khan in FFH 249. *Convolvulus hypocrateriformis* Desr. in Lam. Encycl. 3: 561. 1792.

A woody climber, branchlets silky pubescent. Leaves orbicular, 3–7 × 3.5–8 cm, base cordate, margin entire, tip obtuse-mucronate, glabrous above and densely silky pubescent below. Flowers white, axillary, solitary. Capsules globose, glabrous, 2–4-seeded.

Common in hedges. Fl. & Fr.: July–Nov. *Vern.:* Boddi teega. Gacchibowli, *MSM* 12773.

Ipomoea batatas Lam. (*Vern.:* Chilakadadumpa) is cultivated for its tubers and *I. quamoclit* L. (*Vern.:* Kasiratnam) is planted in gardens for ornamental purpose.

CUSCUTACEAE

CUSCUTA Linnaeus

Cuscuta reflexa Roxb. Pl. Cor. t. 104. 1799. 1832; FBI 4: 225; Gamble 931(654).

A leafless parasitic climbing herb, branchlets glabrous, yellow. Flowers cream in racemes. Capsules globose, 0.5 cm long, glabrous, succulent, 2–4-seeded.

Common on hedges. Fl. & Fr.: June–Dec. Anantasagar beat, *MSM & NV* 10408.

SOLANACEAE

1. Fruit berry:
   2. Stamens free. Fruit with inflated calyx .... *Physalis*
   2. Stamens connivent in a cone. Fruit without inflated calyx .... *Solanum*
1. Fruit capsule .... *Datura*

DATURA Linnaeus

1. Branchlets and flowers purple .... *D. metel*
1. Branchlets and flowers white .... *D. innoxia*

Datura innoxia Miller, Gard. Dict. (ed. 8) 1768. *D. metel* auct. non L.: Sims in Curtis Bot. Mag. t. 1440. 1812; FBI 4: 243; Gamble 941(660).

An annual erect subshrub, up to 70 cm tall, branchlets pubescent. Leaves ovate, 4–15 × 3.5–11 cm, base unequal, margin bluntly dentate, tip acuminate, pubescent on both sides. Flowers white, axillary, solitary. Capsule globose, 4 cm long, prickly, many-seeded. Seeds brown.

Common waste land weed. Fl. & Fr.: All seasons. Puttapahad (Anantasagar), *MSM & NV* 10396.

Datura metel L. Sp. Pl. 179. 1753; Sant. in J. Bombay Nat. Hist. Soc. 47. 657. 1948. *D. fastuosa* auct. non. L. 1759; FBI 4: 242; Gamble 949(660).

A shrub up to 1 m tall, branchlets purple. Leaves ovate, purple, 8–10 × 5–6 cm, base unequal, margin entire, tip acute. Flowers purple, axillary, solitary. Capsules globose, with hard spines.

Common weed in waste places. Fl. & Fr.: Sept.–Jan. *Vern.*: Nalla ummetha, Central University, *MSM* 19217.

PHYSALIS Linnaeus

Physalis minima L. Sp. Pl. 183. 1753; FBI 4: 238; Gamble 939(659).

An annual erect herb, up to 60 cm, branchlets glabrous. Leaves ovate, 2–5 × 1.5–2.5 cm, base obliquely acute-rounded, margin entire, tip acuminate, sparsely strigose above, glabrous below. Flowers yellow, axillary, solitary. Berries globose, 0.7 cm long, over topped by a long inflated calyx, many-seeded.

Common in waste lands. Fl. & Fr.; July–Feb. Kusumasamudram beat, *MSM & NV* 10354; Puttapahad (Anantasagar), *MSM & NV* 10395.

SOLANUM Linnaeus

1. Armed herbs:
   2. Leaves lacerate, berries with green stripes .... ***S. surattense***
   2. Leaves sinuate, berries uniformly yellow or green .... ***S. anguivi***
1. Unarmed herbs .... ***S. nigrum***

Solanum anguivi Lam. Tabl. Encycl. 2: 23. 1794; Hepper in Bot. J. Linn. Soc. 76. 290. 1978. *S. indicum* auct. non L. 1753; FBI 4: 234; Gamble 938(658).

An annual herb, up to 2 m tall, branchlets stellate hairy with yellow prickles. Leaves sinuate ovate, 8–12 × 4–7 cm, base obliquely truncate, stellate hairy on both sides. Flowers violet in extra axillary racemes. Berries 1.5 cm long, glabrous with persistent calyx. Seeds minutely pitted, brown.

Common in waste lands. Fl. & Fr.: Jan.–June. *Vern.*: Mulla wonkai. Anantasagar road, *MSM & NV* 10388.

Solanum nigrum L. Sp. Pl. 186. 1753; FBI 4: 229; Gamble 936(657).

An annual erect herb, up to 1.5 m tall, branchlets glabrous. Leaves ovate-lanceolate, 2–15 × 1.5–9 cm, base truncate-rounded, margin dentate, tip acute-acuminate, glabrous on both sides. Flowers yellow-white in axillary or extra axillary cymes. Berries globose, 0.3 cm long, many-seeded.

Common in waste lands. Fl. & Fr.: All seasons. Venkat Reddipalli, *MSM & KH* 10308, Kusumasamudram road, *MSM & NV* 10351.

Solanum surattense Burm. f. Fl. Ind. 57. 1768. *Solanum xanthocarpum* Schrad. & Wedl. Sert. Hannov. 1: 8. t. 2. 1795; FBI 4: 236; Gamble 938(658).

A perennial armed herb, up to 50 cm, stem and leaves with yellow prickles. Leaves lacerate, 3.5–10 × 1.5–5.5 cm, base attenuate, margin deeply sinuate, tip acute, sparsely stellate pubescent on both sides. Flowers blue in extra axillary cymes. Berries globose, 1.5 cm long, yellow or white with green stripes, many-seeded.

Common waste land weed. Fl. & Fr.: All seasons. Mohammadabad, *MSM & KH* 10231.

Lycopersicon esculentum Mill. (*Vern.:* Tomato), *Capsicum annuum* L., *C. frutescens* L. (*Vern.:* Mirapa), *Solanum melongena* L. (*Vern.:* Vanga) are cultivated for their fruits, *Nicotinana tabaccum* L. (*Vern.:* Pogaku) for leaves and *Cestrum* spp. for ornamental purpose.

SCROPHULARIACEAE

1. Leaves pinnatifid:
   2. Leaves alternate, flowers yellow .... ***Verbascum***
   2. Leaves opposite or whorled, flowers other than yellow:
      3. Calyx 5-partite .... ***Limnophila***
      3. Calyx not 5-partite .... ***Sopubia***

1. Leaves not pinnatifid:
    4. Prostrate herbs .... ***Bacopa***
    4. Erect herbs:
        5. Branchlets scabrid .... ***Striga***
        5. Branchlets glabrous:
            6. Capsules globose .... ***Dopatrium***
            6. Capsules ovoid-globose .... ***Lindernia***

BACOPA Aublet *nom. cons.*

Bacopa monnieri (L.) Penell in Proc. Acad. Nat. Sci. Philadelphia 98: 94. 1946. *Lysimachia monnieri* L. Cent. Pl. II. 9. 1756. *Gratiola monnieri* (L.) L. Syst. Nat. (ed. 10) 851. 1759. *Moniera cuneifolia* Michaux Fl. Bor. Amer. 2: 22. 1803; Gamble 953(669). *Herpestis monnieri* (L.) Kunth. Nov. Gen. Sp. 2. 366. 1818. *nom. illegit.*; FBI 4: 272.

An annual prostrate herb, up to 20 cm, branchlets succulent, glabrous. Leaves oblong-spathulate, 0.5–1.5 × 0.1–0.3 cm, base attenuate, margin entire, tip obtuse-rounded, sessile, glabrous on both sides. Flowers blue, axillary, solitary. Capsules 0.6 cm long, glabrous, many-seeded.

Common in wet places. Fl. & Fr.: All seasons. Mohammadabad, *MSM & KH* 10433.

DOPATARIUM F. Hamilton ex Bentham

Dopatrium junceum (Roxb.) Buch.-Ham. ex Benth. Scroph. Ind. 31. 1835; FBI 4: 274; Gamble 954(670). *Gratiola juncea* Roxb. Pl. Cor. t. 129. 1800.

An annual erect herb, up to 15 cm tall, branchlets sub-succulent, glabrous. Leaves oblong, 1–2 × 0.5–1 cm, gradually smaller upwards, margin entire, tip acute, sessile, glabrous on both sides. Flowers violet, axillary, solitary. Capsules globose, 0.3 cm long, glabrous, many-seeded.

Common in marshy and wet places. Fl. & Fr.: Dec.–Feb. Kothapalli beat, *MSM & KH* 10564.

LIMNOPHILA Robert Brown *nom. cons.*

Limnophila indica (L.) Druce in Bot. Exch. Club. Soc. Brit. Isles 3: 420. 1914. *Hottonia indica* L. Syst. Nat. (ed. 10) 919. 1759. *Limnophila gratioloides* R. Br. Prodr. 442. 1810; FBI 4: 271; Gamble 952(668).

An annual diffuse herb, up to 10 cm tall, branchlets glabrous. Leaves whorled, linear-lanceolate or laciniate, submerged leaves capillaceo-multifid, 0.5–1 × 0.1–0.3 cm, glabrous on both sides. Flowers cream, axillary, solitary. Capsules globose-ovoid, 0.4 cm long, glabrous, many-seeded.

Common in marshy places. Fl. & Fr.: Sept.–Dec. Kondapur *MSM & KH* 10558.

LINDERNIA Allioni

Lindernia parviflora (Roxb.) Haines, Bot. Bihar, Orissa 2: 635(655) 1922. *Gratiola parviflora* Roxb. Pl. Cor. t. 203. 1811. *Ilysanthes parviflora* (Roxb.) Benth. in DC. Prodr. 10: 1846; FBI 4: 283; Gamble 961(675). *Bonnaya hyssopioides* Wight Icon. Pl. Ind. Orient. t. 857. 1844–45, non Benth.

An annual erect herb, up to 12 cm long, branchlets glabrous. Leaves oblong-elliptic, 1–1.5 × 0.3–0.5 cm, base rounded, margin entire, tip acute, upper leaves smaller, glabrous on both sides. Flowers white, solitary or 2 per node. Capsules globose, 0.4 cm long, glabrous, exceeding the calyx, many-seeded.

Common in moist places. Fl. & Fr.: All seasons. Mohammadabad, *MSM & KH* 10559; Rangammagudem, *MSM & KH* 10573; Kondapur, *MSM & KH* 10585.

SOPUBIA F. Hamilton ex D. Don

Sopubia delphinifolia (L.) G. Don. Gen. Hist. 4. 560. 1837; FBI 4: 302; Gamble 970(682). *Gerardia delphinifolia* L. Cent. Pl. 2: 21. 1756.

An erect herb, up to 70 cm tall, branchlets glabrous; leaves pinnatifid below, linear, trifid above, up to 2 cm, lobes filiform, glabrous. Flowers purple in terminal racemes. Capsules 1 cm long, sparsely hairy, many-seeded.

Common in scrub jungles and open forests. Fl. & Fr.: Sept.–Jan. Kanmankalva, *MSM* 10456.

STRIGA Loureiro

1. Flowers pink, calyx 5-ribbed .... ***S. gesnerioides***
1. Flowers yellow or white, calyx 10–15-ribbed:
    2. Flowers white:
        3. Calyx 10-ribbed .... ***S. densiflora***
        3. Calyx 15-ribbed .... ***S. angustifolia***
    2. Flowers yellow .... ***S. asiatica***

Striga angustifolia (D. Don) C.J. Saldanha in Bull. Bot. Surv. Ind. 5: 70. 1963. *Buchnera angustifolia* D. Don Prodr. Fl. Nepal 91. 1825. *Striga euphrasiodes* Benth. in Comp. Bot. Mag. 1: 364. 1835; FBI 4: 299; Gamble 968(680).

An erect herb, up to 35 cm long, branchlets scabrid. Leaves linear, 1–2.5 × 0.2 cm, sessile, margin recurved or revolute, tip acute, scabrid on both sides. Flowers white in terminal spikes, calyx 15-ribbed. Capsules 0.8 cm long, enclosed in persistent calyx, many-seeded.

Common in all places. Fl. & Fr.: All seasons. Mohammadabad, *MSM* 10441.

Striga asiatica (L.) Kuntze. Rev. Gen. Pl. 2: 466. 1891. *Buchnera asiatica* L. Sp. Pl. 630. 1753. *Striga lutea* Lour. Fl. Cochinch. 22. 1790; FBI 4: 299; Gamble 968(680).

An annual erect herb, 50 cm tall, stem 4-angled, scabrid. Leaves linear, 0.5–3 × 0.2–0.3 cm, base attenuate, margin entire, tip acute, glabrous on both sides. Flowers yellow in axillary bracteate spikes. Capsules 0.4 cm long, enclosed in persistent calyx, many-seeded.

Common root parasite on sorghum and sugarcane. Fl. & Fr.: All seasons. Mohammadabad, *MSM & KH* 10219.

Striga densiflora (Benth.) Benth. in Comp. Bot. Mag. 1: 363. 1835; FBI 4: 299; Gamble 967(680). *Buchnera densiflora* Benth. Scroph. Ind. 41. 1835.

An annual erect herb, up to 40 cm tall. Leaves linear, 1–3 × 0.2–0.4 cm, base attenuate, margin entire, tip acute. Flowers white in axillary spikes. Calyx lobes 5, each one ribbed. Capsules globose-ovoid. Seeds oblong.

Occasional in grass lands. Fl. & Fr.: Oct.Jan. Mekavarpalli, *MSM & TP* 11047.

Striga gesnerioides (Willd.) Vatke in Oesterr. Bot. J. 25: 11. 1875. *Buchnera gesnerioides* Willd. Sp. Pl. 3: 338. 1800. *Striga orobanchoides* (Endl.) Benth. in Comp. Bot. Mag. 1: 361. 1836. *nom. illegit.*; FBI 4: 299; Gamble 967(680).

An annual erect herb, up to 25 cm tall. Leaves brown, scale like 0.5–1.5 cm, tip acute, pubescent. Flowers pink in long terminal spikes, calyx 5-ribbed. Capsules sub-globose, enclosed in persistent calyx, many-seeded.

Rare root parasite on *Lepidagathis cristata.* Fl. & Fr.: Aug.–Dec. Anantagiri RF, *MSM & TP* 11085.

VERBASCUM Linnaeus

Verbascum chinense (L.) Santapau. Fl. Purandhar 90. 1958. *Scrophularia chinensis* L. Mant. Pl. 250. 1771. *Celsia coromandeliana* Vahl, Symb. Bot. 3. 79. 1794; FBI 4: 251; Gamble 945(663).

An annual erect herb, up to 1 m tall, stem densely pubescent. Leaves radical and cauline, alternate, oblong, lyrate-pinnatifid, 4–10 × 3–6 cm, base truncate-rounded, margin crenate, tip acute, pubescent on both sides. Flowers yellow in long terminal racemes. Capsules ovoid, 0.6 cm long, pubescent, many-seeded.

Common waste land weed, often in wet places. Fl. & Fr.: June–Dec. Dharmapur, *MSM & KH* 10287.

LENTIBULARIACEAE

UTRICULARIA Linnaeus

Utricularia stellaris L.f. Suppl. Pl. 86. 1781; FBI 4: 328; Gamble 980(689).

An aquatic floating rootless herb, stolons filiform, submerged. Leaves in whorls, pectinate-segmented, submerged. Flowers yellow in long peduncled racemes. spongy floats in a whorl below the inflorescence. Capsules 0.4 cm long, many-seeded.

Occasional in ponds. Fl. & Fr.: Dec.–Mar. Malkalcheruvu, *MSM & KH* 10566.

BIGNONIACEAE

1. Tree up to 20 m tall. Leaves 2–3-pinnate, flowers white, fragrant .... ***Millingtonia***
1. Tree up to 5 m tall, Leaves one-pinnate, flowers yellow or white, not fragrant:
    2. Leaflets oblong-obovate, branchlets pubescent. Calyx spathaceous. Corolla white .... ***Dolichandrone***
    2. Leaflets lanceolate, branchlets glabrous. Calyx cupular. Corolla yellow... ***Tecoma***

DOLICHANDRONE (Fenzl) B. Seemann

1. Leaflets < 3 cm long. Flowers in corymb, fruit falcate .... ***D. falcata***
1. Leaflets to 14 cm long. Flowers in panicles, fruit not falcate .... ***D. atrovirens***

Dolichandrone atrovirens (Roth) Sprague in Bull. Misc. Inform. 1919: 307. 1919; Gamble 996(700); Khan in FFH 256. *Bignonia atrovirens* Heyne ex Roth, Nov. Pl. Spec. 784. 1821. *Dolichandrone crispa* Seem. in J. Bot. 8: 381. 1870; FBI 4: 379.

A deciduous tree, up to 12 m tall, branchlets pubescent. Leaves imparipinnate, leaflets oblong, 8–14 × 2.5–4.5 cm, base obliquely truncate, margin entire, tip acuminate, pubescent on both sides. Flowers white in axillary panicles. Capsules 25 cm long, arcuate with speckles. Seeds winged.

Occasional in forests. Fl. & Fr.: May–Dec. Rangammagudem beat, *MSM & KH* 10316.

Dolichandrone falcata (Wall ex DC.) Seem Fl. Vill. Nov. 151. 1880; FBI 4: 380. 1884; Gamble 996(700). *Spathodea falcata* Wall ex. DC. Prodr. 9: 206. 1845.

A deciduous small tree, up to 5 m tall, branchlets pubescent. Leaves imparipinnate, leaflets obovate-obcordate, 2–3 × 1.5–2.5 cm, thinly pubescent. Flowers white in corymbs. Capsule flat, falcately curved.

Common in deciduous forest. Fl. & Fr. June–May. Central University, *MSM* 15166.

MILLINGTONIA Linnaeus

Millingtonia hortensis L.f. Suppl. Pl. 291. 1781; FBI 4: 377; Gamble 994(669); Khan in FFH 255.

An evergreen tree, up to 20 m tall, branchlets glabrous. Leaves 2–3-pinnate, leaflets ovate, lanceolate, 1–5 × 1–2 cm, base truncate, margin entire, toothed, tip acuminate. Flowers white, fragrant, in corymbose panicles. Capsule linear, compressed, many-seeded.

Common in villages and towns. Fl. & Fr.: July–Jan. Mohammadabad, *MSM* 19240.

TECOMA A.L. Jussieu

Tecoma stans (L.) Kunth in H.B.K. Nov. Gen. Pl. 3: 144. 1819. *Bignonia stans* L. Sp. Pl. (ed. 2) 871. 1763. *Stenolobium stans* (L.) Seem. Ann. Mag. Nat. Hist. (Ser. 3) 10: 30. 1862; Gamble 1000 (703).

A small tree, up to 5 m tall, branchlets glabrous. Leaves imparipinnate, leaflets lanceolate, 4–10 × 1–2.5 cm, base unequal, margin dentate, tip acuminate. Flowers yellow in terminal racemes. Capsule oblong, glabrous.

Cultivated as an ornamental. Fl. & Fr.: Oct.–Mar. Gacchibowli, *MSM* 15190.

Jacaranda mimosifolia D. Don, *Spathodea campanulata* Beauv., *Tabebuia rosea* (Betrol) DC. are planted for ornamental purpose.

PEDALIACEAE

1. Flowers in terminal cymes. Fruits drupes, hooked .... ***Martynia***
1. Flowers solitary. Fruits capsules, beaked .... ***Sesamum***

MARTYNIA Linnaeus

Martynia annua L. Sp. Pl. 618. 1753; Gamble 1003(705). *M. diandra* Gloxin, Observ. Bot. 14 t. 1. 1785; FBI 4: 386; Khan in FFH 262.

Annual erect herb, up to 1 m tall, stem pubescent. Leaves ovate-deltoid, 5–15 x 3–12 cm, base cordate, margin dentate, tip acute, pubescent on both sides. Flowers pinkish-white in terminal cymes. Drupes ovoid, 2.5 cm long, with two sharp anterior hooks, hairy.

Common in waste places on roadsides. Fl. & Fr.: June–Dec. Kommireddipalli, *MSM & NV* 10325.

SESAMUM Linnaeus

Sesamum orientale L. Sp. Pl. 634. 1753. *S. indicum* L. Sp. Pl. 634. 1753; FBI 4: 387; Gamble 1002(704); Khan in FFH 261.

An annual erect herb, up to 1 m tall, branchlets angular, pubescent. Leaves elliptic-lanceolate, 1.5–6 × 0.5–2.5 cm, base rounded, margin entire, tip acuminate. Flowers white-pink, axillary, solitary. Capsule 1.8 cm long, beaked, densely pubescent, many-seeded. Seeds white.

Cultivated, often as an escape. Fl. & Fr.: Sept.–Jan. *Vern.:* Nuvvulu. Kanmankalva beat, *MSM* 10458.

ACANTHACEAE

1. Twining herbs or sub-shurbs .... ***Thunbergia***
1. Erect or prostrate herbs or sub-shrubs:
    2. Armed herbs or sub-shrubs:
        3. Flowers blue .... ***Hygrophila***
        3. Flowers orange .... ***Barleria***
    2. Unarmed herbs or sub-shurbs:
        4. Corolla one lipped .... ***Blepharis***
        4. Corolla two-lipped or with five sub-equal lobes:
            5. Roots tuberous .... ***Ruellia***
            5. Roots not tuberous:
                6. Bracts spine-tipped .... ***Lepidagathis***
                6. Bracts not spine tipped:
                    7. Flowers in panicles:
                        8. Flowers pink .... ***Peristrophe***
                        8 Flowers white:
                            9. Capsules velvety pubescent .... ***Rhinacanthus***
                            9. Capsules glabrous or minutely hairy:
                                10. Capsules bloated .... ***Indoneesiella***
                                10. Capsules not as above .... ***Andrographis***
                    7. Flowers in spikes or clusters
                        11. Flowers in spikes:
                            12. Calyx 4-partite .... ***Rostellularia***
                            12. Calyx 5-partite
                                13. Corolla tube twice as long as corolla lips .... ***Asystasia***
                                13. Corolla tube as long as or slightly larger than corolla lips:
                                    14. Placentas separating from valves .... ***Rungia***
                                    14. Placentas not separating from valves .... ***Justicia***
                        11. Flowers in clusters .... ***Hemigraphis***

ANDROGRAPHIS Wallich

Andrographis paniculata (Burm. f.) Wall. ex Nees in Wall. Pl. As. Rar. 3: 116. 1832; FBI 4: 501; Gamble 1040(734); Khan in FFH 264. *Justicia paniculata* Burm. f. Fl. Ind. 9. 1768. *Andrographis subspathulata* Clarke in Hook. f. Fl. Brit. India 4: 502. 1884.

An erect subshrub, up to 1 m tall, branchlets glabrous, quadrangular, leaves linear-obovate, 3–7.5 × 1–2 cm, base attenuate, margin entire, tip acuminate, glabrous above, appressed pubescent below. Flowers white in terminal and axillary zig-zag branched panicles, calyx stipitate, glandular hairy. Capsule oblong, compressed, minutely hairy.

Common in forests. Fl. & Fr.: Sept.–Jan. Mohammadabad beat, *MSM* 10445.

ASYSTASIA Blume

Asystasia dalzelliana Sant. in Kew. Bull. 1948. *A. violacea* Dalz. ex Clarke in Hook. f. Fl. Brit. India. 4: 494. 1884, non Dalz. 1850; Gamble 1063(745).

An erect herb, branchlets glabrous. Leaves ovate-elliptic, 1.5–3 × 1–2 cm, base truncate, margin entire, tip acute, thinly pubescent above and below. Flowers blue, axillary, solitary. Capsules pubescent, seeds orbicular.

Occasional in forest. Fl. & Fr.: Sept.–Dec. Anantagiri RF, *MSM* 12784.

BARLERIA Linnaeus

Barleria prionitis L. Sp. Pl. 636. 1753; FBI 4: 482; Gamble 1058(741).

An erect armed subshrub, up to 1 m tall, branchlets glabrous. Leaves obovate, 3–8 × 0.7–3 cm, base attenuate, margin entire, tip mucronate, densely pubescent above, glabrous below. Flowers orange, axillary or terminal, solitary or clustered, calyx lobes spine-tipped. Capsules 1.5 cm long, beaked, glabrous, 2-seeded.

Common in scrub jungles and hedges. Fl. & Fr.: Sept.–Dec. *Vern.*: Mullagorinta. Kanmankalva beat, *MSM* 10464.

BLEPHARIS A.L. Jussieu

1. Leaves thin, tip apicalate, flowers white .... ***B. maderaspatensis***
1. Leaves thick, tip obtuse, flowers blue .... ***B. repens***

Blepharis maderaspatensis (L.) Roth Nov. Pl. Spec. 320. 1821. *Acanthus maderaspatensis* L. Sp. Pl. 639. 1753. *Blepharis boerhaviifolia* Pers. Syn. Pl. 2: 180; FBI 4: 478; Gamble 1013(712).

An annual prostrate herb, up to 1 m tall, branchlets hispid. Leaves elliptic-lanceolate, 1.8–8 × 0.5–3 cm, base attenuate, margin distantly toothed, tip apiculate, scabrid on both sides. Flowers white in axillary and terminal clusters, outer calyx lobes unequal, bracts in 4 opposite pairs, dentate with stiff hairs. Capsule 1 cm long, glabrous, 2-seeded.

Occasional in forests. Fl. & Fr.: Sept.–Jan. Anantasagar beat, *MSM* 10508.

Blepharis repens (Vahl) Roth, Nov. Pl. Spec. 321. 1821. *Acanthus repens* Vahl, Symb. Bot. 2: 76. 1791. *Blepharis molluginifolia* Pers. Syn. Pl. 2: 180. 1806; FBI 4: 479; Gamble 1013(712).

An annual prostrate herb, up to 1 m, branchlets hispid. Leaves oblong-obovate, 0.5–2.5 × 0.4–0.8 cm, base cuneate, margin entire, tip obtuse. Flowers blue in axillary and terminal clusters. Capsules oblong, 2-seeded.

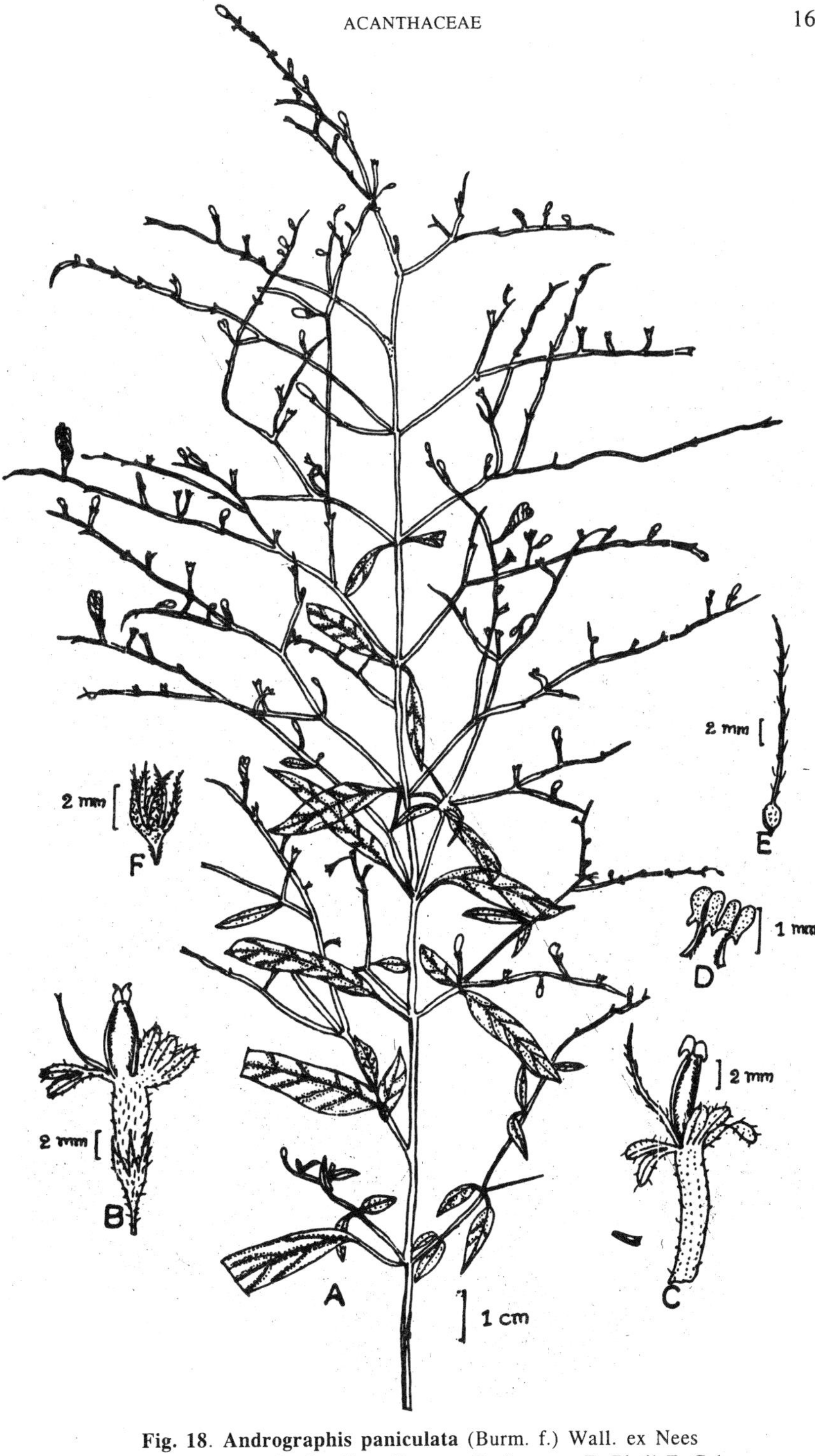

Fig. 18. **Andrographis paniculata** (Burm. f.) Wall. ex Nees
A. Twig B. Flower C. Corolla split open D. Stamens E. Pistil F. Calyx.

Occasional in forest (Seshagiri Rao, 1986). Fl. & Fr.: July–Dec.

HEMIGRAPHIS Nees

Hemigraphis latebrosa (Heyne ex Roth) Roxb. Fl. Ind. 3: 46. 1832; FBI 4: 423; Gamble 1018(715). *Ruellia latebrosa* Heyne ex Roth, Nov. Pl. Spec. 307. 1821.

An annual erect herb, up to 70 cm tall, branchlets quadrangular, hispid. Leaves ovate, 2–5 × 0.5–3 cm, base attenuate-cuneate, margin crenate-serrate, tip acute-acuminate, sparsely hairy on both sides. Flowers blue in axillary and terminal clusters. Capsules 0.7 cm long, sparsely pubescent, 2-seeded.

Common in forest under growth. Fl. & Fr.: Dec.–Mar. Kusumasamudram beat, *MSM & KH* 11028.

HYGROPHILA Robert Brown

Hygrophila auriculata (Schum.) Heine in Kew Bull. 16: 172. 1962. *Barleria auriculata* Schum. & Thonn. Beskr. Guin Pl. 285. 1827. *Asteracantha longifolia* (L.) Nees in Wall. Pl. As. Rar. 3: 90. 1832; Gamble 1015(712). *Barleria longifolia* L. Cent. Pl. 2: 22. 1756. *Hygrophila spinosa* T. Anders. in Thw. Enum. Pl. Zeyl. 225. 1860; FBI 4: 408.

An erect armed subshrub, up to 60 cm tall, stem stout, hispid, leaves 8 and spines 6 in whorls at nodes, oblong-lanceolate, 6–20 × 1.5–4 cm, base attenuate, margin minutely dentate, tip acute, hispid on both sides. Flowers blue in axillary whorls with leafy bracts and bracteoles. Capsules 1 cm long, 4–8-seeded.

Common in wet places, hedges. Fl. & Fr.: Sept.–Jan. Mohammadabad beat, *MSM* 10499.

INDONEESIELLA Sreemadhavan

Indoneesiella echioides (L.) Sreemadh. in Phytologia 15: 271. 1967. *Justicia echioides* L. Sp. Pl. 16. 1753. *Androgrophis echioides* (L.) Nees in Wall. Pl. As. Rar. 3: 117. 1832. *p.p.* FBI 4: 505; Gamble 1051(736).

An annual erect herb, up to 50 cm tall, branchlets hispid. Leaves oblong, 3–5.5 × 1–2.5 cm, base cuneate, margin entire, tip acute, sparsely hairy. Flowers white-pink in axillary, lax racemes. Capsules bloated hairy, many-seeded.

Common in open forest. Fl. & Fr.: Aug.–Jan. Lingampalle, *MSM* 12757.

JUSTICIA Linnaeus

1. Shrubs .... *J. adhathoda*
1. Herbs:
   2. Erect herbs or diffuse herbs:
      3. Bracts white with green nerves .... *J. trinervia*
      3. Bracts green:
         4. Calyx lobes 5 .... *J. glauca*
         4. Calyx lobes 4 .... *J. procumbens*
   2. Prostrate herb .... *J. prostrata*

Justicia adhatoda L. Sp. Pl. 15. 1753. *Adhatoda vasica* Nees in Wall. Pl. As. Rar. 3: 103. 1832; FBI 4: 540; Gamble 1082(758); Khan in FFH 283.

A shrub up to 2 m tall, branchlets angular, pubescent. Leaves ovate-lanceolate, 8–20 × 3–7.5 cm, base acute, margin entire-obscurely crenate, tip acuminate. Flowers white in axillary and terminal spikes. Capsules oblong, pubescent, 4-seeded.

Occasional in hedges. Fl. & Fr.: Dec.–Mar. Mohammadabad, *MSM & PRSR* 15123.

Justicia glauca Rottl. in Ges. Naturf. Frunde Berlin Neue Schriften 4: 219. 1803; FBI 4: 529; Gamble 1079(755).

An annual erect herb, up to 30 cm tall, branchlets angular, pubescent. Leaves ovate, 2–6.5 × 1.5–3 cm, base acute, margin entire, tip acute, pubescent on both sides. Flowers pink in terminal spikes, bracts leafy. Capsules 1 cm long, glabrous, 4-seeded.

Common in all forests. Fl. & Fr.: June–Dec. Kusumasamudram beat, *MSM & NV* 10360; Lingampalli, *MSM & PRSR* 12752.

Justicia procumbens L. Sp. Pl. 15. 1753; FBI 4: 539; Gamble 1080(756).

An annual diffuse herb, up to 25 cm, branchlets pubescent. Leaves obovate, 0.5–2 × 0.5–1 cm, base obtuse, margin enitre, tip obtuse, pubescent on both sides. Flowers pink in terminal spikes. Capsules 0.5 cm long, pubescent except on apex, 4-seeded.

Common in marshy places of forest. Fl. & Fr.: June–Dec. Rangammagudem beat, *MSM & KH* 10576.

Justicia prostrata (Clarke) Gamble Fl. Pres. Madras 1081(757). 1924. *J. diffusa* Willd. var. *prostrata* Clarke in FBI 4: 638. 1885.

An annual prostrate herb, branchlets striate, pubescent. Leaves elliptic 0.5–1 × 0.3–0.5 cm, base rounded, margin entire, tip acute. Flowers pink in long linear spikes. Capsules oblong, sparsely pubescent. 4-seeded.

Common in waste lands and fields (Seshagiri Rao, 1986). Fl. & Fr.: July–Dec.

Justicia trinervia Vahl, Enum. 1: 156. 1834; FBI 4: 536; Gamble 1079(755).

An annual erect herb, up to 75 cm tall, branchlets hairy, leaves elliptic-ovate or oblong, 3–10 × 1.5–4.5 cm, base rounded, margin entire, tip acute-obtuse. Flowers whitish pink in terminal lax spikes, bracts and bracteoles white with green nerves. Capsule oblong, many seeded.

Occasional in forest under growth. Fl. & Fr.: Sept.–Dec. Kanmankalva beat, *MSM* 10457; Anantasagar beat, *MSM* 10503.

LEPIDAGATHIS Willdenow

1. Herbs. Leaves linear-lanceolate. Flowers in pilose spikes .... ***L. cristata***
1. Undershrubs. Leaves oblong-obovate. Flowers in globose heads .... ***L. mitis***

Lepidagathis cristata Willd. Sp. Pl. 3: 400. 1800; FBI 4: 516; Gamble 1066(747).

An annual erect herb, up to 15 cm tall, branchlets quadrangular, glabrous. Leaves linear, lanceolate, 2–5 × 0.5–1 cm, base acute, margin entire, tip acute, pubescent on both sides. Flowers pink in condensed ovoid pilose spikes. Capsules 0.5 cm long, compressed, 2-seeded.

Common in forest. Fl. & Fr.: All seasons. Dharmapur beat, *MSM & KH* 10544.

Lepidagathis mitis Dalz. in Hooker's J. Bot. Kew. Gard. Misc. 3. 226. 1851; FBI 4: 516. 1885; Gamble 2: 1067(747).

A stiff under shrub, branches procumbent from a hard root stock. Leaves oblong-obovate, 4 × 1.5 cm, pubescent above, scabrous below. Flowers white or pale pink in globose heads. Capsule ovoid, 2-seeded.

Common in waste lands (Seshagiri Rao, 1986). Fl. & Fr.: Nov.–Mar.

PERISTROPHE Nees

Peristrophe paniculata (Forssk.) Brumnitt in Kew Bull. 39. 451. 1983. *Dianthera paniculata* Forssk. Fl. Aegypt-Arab. 7. 1775. *Peristrophe bicalyculata* (Retz.) Nees in Wall. Pl. As. Rar. 3: 113. 1832; FBI 4: 554; Gamble 1084(759). *Dianthera bicalyculata* Retz. in Acta. Holm. 1775. 297. t. 9. 1776.

An annual erect herb, up to 1.5 m tall, branchlets quadrangular, glabrous. Leaves ovate-lanceolate, 3–5 × 0.5–1.5 cm, base acute, margin entire, tip acute, pubescent on both sides. Flowers whitish-pink in axillary and terminal lax panicles. Capsules 2.5 cm long, sparsely pubescent, stipitate, 1-seeded.

Common in forests. Fl. & Fr.: Sept.–Jan. Dharmapur beat, *MSM & KH* 10548.

RHINACANTHUS Nees

Rhinacanthus nasutus (L.) Kurz. in J. Asiat. Soc. Bengal 39. t. 2. Nat. Hist. 79. 1870. *Justicia nasutus* L. Sp. Pl. 16. 1753. *Rhinacanthus communis* Nees in Wall. Pl. As. Rar. 3: 109. 1832; FBI 4: 541; Gamble 1083(758).

An annual erect herb, up to 50 cm tall. Leaves elliptic, 4–9 × 2–5 cm, base tapering, margin entire, tip acute. Flowers white in terminal panicles. Capsules narrow, velvety pubescent; seeds tuberculate, glabrous.

Rare in open forest. Fl. & Fr.: Dec.–Feb. Anantagiri RF, *MSM & PRSR* 12785.

ROSTELLULARIA Reichenbach

Note: Bremekamp in Verh. Nederl. Akad. Wet. (2) 45. (2): 61. 1948 separated the species of section 6 of Clarke (*Rostellularia*) and raised this to the rank of a separate genus.

1. Leaves elliptic-ovate. Capsules oblong .... ***R. crinita***
1. Leaves linear-oblong. Capsules ellipsoid .... ***R. simplex***

Rostellularia crinita (Nees) Nees in DC. Prodr. 11: 373. 1847. *Rostellaria crinita* Nees in Wall. Pl. As. Rar. 3: 101. 1832. *Justicia micrantha* Wall. ex Clarke in Hook. f. Fl. Brit. India 4: 536. 1885; Gamble 1080(756).

An annual erect herb, up to 50 cm tall, branchlets hispid. Leaves elliptic-ovate, 1.5–3.5 × 0.8–1.8 cm, base unequal, margin entire, tip acute-obtuse, hispid on both sides. Flowers pink in terminal spikes, spikes up to 2.5 cm long, bracts and bracteoles scarious. Capsule oblong, 0.5 cm long.

Occasional in forest under growth. Fl. & Fr.: Sept.–Jan. Mohammadabad beat, *MSM* 10435.

Rostellularia simplex Wight Ic. t. 1542, 1850. non *Justicia simplex* D. Don 1825. *Justicia notha* Clarke in Hook. f. Fl. Brit. India 4. 537. 1885; Gamble 2: 1080(756).

An annual erect herb, up to 30 cm tall; branchlets striate. Leaves linear oblong, 5–7 × 0.5–0.7 cm, base acute, margin entire, tip acute. Flowers pink in terminal solitary, dense spikes. Capsule ellipsoid.

Occasional in open forest (Seshagiri Rao, 1986). Fl. & Fr.: July–Dec.

RUELLIA Linnaeus emend. Bremekamp

Ruellia tuberosa L. Sp. Pl. 635. 1752; Gamble 1017(714).

An annual erect herb, up to 40 cm tall, stem 4-angled. Leaves oblong-obovate, 6–10 × 3–4 cm, base cuneate, margin entire, tip obtuse. Flowers blue or violet in axillary pairs. Capsules oblong, glabrous.

Occasional as an escape. Fl. & Fr.: All seasons. Central University, *MSM* 15200.

RUNGIA Nees

1. Branchlets densely pubescent .... *R. repens*
1. Branchlets glabrous .... *R. pectinata*

Rungia pectinata (L.) Nees in DC. Prodr. 11: 470. 1841. *Justicia pectinata* L. Amoen. Acad. 4: 299. 1739. *Rungia parvifola* (L.) Nees var. *pectinata* (L.) Clarke in FBI 4: 550. 1885.

An annual erect herb, up to 40 cm tall, branchlets glabrous. Leaves ovate-lanceolate, 1–5 × 0.5–2.5 cm, base attenuate, margin entire, tip acute, pubescent on both sides. Flowers greenish-blue in terminal spikes. Capsules 0.5 cm long, pubescent, 2-seeded. Seeds orbicular.

Occasional in waste lands. Fl. & Fr.: Sept.–May. Kondapur beat, *MSM & KH* 10322.

Rungia repens Nees in Wall. Pl. As. Rar. 3: 110: 1832; FBI 4: 549; Gamble 1070(750), *Justicia repens* L. Sp. Pl. 15. 1753.

An annual procumbent herb, up to 30 cm, branchlets densely pubescent. Leaves elliptic-lanceolate, 2–6 × 0.5–2.5 cm, base rounded, margin serrate, tip acute-acuminate. Flowers pink in long, terminal spikes. Capsules oblong, pubescent, 2–4-seeded.

Occasional weed of forests and waste lands (Seshagiri Rao, 1986). Fl. & Fr.: Sept.–Dec.

THUNBERGIA Retzius *nom. cons.*

Thunbergia fragrans Roxb. Pl. Cor. t. 67. 1795. var. **fragrans;** Clarke in Hook. f. Fl. Brit. India 4; 390. 1884; Gamble 1007(707). *T. fragrans* Roxb. var. *laevis* (Nees) Clarke in Hook. f. Fl. Brit. India 4; 391. 1884; Gamble 1007(707). *T. laevis* Nees in Wall. Pl. As. Rar. 3: 77. 1832.

A slender twiner. Leaves obovate, 4–12 × 3–6 cm, base hastate, margin entire, tip acute. Flowers white, axillary, solitary. Capsules glabrous, 4-seeded.

Occasional on hedges and bushes. Fl. & Fr.: Sept.–Feb. Naskal, *MSM* 15108.

Crossandra infundibuliformis (L.) Nees (*Vern.:* Kanakambaram) is planted for its ornamental flowers.

VERBENACEAE

1. Prostrate herbs .... *Phyla*
1. Erect herbs, shurbs or trees:
   2. Plants armed .... *Lantana*
   2. Plants unarmed:
      3. Leaves compound .... *Vitex*
      3. Leaves simple:
         4. Herbs or shrubs:

5. Flowers in spikes .... *Stachytarpheta*
5. Flowers in cymes:
   6. Flowers in aggregate capitate cyme .... *Symphorema*
   6. Flowers in lax panicle .... *Clerodendrum*

4. Trees:
   7. Young branchlets tomentose .... *Premna*
   7. Young branchlets not as above:
      8. Drupes 4-lobed, clothed with peltate hairs .... *Tectona*
      8. Drupes pyriform, smooth .... *Gmelina*

CLERODENDRUM Linnaeus

1. Straggling shrubs .... *C. inerme*
1. Erect shrubs:
   2. Flowers blue .... *C. serratum*
   2. Flowers white:
      3. Branchlets glabrous, leaves broad ovate .... *C. philippinum*
      3. Branchlets pubescent, leaves ovate-deltoid .... *C. phlomides*

Clerodendrum inerme (L.) Gaertn. Fruct. 1: 271. 1788; FBI 4: 589; Gamble 1099(769). *Volkameria inermis* L. Sp. Pl. 637. 1753.

A straggling shrub, up to 2 m, branchlets pubescent. Leaves elliptic-ovate, 4–7 × 2–4 cm, base cuneate, margin entire, tip acute-obtuse, glabrous on both sides. Flowers white in axillary monochasial cyme. Drupes pyriform, glabrous, brown when ripe.

A common hedge plant. Fl. & Fr.: All seasons. Central University, *MSM* 15197.

Clerodendrum philippinum Schauer in DC. Prodr. 11. 667. 1847 (*Clerodendron*); Howard & Powell in Taxon 1754. 1968. *C. fragrans* Hort. ex Vent. Jard. Mal. 2: t. 70: 1804 pro syn. Gamble 1100(770).

A small shrub, branchlets glabrous. Leaves ovate, 8 × 5 cm, base rounded, margin coarsely toothed, tip acute. Flowers white in compact corymbose panicles. Fruits not seen.

Rare on roadsides. It is often met in garden. Fl. & Fr.: Throughout the year. Vikarabad, *MSM & TP* 11089.

Clerodendrum phlomides L.f. Suppl. Pl. 292. 1781; FBI 4: 590; Gamble 1099(769); Khan in FFH 275.

A shrub, up to 4 m tall, branchlets pubescent. Leaves ovate-deltoid, 3–7 × 2.5–4 cm, base truncate, margin entire, tip acute. Flowers creamish white in axillary and terminal panicles. Drupes globose, black when ripe.

Occasional in forest. Fl. & Fr.: June–Aug. Mekavarpalli *MSM & TP* 11049.

Clerodendrum serratum (L.) Moon. Cat. Pl. Ceylon 46. 1824; FBI 4: 592; Gamble 1100(770). *Volkameria serrata* L. Mant. Pl. 90. 1767.

A shrub, up to 1 m tall, branchlets angular. Leaves ovate-elliptic, 2–20 × 1.5–12 cm, base attenuate, margin serrate, tip acuminate. Flowers pale blue in terminal long panicles. Drupes obovoid.

Occasional in deciduous forest. Fl. & Fr.: Aug.–Dec. Rudraram RF, *MSM & TP* 12731.

GMELINA Linnaeus

Gmelina arborea Roxb. Pl. Cor. t. t. 246. 246. 1815; FBI 4: 581; Gamble 1097(768).

A deciduous tree, up to 10 m tall, bark greyish-yellow, branchlets quadrangular, pubescent. Leaves ovate, 10–20 × 8–15 cm, base truncate, cordate, margin entire, tip obtuse-acuminate, glabrous above, tomentose below. Flowers brownish-yellow in long fulvous, hairy panicles. Drupes pyriform, smooth, orange-yellow when ripe.

Occasional in deciduous forest. Fl. & Fr.: Feb.–Apr. *Vern.:* Gommerteku. Tandur, *MSM* 19241.

LANTANA Linnaeus

Lantana camara L. Sp. Pl. 627. 1753; FBI 4: 562. *L. aculeata* L. Sp. Pl. 627. 1753; Gamble 1087(761); Khan in FFH 266; *L. camara* L. var. *aculeata* (L.) Mold. in Torreya 34: 9. 1934.

An annual straggling shrub, up to 2 m, branchlets quadrangular, pubescent. Leaves ovate, 2–5.5 × 2–4 cm, base truncate, margin crenate-serrate, tip acute, sparsely pubescent above, densely pubescent below. Flowers red in axillary spikes. Drupes globose, 0.4 cm long, succulent, 1-seeded.

Common in scrub jungles in all forests. Fl. & Fr.: All seasons. Mohammadabad beat, *MSM & KH* 10207.

PHYLA Loureiro

Phyla nodiflora (L.) E. Greene in Pittonia 4: 46. 1899. *Verbena nodiflora* L. Sp. Pl. 20. 1753. *Lippia nodiflora* (L.) A. Rich in Michaux, Fl. Bor. Amer. 2: 15. 1803; FBI 4: 563; Gamble 1098(762).

An annual prostrate herb, up to 20 cm, stem glabrous. Leaves spathulate, 1–2 × 0.4–1 cm, base attenuate, margin serrate, tip rounded, sessile, pubescent on both sides. Flowers white in axillary long stalked globose heads. Drupes 0.2 cm long, globose, 1-seeded.

Common at ponds. Fl. & Fr.: All seasons. Gadirayal beat, *MSM* 10524.

PREMNA Linnaeus

1. Branchlets and leaves tomentose .... *P. tomentosa*
1. Branchlets glabrous, leaves pubescent .... *P. latifolia*

Premna latifolia Roxb. Fl. Ind. 3: 76. 1832; var. **mollissima** C.B. Clarke in Hook. f. Fl. Brit. India. 4: 578. 1885; Gamble 1096(767).

A medium sized tree, up to 3 m tall, branchlets glabrous. Leaves ovate, 4–6 × 3–5 cm, base rounded, margin serrate, tip obtuse, pubescent on both sides. Flowers greenish white in terminal corymbs. Drupes globose, black.

Occasional in forest. Fl. & Fr.: Mar.–Oct. Rudraram RF, *MSM & TP* 12710.

Premna tomentosa Willd. Sp. Pl. 3: 314. 1800; FBI 4: 576; Gamble 1095(767); Khan in FFH 271. *Cornutia corymbosa* Lam. Encycl. 1: 54. 1783; non Burm. f. 1768.

A deciduous tree, up to 6 m tall, young branchlets yellow-tomentose, bark yellow. Leaves ovate, 3–10 × 3–8 cm, base cordate, margin entire, tip acuminate, glabrous above, fulvous stellate-tomentose below. Flowers white-cream in terminal corymbs. Drupes 0.5 cm long, glabrous, pyrenes 1.

Common in dry deciduous forests. Fl. & Fr.: Jan.–May. *Vern.*: Nagur. Mohammadabad beat, *MSM & KH* 10212.

STACHYTARPHETA Vahl *nom. cons.*

Stachytarpheta jamaicensis (L.) Vahl, Enum. Pl. 1: 206. 1804. *Verbena jamaicensis* L. Sp. Pl. 19. 1753. *Stachytarpheta indica* auct. non Vahl; FBI 4: 564; Gamble 1090(763).

An annual erect herb, up to 70 cm tall, branchlets rigid. Leaves elliptic 2–10 × 1–5 cm, base tapering, margin crente-serrate, tip acute, glabrous on both sides. Flowers blue in terminal, long spikes. Fruits oblong, 1-seeded pyrenes.

Common in wet waste lands and roadsides. Fl. & Fr.: Aug.–Dec. Central University, *MSM* 19227.

SYMPHOREMA Roxburgh

Symphorema involucratum Roxb. Pl. Cor. t. 186. 1805; FBI 4: 599; Gamble 1104(773).

A straggling shrub, up to 5 m, young branchlets grey tomentose. Leaves ovate, 3–7.5 × 2.5–5 cm, base rounded-truncate, margin crenate-serrate, tip acute, grey tomentose above, grey stellate tomentose below. Flowers white, aggregated in capitate, pedunculate cymes, surrounded by obovate spathulate involucral bracts. Drupes globose.

Occasional in forests. Fl. & Fr.: May–Jan. Kothapalli beat, *MSM & KH* 10299.

TECTONA Linnaeus f.

Tectona grandis L.f. Suppl. Pl. 151. 1781; FBI 4: 570; Gamble 1092(765); Khan in FFH 266.

A deciduous tree, up to 20 m tall, branchlets tetragonous, tomentose. Leaves broadly elliptic, 15–30 × 8–20 cm, base cuneate, margin entire, tip obtuse or acute, rough, glabrous above, rusty-tomentose below. Flowers light yellow in terminal cymose panicles. Drupes 4-lobed, clothed with peltate stellate hairs.

Common in dry deciduous forests. Fl. & Fr.: June–Mar. *Vern.*: Teku. Anantasagar beat, *MSM & KH* 10407.

VITEX Linnaeus

Vitex negundo L. Sp. Pl. 638. 1753; FBI 4: 583; Gamble 1102(771); Khan in FFH 273.

A shrub, up to 5 m tall, branchlets tetragonous, whitish grey pubescent, bark thin, grey. Leaves 3–5-foliolate, leaflets oblong-lanceolate, terminal leaflet larger than laterals, 5–15 × 1–3 cm, base acute, margin entire, tip acute, glabrous above, grey tomentose below. Flowers bluish-purple in terminal panicles. Drupes globose, 0.5 cm long, glabrous, black.

Common in hedges, on banks of rivers and streams. Fl. & Fr.: All seasons. *Vern.*: Vavili. *Uses*: Leaves and fruit juice as medicine. Mohammadabad beat, *MSM & KH* 10232.

Citherexylum subserratum Sw. and *Duranta repens* L. are planted for ornamental purpose.

LAMIACEAE (*nom. alter.* LABIATAE)

1\. Calyx 8–10-toothed:
  2\. Shrubs up to 1.5 m tall, flowers orange .... ***Leonotis***
  2\. Herbs up to 50 cm tall, flowers white .... ***Leucas***
1\. Calyx 5-toothed:
  3\. Calyx campanulate:
    4\. Calyx 5-toothed .... ***Hyptis***
    4\. Calyx not as above:
      5\. Thrysus 1 cm across .... ***Anisomeles***
      5\. Thrysus 2.5 cm across .... ***Basilicum***
  3\. Calyx 2-lipped:
    6\. Flowers in dense heads:
      7\. Lower lip of calyx truncate .... ***Anisochilus***
      7\. Lower lip of calyx 2 or 4-lipped .... ***Ocimum***
    6\. Flowers in whorls or panicles:

8. Lower lip of calyx 2-toothed, mucronate .... ***Becium***
8. Lower lip of calyx 4-toothed, 2 lobed:
   9. Leaves pinnatisect, lobes linear .... ***Lavandula***
   9. Leaves not pinnatisect, not lobed:
      10. Flowers blue .... ***Plectranthus***
      10. Flowers white lilac .... ***Orthosiphon***

ANISOCHILUS Wallich ex Bentham

Anisochilus carnosus (L.f.) Wall. Pl. As. Rar. 2: 18. 1831; FBI 4: 627; Gamble 1126(788). *Lavandula carnosa* L.f. Suppl. Pl. 273. 1781.

An annual erect herb, up to 70 cm tall, branchlets angular, pubescent. Leaves ovate, 1.5–7.5 × 1–4.5 cm, base cordate, margin serrate, tip acuminate, glandular above, densely pubescent below. Flowers blue in terminal, long spikes. Nutlets 0.2 cm long, brown.

Rare in dry deciduous forests on hill top. Fl. & Fr.: Sept.–Jan. Rangammagudem beat, *MSM* 10473.

ANISOMELES Robert Brown

Anisomeles indica (L.) O. Kuntze Rev. Gen. Pl. 2: 512. 1891; Gamble 1140(797). *Nepeta indica* L. Sp. Pl. 571. 1753. *Anisomeles ovata* R.Br. in W.T. Aiton, Hortus Kew. 3: 364.1811; FBI 4; 672.

An erect under shrub, up to 1.5 m tall, branchlets quadrangular, pubescent. Leaves ovate, 4–9 × 1.5–5.5 cm, base cuneate, margin crenate-dentate, densely pubescent on both sides. Flowers violet in dense axillary whorls on long spikes, calyx lobes lanceolate. Nutlets 0.5 cm long, black.

Common in forests and along stream banks and fences. Fl. & Fr.: Sept.–Dec. Kanmankalva beat, *MSM* 10460.

BASILICUM Moench

Basilicum polystachyon (L.) Monench. Suppl. Meth. 143. 1802. *Ocimum polystachyum* L. Mant. Pl. 567. 1771. *Moschosma polystachyum (L.) Benth.* in Wall. Pl. As. Rar. 2: 13. 1830–31; FBI 4: 612; Gamble 1117(782).

An erect under shrub, up to 70 cm tall, branchlets glabrous, quadrangular. Leaves ovate, 3–10 × 1–3.5 cm, base truncate, margin serrate, tip acuminate, glabrous on both sides. Flowers blue or pink in axillary racemes. Nutlets ellipsoid, 0.2 cm, smooth, 4–5-seeded.

Common in forests. Fl. & Fr.: Sept.–Dec. Dharampur, *MSM* 10484.

BECIUM Lindley

Note: Harley (1983) transferred *Ocimum adscendens* Willd. to the African genus *Becium* Lindl. based on calyx character. The genus *Ocimum* is distinguished from *Becium* by having shorter, but distinct teeth of lateral calyx which are obsolete and replaced by a row of minute teeth or by a row of dense cilia in the latter.

Becium filamentosum (Forssk.) Chiov. Nuovo Giorn. Bot. Ital. 26. 162. 1919. *Ocimum filamentosum* Forssk. Fl. Aegypt. Arab. 108. 1775. *O. adscendens* Willd. Sp. Pl. 3: 166. 1800; FBI 4: 609; Gamble 1112(775); Khan in FFH 276.

An annual erect herb, up to 50 cm tall, branchlets angular, pubescent. Leaves ovate-elliptic, 2–4 × 1.5–2 cm, base cuneate, margin serrulate, tip acute. Flowers white in terminal racemes. Nutlets trigonous.

Occasional in waste lands. Fl. & Fr.: Sept.–Dec. Gacchibowli, *MSM* 12769.

HYPTIS N.J. Jacquin *nom. cons.*

Hyptis suaveolens (L.) Poit. in Ann. Mus. Natl. Hist. Nat. 7: 472. t. 29. t. 2. 1806; FBI 4: 630; Gamble 1129(789). *Ballota suaveolens* L. Syst. Nat. ed. 10. 1110. 1759.

An annual erect sub shrub, up to 1.5 cm tall, branchlets angular, hispid. Leaves ovate, 1.5–4.5 × 1–2 cm, base truncate rounded, margin irregularly serrate, tip acute, strigose on both sides. Flowers blue in whorls or in long cymes at nodes. Calyx 10-ribbed with 5 aristate teeth. Nutlets ovoid.

Common in forest undergrowth. Fl. & Fr.: Dec.–Mar. Mohammadabad beat, *MSM & KH* 10556.

LAVANDULA Linnaeus

Lavandula bipinnata O. Kuntze Rev. Gen. Pl. 2: 521. 1891; Gamble 2: 1129(791). *L. burmanni* Benth. Lab. Gen. et. Sp. 151. 1861; FBI 4: 631.

A slender erect herb, up to 1 m tall, stem 4-angled. Leaves pinnatisect, lobes linear, whorled, glabrescent. Flowers violet in panicled spikes.

Rare in hedges and foot hills. Fl. & Fr.: Nov.–Feb. Anantagiri RF, *MSM & PRSR* 12793.

LEONOTIS (Persoon) Linnaeus

Leonotis nepetifolia (L.) R. Br. Prodr. 504. 1810; FBI 4: 691; Gamble 1155(807).

An annual erect herb up to 1.5 cm tall. Stem quadrangular. Leaves ovate, 1.5–4 × 1–2 cm, base truncate, margin crenate, tip acute, pubescent on both sides. Flowers saffron in axillary globose whorls. Nutlets oblong-obovoid.

Common in waste lands and roadsides. Fl. & Fr.: Sept.–Jan. Pargi, *MSM & PRSR* 15101.

LEUCAS Robert Brown

1. Flowers in dense globose verticils .... ***L. urticaefolia***
1. Flowers in globose heads:
   2. Inflorescence terminal:
      3. Globose heads large, leaves ovate .... ***L. cephalotes***
      3. Globose heads small, leaves oblong .... ***L. nutans***
   2. Inflorescence both axillary and terminal .... ***L. aspera***

Leucas aspera (Willd.) Link, Enum. Hort. Berol. Act. 2: 113. 1822; FBI 4: 890; Gamble 1150(803). *Phlomis aspera* Willd. Enum. Hort. Berol. 621. 1809.

An annual erect herb, up to 50 cm tall, branchlets quadrangular, hispid. Leaves linear-lanceolate, 1.5–5 × 0.5–1 cm, base cuneate, serrate, tip acute, puberulous above and glabrous below. Flowers white in terminal heads. Nutlets 3-gonous.

Common in roadsides. Fl. & Fr.: All seasons. Kusumasamudram beat, *MSM & KH* 11026.

Leucas cephalotes (Roth) Spreng. in L. Syst. Veg. 2: 743. 1825; FBI 4: 689; Gamble (804). *Phlomis cephalotes* Roth, Nov. Pl. Spec. 262. 1821.

An annual erect pubescent herb, up to 50 cm tall. Leaves ovate-lanceolate, 3–6 × 1–3 cm, base truncate, margin crenate-serrate, tip acute, pubescent on both sides. Flowers white in terminal globose heads. Nutlets smooth, brown.

Occasional in cultivated fields. Fl. & Fr.: July–Jan. Mekavarpalli, *MSM & TP* 11048.

Leucas nutans Spreng in L. Syst. Veg. ed. 16. 2: 743. 1825; FBI 4: 688; Gamble 1149(803).

An erect herb, up to 40 cm tall. Leaves oblong-lanceolate, 1.5–3.5 × 0.5–1 cm, base cuneate, margin serrate, tip acute, hispid on both sides. Flowers white in axillary and terminal globose heads. Nutlets ovate.

Occasional in forest. Fl. & Fr.: June–Dec. Lingampalle, *MSM* 12765.

Leucas urticaefolia (Vahl) R. Br. Prodr. 504. 1810; FBI 4: 680. Gamble 1149(803). *Phlomis urticifolia* Vahl, Symb. Bot. 2: 76. 1794.

An annual erect herb, up to 50 cm tall. Leaves ovate-elliptic, 3–5 × 2–3 cm, base rounded, cuneate, margin serrate, tip acute. Flowers white in axillary dense verticils. Nutlets trigonous.

Occasional in waste lands (Seshagiri Rao, 1986). Fl. & Fr.: July–Nov.

OCIMUM Linnaeus

1. Sub shrubs, nutlets large .... ***O. gratissimum***
1. Herbs, nutlets small:
    2. Flowers purple, nutlets smooth .... ***O. tenuiflorum***
    2. Flowers white, nutlets pitted .... ***O. americanum***

Ocimum americanum L. Cent. Pl. 1: 15. 1755. *O. canum* Sims in Curtis Bot. Mag. 51. Pl. 2452. 1823; FBI 4: 607; Khan in FFH 276. *O. africanum* Lour. Fl. Cochinch. 370. 1790; Gamble (1111) 777.

An annual erect herb, up to 50 cm tall, branchlets 4-gonous, pubescent. Leaves elliptic-lanceolate, 1–3.5 × 0.5–2.5 cm, base truncate, margin sparingly serrate, tip acute, pubescent on both sides. Flowers white; about 6 in a whorl in terminal spicate racemes. Nutlets 0.1 cm long, pitted, black. Seeds mucilaginous when wet.

Gregarious in waste lands. Fl. & Fr.: All seasons. *Vern.:* Kukka tulasi. Mohammabad beat, *MSM & KH* 10229.

Ocimum gratissimum L. Sp. Pl. 1197. 1753; FBI 4: 608; Gamble 1111(777).

A perennial sub-shrub, up to 1 m tall, branchlets 4-gonous, glabrescent. Leaves ovate, 4–15 × 1.5–5.5 cm, base cuneate, coarsely crenate-serrate, tip acuminate, glabrescent on both sides. Flowers cream in terminal thyrses. Nutlets 4, 0.1 cm long, rugose-brown.

Occasional in hedges. Fl. & Fr.: June–Dec. Anantasagar beat, *MSM & NV* 10406.

Ocimum tenuiflorum L. Sp. Pl. 597. 1753. *O. sanctum* L. Mant. Pl. 85. 1767; FBI 4: 609; Gamble 1111(778); Khan in FFH.

An annual erect herb, up to 1 m tall. Leaves elliptic-oblong, 2–4 × 0.5–1.5 cm, base cuneate, margin entire-serrate, tip acute, pubescent on both sides. Flowers blue in close whorls on long trichotomous racemes, calyx lower lip 4-lobed. Nutlets smooth.

Common in villages. Fl. & Fr.: All seasons. *Vern.:* Tulsi. Central University, *MSM* 19242.

ORTHOSIPHON Bentham

Orthosiphon rubicundus (D. Don) Benth. in. Wall; Pl. As. Rar. 2: 41. 1830–31. var. *rubicundus;* FBI 4: 614; Gamble 1114(779). *Plectranthus rubicundus* D. Don, Prodr. Fl. Nepal 116. 1825 '*rubicunda*'.

An annual erect herb, up to 30 cm tall. Leaves ovate-oblong, 4.5–7.5 × 1–3 cm, base cuneate, margin serrate-dentate, tip obtuse. Flowers purple in distant whorls in slender racemes. Nutlets smooth, punctate.

Occasional in forest undergrowth. Fl. & Fr.: Aug.–Oct. Anantagiri RF, *MSM & TP* 11070.

PLECTRANTHUS L. Heritier *nom. cons.*

1. Flowers blue in trichotomous racemes .... ***P. mollis***
1. Flowers purple in dense crowded whorls .... ***P. amboinicus***

Plectranthus amboinicus (Lour.) Spreng. Syst. Veg. 2. 690. 1825; Willemse in Blumea 25. 509. 1979. *Coleus amboinicus* Lour. Fl. Cochinchi 2: 372. 1790; Gamble 1124(786). *C. aromaticus* Benth. in Wall. Pl. As. Rar. 2: 15. 1830–1831; FBI 4: 625.

An aromatic herb, up to 80 cm tall. Leaves fleshy, ovate, 3–6 cm long, base cordate, margin crenate. Flowers purple in densely crowded whorls.

Cultivated or as an escape (Seshagiri Rao, 1986). Fl. & Fr.: Sept.–Jan.

Plectranthus mollis (Ait.) Spreng. in L. Syst. Veg. ed. 16. 2: 690. 1825. *Ocimum molle* Ait. Hort. Kew 2: 322. 1789. *Plectranthus incanus* Link Enum. Hort. Berol. Alt. 2: 1822; FBI 4: 621; Gamble 1121(784).

An annual erect herb, up to 70 cm tall, branchlets quadrangular, puberulous. Leaves ovate, 3–10 × 2–9 cm, base cordate, margin crenate, tip obtuse, glabrous above and puberulous below. Flowers blue in trichotomous racemes. Nutlets sub-globose, light brown.

Rare in waste land (Seshagiri Rao, 1986). Fl. & Fr.: June–Dec.

Mentha piperata L. is cultivated for its leaves which are used as spice. *Coleus* and *Salvia* spp. are cultivated in gardens for ornamental purpose.

NYCTAGINACEAE

1. Flowers in umbels, anthocarp club shaped .... ***Boerhavia***
1. Flowers in panicles, anthocarp fusiform .... ***Commicarpus***

BOERHAVIA Linnaeus

Note: The genus was named after Boerhave. Linnaeus latinized the name as Boerhavius and named the genus *Boerhavia* in Sp. Pl. and Gen. Pl. But in Philos. Botanica gives *Boerhaavia*. Later authors changed it into *Boerhaavia* Mill. 1754, *Boerhavia* Neck 1790, *Boerhavia* D. Kuntze 1891. Linnaeus spelling should be retained.

Boerhavia diffusa L. Sp. Pl. 3. 1753; Gamble 1162(814); Khan in FFH 277. *B. repens* L. Sp. Pl. 3. 1753; FBI 4: 709.

An annual prostrate herb, branchlets pubescent. Leaves ovate-orbicular, 0.5–3 × 0.5–2.5 cm, base rounded-truncate, margin entire-undulate, tip acute-apiculate. Flowers violet in axillary umbel. Anthocarps club-shaped, glandular 1-seeded.

Common weed of all places. Fl. & Fr.: All seasons. Mohammadabad, *MSM & PRSR* 15124.

COMMICARPUS Standley

Commicarpus chinensis (L.) Heimerl in Engl. Nat. Pflanzen Fam (ed. 2) 16c: 117. 1934. *Valeriana chinensis* L. Sp. Pl. 33. 1753. *Boerhavia chinensis* (L.) Asch. & Schweinf. Beitr. Fl. Aethip: 167. 1867. *B. repanda* Willd. Sp. Pl. 1: 22. 1797; FBI 4: 709; Gamble 1163(814).

A diffuse scandent herb, up to 75 cm, branchlets glabrous. Leaves triangular-ovate, 1–3.5 × 0.5–2 cm, base cordate, margin shallowly-lobed, tip acute, glabrous above, pubescent below. Flowers pink in axillary and terminal panicles. Anthocarp fusiform, 0.9 cm long, 10-ribbed, wart like appendanged.

Common in hedges and on thickets. Fl. & Fr.: May–Sept. Venkat Reddipalli, *MSM* 10310.

Bougainvillea glabra Choisy, *B. spectabilis* Willd. (*Vern.:* Kagita puvvu) and *Mirabilis jalapa* L. (*Vern.:* Chandrakanta) are planted for ornamental purpose.

AMARANTHACEAE

1. Leaves linear, needle like .... ***Trichurus***
1. Leaves not needle like:
    2. Leaves opposite:
        3. Flowers in long spikes:
            4. Fruits with bristles .... ***Pupalia***
            4. Fruit without bristles .... ***Achyranthes***
        3. Flowers not in spikes:
            5. Branchlets pubescent, flowers in silvery globose heads .... ***Gomphrena***
            5. Branchlets almost glabrous, flowers in globose straw colour heads .... ***Alternanthera***
    2. Leaves alternate:
        6. Branchlets wiry and woolly .... ***Aerva***
        6. Branchlets not as above:
            7. Branchlets pink and flowers rose .... ***Digera***
            7. Branchlets not pink, flowers other than rose:
                8. Spikes capitate .... ***Allmania***
                8. Spikes not capitate:
                    9. Flowers white, bisexual .... ***Celosia***
                    9. Flowers green, unisexual .... ***Amaranthus***

ACHYRANTHES Linnaeus

Achyranthes aspera L. Sp. Pl. 204. 1753; FBI 4: 730. Gamble 1176(823).

An erect subshrub up to 60 cm tall. Leaves obovate-orbicular, 2–6 × 1–3 cm, base attenuate, margin entire, tip acute, pubescent on both sides. Flowers green in terminal spikes. Utricles ovate, tip spiny.

Common in waste lands and roadsides. Fl. & Fr.: All seasons. *Vern.*: Uttareni, Central University, *MSM & TP* 15128.

AERVA Forsskal

Aerva lanata (L.) Juss. Ann. Mus. Nat. Hist. Par. 2. 131. 1803; FBI 4: 728; Gamble 1178(825). *Achyranthes lanata* L. Sp. Pl. 204. 1753.

An annual erect herb, up to 60 cm tall, stem grey-pubescent. Leaves obovate, 1–3 × 0.5–1.5 cm, base cuneate, margin entire, tip acute, pubescent above, white woolly below. Flowers white in axillary spikes. Seeds smooth, black.

Common weed of waste places and roadsides. Fl. & Fr.: All seasons. Mohammadabad beat, *MSM & TP* 10419.

ALLMANIA Robert Brown ex R. Wight

Allmania nodiflora (L.) R. Br. ex Wight

1. Erect herb not much branched. Flowers orange in globose heads .... var. **nodiflora**
1. Straggling herb, many branched. Flowers white in sessile heads ...... var. **roxburghii**

Allmania nodiflora (L.) R. Br. ex Wight in Hook. J. Bot. 1: 227 t. 128. 1834; var. *nodiflora* FBI 4: 716; Gamble 1167(817). *Celosia nodiflora* L. Sp. Pl. 205. 1753.

An annual erect herb, up to 50 cm tall, stem glabrous. Leaves oblong-spathulate, 2–5 × 0.5–1.5 cm, base cuneate, margin entire, tip obtuse, glabrous on both sides. Flowers orange or brown in terminal globose heads. Utricles 3 cm long, membranous, seed compressed, with cupular aril.

Occasional in open forests. Fl. & Fr.: June–Jan. Kusumasamudram beat, *MSM & NV* 10364; Central University, *MSM* 15185.

Allmania nodiflora (L.) R. Br. ex Wight *var.* **roxburghii** Wight in Hook. J. Bot. 1. 226. t. 128. 1834. *Achyranthes nodiflora* Roxb. Fl. Ind. 678. *Celosia nodiflora* L. Sp. Pl. 205. 1753; Gamble 1167(819).

A straggling many branched herb with stout root-stock. Leaves orbicular, often apiculate. Heads axillary, sessile, white.

Rare in open forest. Fl. & Fr.: July–Dec. Central University, *MSM* 15146.

ALTERNANTHERA Forsskal

1. Bracts spinescent .... *A. pungens*
1. Bracts not spinescent:
   2. Leaves lanceolate-spathulate .... *A. sessilis*
   2. Leaves elliptic-lanceolate .... *A. tenella*

Alternanthera pungens Kunth, Nov. Gen. Sp. 2: 206. 1817. *A. echinata* auct. non Smith; Tadulingam & Venkatanaryana. Handb. S. Ind. Weeds. 274 t. 114. 1932.

A perennial prostrate herb, branchlets pubescent. Leaves orbicular-elliptic, 1–3 × 0.5–2 cm, base attenuate, margin entire, tip apiculate, pubescent on both sides. Flowers yellow in axillary spikes, bracts spinescent, oblanceolate. Seeds discoid.

Common in waste places. Fl. & Fr.: May–Oct. Mokarlabad, *MSM & KH* 10263.

Alternanthera sessilis (L.) R. Br. ex DC. Cat. Pl. Horti Monsp. 4: 77. 1813; FBI 4: 731. *Gomphrena sessilis* L. Sp. Pl. 225. 1753. *Alternanthera triandra* Lam. Encycl. 1: 95. 1783; Gamble 1179(825).

An aquatic prostrate herb, branchlets glabrous. Leaves lanceolate, spathulate, 1.5–5.5 × 0.5–1 cm, base cuneate, margin entire, tip acute, glabrous on both sides. Flowers white in axillary sessile heads. Utricles compressed, 1.2 cm long. Seeds orbicular.

Common in damp places. Fl. & Fr.: May–Dec. Dharmapur, *MSM & KH* 10283.

Alternanthera tenella Colla, Mem. Acad. Sci. Terime 33: 131. t. 9. 1820; Mears in Proc. Acad. Nat. Sci. Philadelphia 129: 19. 1977.

An annual decumbent herb up to 30 cm tall, branchlets striate. Leaves opposite, green, petiolate, elliptic-lanceolate, 2–5 x 1–7.5 cm, base cuneate, margin entire, tip acute. Flowers white in axillary nodal heads.

Occasional in waste lands and roadisdes. Fl. & Fr.: Oct.–Jan. Central University, *MSM* 15149.

AMARANTHUS Linnaeus

Note: The generic name is derived from the Amaranth of ancient Greeks. Its latin form is *Amarantus.* As a generic name it was spelt by some pre Linnaen botanists as *Amarantus* and by most as *Amaranthus*. Linnaeus consistently spelt it as *Amaranthus*. He knew both forms and seems to have chosen delibrately the form *Amaranthus* but later on he wavered and gave philologically *Amarantus* in Syst. Nat. (ed. 10) 1268.–1759.

1. Armed herbs .... ***A. spinosus***
1. Unarmed herbs:
    2. Flowers in axillary clusters .... ***A. tricolor***
    2. Flowers in axillary and terminal spikes .... ***A. viridis***

Amaranthus spinosus L. Sp. Pl. 991. 1753; FBI 4: 718; Gamble 1170(819); Khan in FFH 279.

An erect herb up to 30 cm tall. Leaves elliptic-lanceolate, 1–3 × 0.5–1.5 cm, base acute, margin entire, tip acute, glabrous on both sides. Flowers green in axillary and terminal panicles. Utricles ovate.

Common on roadsides. Fl. & Fr.: All seasons. Vikarabad, *MSM & TP* 11090.

Amaranthus tricolor L. Sp. Pl. 989. 1752. *A. mangostanus* L. Cent. Pl. 1: 32. 1755; FBI 4: 720; Gamble 1171 (820). *A. gangeticus* L. Syst. Nat. ed. 10. 1268. 1759 incl. var. *tristis*; FBI 4: 719; Gamble 1171 (810).

Erect herb, to 90 cm tall. Leaves ovate, 3–5 × 2–4.5 cm, base cuneate, margin entire, tip retuse, glabrous. Inflorescence axillary fascicles or aggregated into terminal spiciform racemes. Flowers unisexual, male and female intermixed. Utricle flask shaped, circumscissile. Seeds orbicular, 1.5 mm across, black.

Occasional weed in waste places, fields and along road sides. *Vern.:* Perugu thota kura. Fl. & Fr.: All seasons. Naskal, *MSM & PRSR* 115111.

Amaranthus viridis L. Sp. Pl. ed. 2: 1405. 1763; FBI 4: 720; Gamble 1171 (820).

An annual erect herb, up to 70 cm tall, branchlets striate, glabrous. Leaves ovate-rhomboid, 1–4 × 0.5–3 cm, base cuneate, margin entire, tip notched. Flowers greenish-yellow in axillary and terminal spikes. Utricles ovoid, 0.3 cm long, circumscissile, 1-seeded.

Common in marshy places. Fl. & Fr.: All seasons. Anantasagar beat, *MSM & KH* 11001.

CELOSIA Linnaeus

Celosia argentea L. Sp. Pl. 205. 1753; FBI 4: 714; Gamble 1166(816); Khan in FFH 278.

An erect herb up to 1 m tall. Leaves ovate-linear lanceolate, 3–5 × 0.5–2 cm, base cuneate, margin entire, tip acute. Flowers white in terminal spikes. Utricles globose.

Common weed in cultivated fields. Fl. & Fr.: All seasons. Lingampalle, *MSM* 12745.

DIGERA Forsskal

Digera muricata (L.) C. Martius, Beitr. Amar. 77. 1825. *Achyranthes muricata* L. Sp. Pl. ed. 2: 295. 1762. *Digera arvensis* Forssk., Fl. Aegypt-Arab. 65. 1775; FBI 4: 717; Gamble 1169(818).

An annual erect herb up to 50 cm tall, stem glabrous. Leaves ovate-elliptic, 2–6 × 1–4 cm, base truncate, margin entire, tip acute. Flowers rose in axillary and terminal spikes. Nutlets ovoid, rugose, 1-seeded.

Common weed in cultivated fields. Fl. & Fr.: Aug.–Nov. *Vern.*: Chenchalikura. Mekavarpalli, *MSM & TP* 11038.

GOMPHRENA Linnaeus

Gomphrena serrata L. Sp. Pl. 224. 1753. *G. decumbens* Jacq. Hort. Schoenbr. t. 482. 1805; Gamble 1179 (825). *G. celosioides* auct. non Mart. Bennet. Fl. Howrah 142. 1979.

An annual erect herb, up to 30 cm tall, branchlets pubescent. Leaves ovate-lanceolate, 1.5–2 × 0.5–2 cm, base attenuate, margin entire, tip obtusely apiculate, pubescent on both sides. Flowers white in axillary or terminal spikes. Utricles 0.2 cm long, 1-seeded. Seed reticulate, shiny.

Occasional in open forests and roadsides. Fl. & Fr.: May–Dec. Gadirayal beat, *MSM & KH* 10266, Central University, *MSM* 15143.

Note: According to Mears (loc. cit.) the species to which the name *Gomphrena serrata* L. applies was renamed by Jacquin (1805) as *G. decumbens* without referring Linnaeus taxon. Hence the former name is retained.

PUPALIA A.L. Jussieu *nom. cons.*

Pupalia lappacea (L.) A.L. Juss. in Ann. Mus. Natl. Hist. Nat. (Paris) 2: 132. 1803; FBI 4: 724; Gamble 1173(821) var. **lappacea.** *Achyranthes lappacea* L. .Sp. Pl. 204. 1753. *Pupalia atropurpurea* (Lam.) Moq. in DC. Prodr. 13(2): 331. 1849; FBI 4: 723; Gamble 1173(821). *Achyranthes atropurpurea* Lam. Encycl. 1: 546. 1785.

An erect under shrub, up to 1 m tall, stem pubescent. Leaves elliptic-ovate, 1.5–8.5 × 1–4.5 cm, base cuneate, margin entire, tip shortly acuminate, sparsely pubescent on both sides. Flowers white or yellow in axillary and terminal spikes, bracts and bracteoles awned. Utricles 0.3 cm long, surrounded by 10–20 yellow bristles, 1-seeded.

Common in hedges and open forests. Fl. & Fr.: Sept.–Dec. Rangammagudem beat, *MSM* 10474.

TRICHURUS C. Townsend

Trichurus monsoniae (L.f.) Towns. in Kew Bull. 29: 466. 1974. *Illecebrum monsoniae* L.f. Suppl. Pl. 161. 1781. *Aerva monsoniae* (L.f.) C. Martius in Nov. Act. Phys. Med. Acad. Caes. Leop. Carol. Nat. Cur. 13(1): 291. 1826; FBI 4: 729; Gamble 178(825). *Celosia monsoniae* Retz. Obs. Bot. 2: 13. 1781.

A perennial prostrate herb, up to 50 cm, branchlets white, tomentose, leaves linear, 1–3 × 0.1 cm, fascicled at nodes, base acute, margin entire, tip spine tipped, glabrous on both sides. Flowers whitish-pink in terminal spikes. Utricles ovoid, 0.3 cm long, circumcissile, 1-seeded.

Common in open forest. Fl. & Fr.: Sept.–Dec. Kothapalli beat, *MSM & KH* 10565.

Celosia argentea L. var. *cristata* (L.) O. Ktze. and *Gomphrena globosa* L. are cultivated for ornamental purpose.

POLYGONACEAE

1. A climbing shrub .... ***Antigonon***
1. An erect herb .... ***Polygonum***

ANTIGONON Endlicher

Antigonon leptopus Hook. & Arn. Bot. Beechey Voy. 308. t. 69. 1841; Gamble 1192(835).

A climbing shrub, stem angled, pubescent, tendrils simple. Leaves hastate, 3–15 × 2–8 cm, base cordate, margin entire, tip acute. Flowers rose in terminal panicles. Nutlets trigonous, glabrous.

Occasional in villages. Fl. & Fr.: All seasons. Vikarabad, *MSM & TP* 11091.

POLYGONUM Linnaeus

1. Erect herbs. Nutlets glabrous .... ***P. glabrum***
1. Prostrate herbs. Nutlets pubescent .... ***P. plebeium***

Polygonum glabrum Willd. Sp. Pl. 2: 447. 799; FBI 5: 34. *p.p.* Gamble 1188(832).

A perennial erect herb, up to 1.5 m tall, stem reddish-brown, glabrous. Leaves lanceolate, 9–20 × 1.5–3 cm, base attenuate, margin entire, tip acuminate, ocreae tubular, glabrous on both sides. Flowers pink in axillary-terminal racemes. Nutlets 0.3 cm long, biconvex, glabrous.

Occasional on bank of streams and ponds. Fl. & Fr.: May–Dec. Venkatamma cheruvu, *MSM & KH* 10259.

Polygonum plebeium R. Br. Prodr. 420 1810; FBI 5: 27; Gamble 1188(832).

P. indicum Heyne ex Roth Nov. Pl. Spec. 208. 1821.

A perennial prostrate herb, up to 30 cm, stem striate, glabrous. Leaves oblong-lanceolate,0.8–1 × 0.1–0.3 cm, margin entire, tip acute, ocreae scarious, glabrous on both sides. Flowers red or pink, axillary, solitary or 2–3 in fascicles. Nutlets 3-gonous, 0.2 cm long, pubescent.

Occasional in dry ponds and lakes. Fl. & Fr.: July–Jan. Peddakunta, *MSM* & NV 10422.

Rumex vesicarius L. (*Vern.:* Chukkakura) is cultivated as leafy vegetable.

ARISTOLOCHIACEAE

ARISTOLOCHIA Linnaeus

Aristolochia indica L. Sp. Pl. 960. 1753; FBI 5: 75; Gamble 1202(841).

A perennial climbing shrub, stem grooved, glabrous. Leaves obovate-oblong, 4–8 × 3–5 cm, base cuneate-rounded, margin entire, tip acuminate, glabrous on both sides. Flowers cream in axillary racemes. Capsules septicidal, 5 cm long, 6-valved, many-seeded.

Occasional in forests. Fl. & Fr.: July–Apr. *Vern.*: Gadida gadapa. *Uses:* Antidote for snake bite. Dharmapur, *MSM & KH* 10288; Kusumasamudram beat, *MSM & NV* 10367.

LAURACEAE

CASSYTHA Linnaeus

Cassytha filiformis L. Sp. Pl. 35. 1753; FBI 5: 188; Gamble 1241(868).

Leafless twining parasite, stem dark green, glabrous. Flowers yellow in long spikes. Drupes 0.6 cm long, enclosed in a perianth tube, crowned, glabrous, 1-seeded.

Common parasite on trees and bushes. Fl. & Fr.: June–Jan. Mohammadabad beat, *MSM & KH* 10599.

HERNANDIACEAE

GYROCARPUS N.J. Jacquin

Gyrocarpus americanus Jacq. Select. Strip. Amer. Hist. 282. t. 178. f. 80. 1763; Gamble 1242(869). *G. jacquini* Roxb. Pl. Cor. t. 1. 1795; FBI 2:471; Khan in FFH 174.

A deciduous tree, up to 10 m tall, branchlets pubescent, bark greyish white. Leaves ovate, 8–15 × 10–15 cm, base cordate, margin entire, tip acuminate, pubescent above and villous below. Flowers pale yellow in terminal panicles, unisexual. Samaras 9 cm long, 2-winged, glabrous. Seed solitary.

Common in forests. Fl. & Fr.: Jan.–May. Gadirayal beat, *MSM & KH* 10260.

LORANTHACEAE

1. Flowers actinomorphic, bract partly enclosing calyx .... ***Dendrophthoe***
1. Flowers zygomorphic, bracts free from calyx:

2. Inflorescence racemes, corolla 4-lobed .... *Scurrula*
2. Inflorescence fascicles, corolla 5-lobed .... *Taxillus*

DENDROPHTHOE C.F.P. Martius

Dendrophthoe falcata (L.f.) Ettingsh. in Akad. Wiss. Wien. Math Naturwiss. Kl. Denkschr. 32: 68 t. 13. ff. 14 & 15. 1872. *Loranthus falcatus* L.f. Suppl. Pl. 211. 1781. *L. longiflorus* Desr. in Lam. Encycl. 3: 598. 1792; FBI 5: 214; Gamble 1253(877); Khan in FFH 284. *L. longiflorus* Desr. var. *falcata* (L.f.) Kurz. For. Fl. 2: 321. 1877; FBI 5: 214; Gamble 1253(877). *L. longiflorus* Desr. var. *amplexifolia* (DC.) Thw. Enum. Pl. Zeyl. 134. 1859; Gamble 1253(877).

A much branched partial stem parasite, stem glabrous, woody. Leaves elliptic-lanceolate, 6–15 × 1–4.5 cm, base cuneate, margin entire, tip acute or rounded, glabrous on both sides. Flowers yellow-orange in axillary racemes. Berries 1.5 cm long, glabrous, crowned.

Common in forests. Fl. & Fr.: Dec.–Mar. *Vern.:* Bhadanika, Lothuvagu, *MSM & KH* 10296; Kothapalli beat, *MSM* 10448; Kusumasamudram beat, *MSM & KH* 11022.

SCURRULA Linnaeus

Scurrula parasitica L. Sp. Pl. 110. 1753. *Loranthus buddleioides* Desr. in Lam. Encycl. 3: 600. 1792; Gamble 1251(876). *L. scurrula* (L.) L. Sp. Pl. (ed. 2) 472. 1762; Khan in FFH 265.

A perennial much-branched stem parasite, branchlets rusty tomentose. Leaves elliptic-ovate, 3.5–8 × 2–5 cm, base rounded, margin entire, tip rounded, grey-tomentose on both sides. Flowers scarlet in axillary fascicles. Berries pyriform, 1 cm long, tomentose.

Occasional in forests. Fl. & Fr.: May–Dec. Kothapalli beat, *MSM & KH*10298.

TAXILLUS Van Tieghem

Taxillus bracteatus (Wall.) Tieghem in Bull. Soc. Bot. France 42: 256. 1895. *Loranthus bracteatus* Heyne ex Wall. in Roxb. Fl. Ind. 2: 220. 1824; FBI 5: 213; Gamble 1252(876).

A much branched stem parasite, branchlets tawny-tomentose. Leaves orbicular, 1–4 × 1–4 cm, 3–5-nerved from base, base cuneate, margin entire, tip obtuse, pilose above, pubescent below. Flowers greenish-yellow in axillary fascicles.

Occasional in forests. Fl. & Fr.: Sept.–Dec. Kothapalli beat, *MSM & KH* 10298.

VISCACEAE

VISCUM Linnaeus

Viscum nepalense Spreng. Syst. Curi-post. 47. 1827. *V. articulatum* auct. non Burm. f.: FBI 5: 226; Gamble 1259(881); Khan in FFH 285.

A leafless dichotomously branched stem parasite, branchlets articulate. Internodes 3–8 × 0.2–0.6 cm, flattened, yellow. Flowers purple in fascicles at nodes, bracts cup-shaped. Berries 0.3 cm long, glabrous, 1-seeded.

Rare in forests. Fl. & Fr.: May–Sept. Rangammagudem beat, *MSM & KH* 10315.

EUPHORBIACEAE

1. Inflorecence a cyathium:
   2. Stem succulent, branchlets zig-zag .... ***Pedilanthus***
   2. Stem not succulent, branchlets straight .... ***Euphorbia***
1. Inflorescence not a cyathium:
   3. Branchlets and leaves with stinging hairs .... ***Tragia***
   3. Branchlets and leaves not as above:
      4. Trees:
         5. Branchlets fulvous tomentose, cells of ovary 2-ovuled .... ***Givotia***
         5. Branchlets glabrous or pubescent or puberlous:
            6. Flowers purple, in spikes .... ***Bridelia (retusa)***
            6. Flowers green, solitary or in clusters:
               7. Leaves 5–10 × 2–4 cm .... ***Drypetes***
               7. Leaves 1–1.5 × 0.2–0.4 cm .... ***Phyllanthus***
      4. Herbs or shrubs:
         8. Fruit with prickles .... ***Sebastiania***
         8. Fruit without prickles:
            9. Leaf margin serrate-crenate or dentate or toothed:
               10. Stem glaucous .... ***Croton***
               10. Stem not glaucous:
                  11. Leaves irregularly toothed .... ***Baliospermum***
                  11. Leaves crenate-serrate .... ***Acalypha***
            9. Leaf margin entire:
               12. Branchlets brown tomentose .... ***Chrozophora***
               12. Branchlets green glabrous or pubescent:
                  13. Leaves cordate at base .... ***Jatropha***
                  13. Leaves not cordate at base .... ***Bridelia (montana)***

ACALYPHA Linnaeus

1. Capsules glabrous .... ***A. ciliata***
1. Capsules hispid:
   2. Branchlets pubescent .... ***A. lanceolate***
   2. Branchlets hispid .... ***A. indica***

Acalypha ciliata Forssk., Fl. Aegypt-Arab. 162. 1775; FBI 5: 417; Gamble 1330(930).

An annual erect herb up to 60 cm tall, branchlets striate, hispid. Leaves ovate, 4–6 × 2–4 cm, base rounded, margin serrate, tip acuminate. Flowers green in axillary spikes, male ones in upper half, female in lower half. Capsules globose, glabrous.

Occasional in waste land. Fl. & Fr.: July–Nov. Anantagiri RF, MSM & TP 11063.

Acalypha indica L. Sp. Pl. 1003. 1753; FBI 5: 416; Gamble 1330(930).

An annual erect herb up to 70 cm tall, branchlets striate, hispid. Leaves ovate-rhomboid, 1.5–4 × 1–2.5 cm, base cuneate, margin serrate, tip acute. Flowers green in axillary long spikes, male ones in upper half and female one in lower half. Capsules globose-hispid.

Common in waste lands as weed. Fl. & Fr.: All seasons: Mekavarpalli, MSM & TP 11041.

Acalypha lanceolata Willd. Sp. Pl. 4. 524. 1805; Gamble 1331(931). A. fallax Muell.-Arg. in Linnaea 34: 43. 1865; FBI 5; 416.

An annual erect herb, up to 60 cm tall, branchlets pubescent. Leaves ovate, 2–6 × 1–2.5 cm, base acute, margin serrate, crenate-serrate, tip acute, glabrous above and pubescent below. Flowers creamish-white in axillary spikes. Capsules 0.2 cm long, glandular ciliate. Seeds globose, smooth.

Occasional in waste places, open forests. Fl. & Fr.: All seasons. Kusumasamudram, MSM & NV 10348.

BALIOSPERMUM Blume

Baliospermum montanum (Willd.) Muell-Arg. in DC. Prodr. 15(2): 1125. 1866; Gamble 1343(939). *Jatropha montana* Willd. Sp. Pl. 4: 563. 1805. *Baliospermum axillare* Bl., Bijdr. 604. 1825; FBI 5: 461.

An annual erect herb up to 70 cm tall, branchlets glabrous. Leaves ovate-lanceolate, base rounded, margin-dentate, tip acuminate. Flowers green in axillary long racemes, monoecious or dioecious. Capsules ovoid, 3-lobed, hairy.

Common weed of cultivated fields. Fl. & Fr.: All seasons. *Vern.:* Nelajidi. Pargi, *MSM & PRSR* 15103.

BRIDELIA Willdenow

1. Shrub up to 3 m tall, branchlets glabrous. Secondary nerves 5–6 pairs ... . ***B. montana***
1. Tree up to 12 m tall, branchlets tomentose. Secondary nerves numerous ... ***B. retusa***

Bridelia montana Willd. Sp. Pl. 4: 978. 1806; FBI 5: 269. *p.p.* Gamble 1290(896). *B. hamiltoniana* Wall. ex Muell-Arg. in Linnaea 34: 77. 1865–66; FBI 5: 271; Khan in FFH 292.

A deciduous shrub, up to 4 m tall, branchlets glabrous. Leaves obovate, 3–6 × 1–3 cm, base cuneate, margin entire, tip obtuse or rounded, glabrous above, glaucous below. Flowers cream in axillary spikes. Drupes 0.3 cm long, glabrous, 1–2-seeded.

Common in forests. Fl. & Fr.: Sept.–Jan. Gadirayal beat, *MSM* 10518; Central University, *MSM* 15127.

Bridelia retusa (L.) Spreng in L. Syst. Veg. (ed. 16)3: 48. 1826; FBI 5: 268; *p.p.* Gamble 1280(896); Khan in FFH 292. *Cluytia retusa* L. Sp. Pl. 1042. 1753.

A deciduous tree, up to 8 m tall, bark grey with cracks, branchlets tomentose. Leaves oblong-obovate, 5–15 × 3–8 cm, base rounded, margin entire, obscurely serrate, tip obtuse, retuse, glabrous above, pubescent below. Flowers purple in axillary or terminal spikes. Drupes globose, glabrous, black when ripe. 2-seeded.

Common in all forests. Fl. & Fr.: June–Dec. Anantagiri RF, *MSM & PRSR* 12782.

CHROZOPHORA A. Jussieu

Chrozophora rottleri (Geis.) Adr. Juss. Euphorb. Gen. 28. 1824; Gamble 1317(921). *Croton rottleri* Geis. Crot. Monogr. 54. 1807.

An annual erect monoecious herb, up to 50 cm tall, branchlets tomentose. Leaves ovate, 2–6 × 1–3 cm, base cuneate, margin sinuate, undulate, densely stellate tomentose below. Flowers yellow in terminal racemes. Capsule 0.5 cm long, 3-lobed, densely stellate tomentose. Seeds globose, smooth.

Rare in open forests in dried ponds. Fl. & Fr.: Jan.–May. Mohammadabad beat, *MSM & KH* 10234, *MSM & NV* 10426.

CROTON Linnaeus

Croton bonplandianum Baill. in Adansonia 4: 339. 1864. *C. sparsiflorus* Morong in Ann. New York Acad. Sci. 7: 221. 1893; Gamble 1316(920).

An annual erect herb, up to 60 cm tall. Leaves ovate-lanceolate, 2–7 × 1.5–3 cm, base acute, margin serrate, tip acuminate. Flowers greenish-yellow in long terminal racemes. Male above, female below. Capsules oblong, glandular, 3-valved.

Common in waste lands. Fl. & Fr.: All seasons. *Vern.*: Kukka mirapa. Central University, *MSM* 19243.

DRYPETES Vahl

Drypetes sepiaria (Wight & Arn.) Pax & Hoffm. in Engler Pflanzenr. 81: 27. 1922. *Hemicyclia sepiara* Wight & Arn. in Edinburgh New Philos J. 14: 297. 1833;. FBI 5: 337. Gamble 1299(909).

A small tree up to 5 m tall, branchlets puberulous. Leaves elliptic-oblong, base rounded, margin entire, tip obtuse-retuse. Flowers axillary, pubescent. Drupe globose. Seed arillate.

Rare in scrub jungle. Fl. & Fr.: Dec.–Apr. Lingampalle, *MSM* 12755.

EUPHORBIA Linnaeus

1. Plants armed:
    2. Leaves linear-oblong .... ***E. nivulia***
    2. Leaves modified into spines .... ***E. antiquorum***
1. Plants unarmed:
    3. Stems succulent, cylindirc .... ***E. tirucalli***
    3. Stems not succulent, other than cylindric:
        4. Prostrate herbs:
            5. Leaves serrate .... ***E. thymifolia***
            5. Leaves entire or crenulate:
                6. Branches hispid .... ***E. chamaesyce***
                6. Branches glabrous .... ***E. heyneana***
        4. Erect or ascending herbs:
            7. Leaves whorled .... ***E. heterophylla***
            7. Leaves opposite:
                8. Hispid herbs, cymes not capitate .... ***E. hirta***
                8. Glabrous herbs, cymes capitate .... ***E. indica***

Euphorbia antiquorum L. Sp. Pl. 450. 1753; FBI 5: 255. Gamble 1277(894).

A succulent shrub, up to 4 m tall, branchlets winged with persistent spines. Leaves obovate-oblong, base cuneate, margin enitre, tip rounded or subtruncate. Flowers in axillary, lax cymes. Capsules trigonous. Seeds globose.

Occasional in scrub jungles (Seshagiri Rao, 1986). Fl. & Fr.: Dec.–Feb.

Euphorbia chamaesyce L. Sp. Pl. 455. 1753; Rechingee & Schiman, Czeika in Rochinger, Fl. Iranica 6: 17. 1964. *E. prostrata* Ait. Hort. Kew. ed. 1. 2. 139. 1789; FBI 5: 226. 1887 in adnot. Gamble 1276(893).

A spreading herb, branchlets pubescent. Leaves elliptic-obovate, 0.4–1.5 × 0.3–0.6 cm base truncate, margin crenulate, tip obtuse. Flowers in axillary cyathia. Capsules beaked. Seed 4-angled, furrowed.

Common weed in waste lands (Seshagiri Rao, 1986). Fl. & Fr.: July–Jan.

Euphorbia heterophylla L. Sp. Pl. 453. 1753. *E. geniculata* Orteg., Nov. Rar. Pl. Matrit. 18. 1797; FBI 5: 239; Gamble 1884(1303).

An annual erect herb, up to 70 cm tall, stem striate, glabrous. Leaves elliptic, 4–6 × 0.5–3 cm, base attenuate, margin entire-serrate, tip acute-acuminate. Flowers in cyathia in terminal clusters. Capsules globose, nearly-glabrous. Seeds angled, tuberculate.

Weed of waste lands and cultivated fields. Fl. & Fr.: All seasons. Mekavarpalli, *MSM & TP* 11042.

Euphorbia heyneana Spreng. in L. Syst. Veg. (ed. 16) 3: 791. 1826; ssp. **heyneana**; Panigr. in Kew Bull. 29: 695. 1974. *E. microphylla* Heyne ex Roth, Nov. Pl. Spec. 229. 1821, non Lam. 1788; FBI 5: 252; Gamble 1276(893).

An annual prostrate herb, branchlets thin, glabrous. Leaves obovate, oblanceolate, 0.2–0.4 × 0.1–0.2 cm, base oblique, margin entire, tip obtuse, glabrous on both sides. Flowers pink in axillary cyathia. Capsules 0.1 cm long, glabrous, 3-lobed. Seed 4-angled.

Occasional in open forests, ponds and dried tanks. Fl. & Fr.: All seasons. Gadirayal beat, *MSM & KH*10265.

Euphorbia hirta L. Sp. Pl. 454. 1753; Gamble 1275(892). *Chamaesyce hirta* (L.) Millsp. Publ. Field Columbian Mus. Bot. Ser. 2: 303. 1909. *Euphorbia pilulifera* auct. non L.: FBI 5: 250.

An annual erect herb, up to 50 cm tall. Stem hispid. Leaves elliptic-lanceolate, 1.5–3.5 × 0.5–1.5 cm, base obliquely truncate, margin serrate, tip acute, glabrous above and grey-pubescent below. Flowers green in axillary cyathia. Capsules 0.2 cm long, pubescent, seeds 4-angled.

Common weed of roadsides, open forests and waste places. Fl. & Fr.: All seasons. Gadirayal beat *MSM & KH* 10264.

Euphorbia indica Lam. Encycl. 2: 423. 1786. *E. hypericifolia* auct. non L.; FBI 5: 249; Gamble 1275(892).

An annual erect herb, up to 60 cm tall, stem glabrescent. Leaves oblong-obovate, 1–4 × 0.5–1.5 cm, base rounded, margin serrate, tip obtuse, glabrous above and glaucous below. Flowers pink in axillary cyathia. Capsules 0.1 cm long, 3-lobed, glabrous. Seeds angled, smooth.

Common in open forests. Fl. & Fr. All seasons. Mohammadabad, *MSM & KH* 10220.

Euphorbia nivulia Buch.-Ham. in Trans. Linn. Soc. London 14: 286. 1824; FBI 5: 255; Gamble 1277(893).

An armed deciduous tree, 5 m tall, branchlets whorled, succulent, glabrous, bark fissured. Spines short, paired. Leaves lanceolate, 10–15 × 3–7 cm, base cuneate, tip obtuse. Flowers in sub terminal cyathia. Capsules 0.5 cm long, glabrous. Seed 4-angular, smooth.

Common in scrub jungles and open forests. Fl. & Fr.: Dec.–May. Kanmankalva beat, *MSM & KH* 10302.

Euphorbia thymifolia L. Sp. Pl. 454. 1753; FBI 5: 252. Gamble 1276(893). *Chamaesyce thymifolia* (L.) Millsp. in Pub. Field Columb. Bot. Mus. Ser. 2: 412. 1916.

A prostrate herb, branchlets pubescent. Leaves elliptic-oblong, 3–6 × 1.5–3 mm, base obliquely cordate, margin-serrate-crenate, tip obtuse-rounded. Flowers in solitary or 2–3 cyathia. Capsule obtusely keeled. Seed 4-angled.

Common in rocky areas (Seshagiri Rao, 1986). Fl. & Fr.: July–Mar.

Euphorbia tirucalli L. Sp. Pl. 452. 753; FBI 5: 254; Gamble 1276(893); Khan in FFH 290.

A shrub, up to 2 m, stem cylindric, succulent, glabrous. Leaves linear, 1.5 × 0.3 cm, base cuneate, margin entire, tip sub-acute. Cyathia in terminal clusters and at forks of branches. Capsules globose, 3-lobed, villous, seeds globose, smooth.

Occasional in hedges. Fl. & Fr.: Lingampalle, *MSM* 19225.

GIVOTIA Griffith

Givotia rottleriformis Griff. in Calcutta J. Nat. Hist. 4: 388. 1844; FBI 5:395; Gamble 1342(938); Khan in FFH 300.

A deciduous tree, up to 20 m tall, young branchlets fulvous-tomentose, bark whitish. Leaves ovate, 6–15 × 3.5–10 cm, base cordate, margin dentate, tip acuminate, grey tomentose above, villous below. Flowers dioecious, light yellow, in terminal panicled cymes. Drupes 2.5 cm long, glabrous, 1-seeded. Seed ellipsoid, testa bony.

Common in all forests. Fl. & Fr.: May–Jan. *Vern.*: Polliki, Tella Poliki. Mohammadabad beat, *MSM & KH* 10201.

JATROPHA Linnaeus

1. Shrubs more than 1 m tall:
   2. Peduncles and fruits pubescent .... ***J. villosa***
   2. Peduncles and fruits not pubescent .... ***J. curcas***
1. Shurbs less than 1 m tall:
   3. Flowers red .... ***J. gossypifolia***
   3. Flowers yellow .... ***J. glandulifera***

Jatropha curcas L. Sp. Pl. 1006. 1753; FBI 5: 383; Gamble 1340(937); Khan in FFH 299.

A shrub, up to 2 m tall, branchlets glabrous, bark peeling off in white flakes. Leaves cordate, 8–15 × 6–10 cm, base cordate, margin entire, tip acute, glabrous on both sides. Flowers green in terminal dichasial cymes. Capsules globose, 3 cm long, glabrous, seeds ellipsoid.

Common along the hedges. Fl. & Fr.: May–August. Lankalakunta, *MSM & KH* 10273.

Jatropha glandulifera Roxb. Fl. Ind. 3: 688. 1832; FBI 5: 382; Gamble 1340(937).

A shrub, up to 1 m tall, stem grey white. Leaves deeply 3–5-lobed, 4–8 × 6–9 cm, lobes obovate, 3–6 × 1–5 cm, margin serrate, tip acuminate, gland tipped. Flowers yellowish-green in terminal cymes.

Common in waste lands and roadsides. Fl. & Fr.: All seasons. Central University, *MSM* 19211.

Jatropha gossypifolia L. Sp. Pl. 1006. 1753; FBI 5: 383; Gamble 1340(937); Khan in FFH 299.

A shrub, up to 70 cm tall, branchlets glabrous. Leaves palmately 3-lobed, 5–10 × 8–12 cm, base cordate, lobes oblong, elliptic, tip acute, petiole glandular. Flowers red in axillary and terminal paniculate cymes. Capsules ellipsoid, 3-lobed, glabrous, red when ripe.

Common on roadsides. Fl. & Fr.: Aug.–Oct. Dharur, *MSM & TP* 11099.

Jatropha villosa Wight Ic. 4(1): 1. 1848. *J. wightiana* Muell.-Arg. in DC. Prodr. 15(2): 1080. 1866; FBI 5: 383. Gamble 1340(937).

A shrub, up to 2 m tall, monoecious, branchlets striate, pubescent. Leaves cordiform, 4–8 × 4–6 cm, base cordate, margin entire, tip acute, pubescent along nerves. Flowers green in terminal cymes, peduncles pubescent. Capsule globose, pubescent.

Occasional in scrub jungle. Fl. & Fr.: Aug.–Dec. Tandur, *MSM* 19244.

PEDILANTHUS Necker ex Poiteau *nom. cons.*

Pedilanthus tithymaloides (L.) Poit. in Ann. Mus. Natl. Hist. Nat. 19: 390. t. 19. 1812; FBI 5: 239; Gamble 1349(942). *Euphorbia tithymaloides* L. Sp. Pl. 453. 1753.

A perennial succulent subshrub, up to 1 m tall, branchlets zig-zag, glaucous. Leaves ellipitc-ovate, 4–6 × 3–5 cm, base attenuate, margin entire, tip acute. Cyathia in terminal cymes subtended by scarlet or orange lanceolate bracts. Involucre 4-angular.

Cultivated in gardens, rarely as an escape (Seshagiri Rao, 1986). Fl. & Fr.: July–Jan.

PHYLLANTHUS Linnaeus

1. Trees or shrubs:
    2. Trees. Fruit a drupe, yellowish green .... *P. emblica*
    2. Shrubs. Fruit a berry, red-black when ripe .... *P. reticulatus*
1. Herbs:
    3. Branchlets thinly puberulous .... *P. urinaria*
    3. Branchlets glabrous:
        4. Leaves elliptic-ovate .... *P. debilis*
        4. Leaves oblong-obovate:

5. Leaves obovate .... *P. maderaspatensis*
5. Leaves oblong:
   6. Erect herbs .... *P. amarus*
   6. Prostrate herbs .... *P. virgatus*

Phyllanthus amarus Schum. & Thonn. Kongl. Danske Vidensk. Selsk. Skr. 4: 195. 1829. *Phyllanthus niruri* auct. non L. 1753; FBI 5: 298; Gamble 1290(903).

An annual erect herb, up to 20 cm tall, stem glabrous. Leaves oblong, 0.4–1 × 0.3–0.5 cm, base rounded, margin entire, tip acute, glabrous above and glaucous below. Male flowers few, terminal; female flowers numerous. Seeds 3-gonous, vertically muricate.

Common in wet places. Fl. & Fr.: May–Jan. *Vern.:* Nela usiri. Gadirayal beat, *MSM & KH* 10267.

Note: This taxon was previously reported as *P. niruri* L. Webster (1975) has shown that *P. niruri* is an American taxon endemic to West Indies and not represented in India. *P. amarus* can easily be distinguished by having equilateral leaf bases, bisexual cymules, ribbed seeds from *P. niruri.*

Phyllanthus debilis Klein ex Willd. Sp. Pl. 4: 582. 1805; FBI 5: 299; Gamble 1290(903).

Erect herbs, up to 60 cm tall. Leaves elliptic-ovate, 1–2 × 0.5–1 cm, base truncate, margin entire, tip obtuse. Flowers white, axillary, solitary. Capsules globose, 3-valved, seeds trigonous.

Occasional in forest. Fl. & Fr.: June–Dec. Anantagiri RF, *MSM & TP* 11058.

Phyllanthus emblica L. Sp. Pl. 982. 1753; FBI 5: 289; Khan in FFH 295. *Emblica officinalis* Gaertn., Fruct. 2: 122. t. 108. 1790; Gamble 1295(906).

A deciduous tree, up to 10 m tall, bark grey, branchlets glabrous. Leaves oblong, 1–2 × 0.2–0.4 cm, base truncate, margin entire, tip apiculate, glabrous above, glaucous below. Flowers green, axillary, solitary, or in clusters. Drupes globose, yellowish green, glabrous. Seeds trigonous, green.

Occasional in deciduous forest. Fl. & Fr.: Nov.–Mar. *Vern.:* Usiri. Anantagiri RF, *MSM & PRSR* 12783.

Phyllanthus maderaspatensis L. Sp. Pl. 982. 1753; FBI 5: 292; Gamble 1289(902).

An annual erect herb, up to 40 cm tall, branchlets glabrous. Leaves obovate, spathulate, 1–2 × 0.3–0.7 cm, base cuneate, margin entire, tip rounded or mucronate. Flowers green, axillary, solitary or in clusters. Capsules glabrous. Seeds wedge shaped.

Common weed of cultivated fields. Fl. & Fr.: All seasons. Mekavarpalli, *MSM & TP* 11040.

Phyllanthus reticulatus Poir. in Lam. Encycl. 5: 298. 1804; FBI 5: 288. *Kirganelia reticulata* (Poir.) Baill. Etude Euphorb. 613. 1874; Gamble 1294(905). *Anisonema multiflora* Wight, Icon Pl. Ind. Orient. t. 1899. 1852; Khan in FFH 294.

A woody shrub, up to 3 m tall, branchlets tomentose, leaves oblong-elliptic, 0.8–2.5 × 0.5–1.5 cm, base obtuse, margin entire, tip rounded, glabrous on both sides. Flowers purple, dioecious, male in axillary fascicles, female solitary. Berries red, black when ripe, 0.5 cm long, glabrous. Seeds triquetrous.

Common in forests, along streams, hedges. Fl. & Fr.: All seasons. *Vern.:* Pulcheri. Gadirayal beat, *MSM & KH* 10249.

Phyllanthus urinaria L. Sp. Pl. 982. 1753; FBI 5: 293; Gamble 1289(902).

An annual erect herb up to 30 cm tall, branchlets thinly puberulous. Leaves linear-oblong, 0.5–1.5 × 0.2–0.4 cm, base rounded, margin serrulate, tip apiculate, stipules lanceolate. Flowers cream, solitary. Capsule globose, seeds 3-gonous, furrowed.

Occasional in forest undergrowth. Fl. & Fr.: Oct.–Dec. Vikarabad, *MSM* 19245.

Phyllanthus virgatus Forst. f. Fl. Ins. Austr. 65. 1786. *P. simplex* Retz. Obs. Bot. 5: 29. 1788; FBI 5: 295; Gamble 1289(902).

An annual prostrate herb, up to 50 cm tall, branchlets glabrous. Leaves oblong-lanceolate, 1–2.5 × 0.3–0.8 cm, base truncate, margin entire, tip mucronate, glabrous above, glaucous below. Flowers greenish, axillary, solitary or in fascicles. Capsules globose, 0.3 cm long, 3-lobed, glabrous. Seeds 3-gonous, covered with minute tubercles.

Common along stream and open forests. Fl. & Fr.; June–Dec. Kusumasamudram beat, *MSM & NV* 10359; Tandur, *MSM & TP* 12703.

SEBASTIANIA Sprengel

Sebastiania chamaelea (L.) Muell-Arg. in DC. Prodr. 15: 1175. 1866; FBI 5: 475; Gamble 1344(940). *Tragia chamaelea* L. Sp. Pl. 981. 1753.

An annual erect herb up to 70 cm tall, branchlets angled, glabrous. Leaves oblong, 3–6 × 0.2–0.5 cm, base rounded, margin revolute-serrate, tip acute. Flowers in axillary spikes, male in clusters of 3, female solitary. Capsules oblong, 3-lobed, each with 2 rows of prickles.

Occasional in waste lands. Fl. & Fr.: All seasons. Lingampalle, *MSM* 12766.

TRAGIA Linnaeus

Tragia plukenetii Radclif Smitth in Kew Bull. 37: 688. 1983. *T. cannabina* L.f. Suppl. Pl. 415. 1781. *nom. illegit.:* Gamble 1333(932). *T. involucrata*

L. var. *cannabina* (L.f) Muell-Arg. in DC. Prodr. 15(2) 944. 1862; FBI 5: 465.

An annual erect herb up to 50 cm tall. Stem with stinging hairs. Leaves palmatifid, 3-lobed, 4–9 × 2–5 cm, lobes lanceolate, sharply serrate, tip-acute, pubescent. Flowers cream or green in long spikes. Tepals 3 in male, 6 in female. Capsules 3-lobed, hairy.

Occasional on roadsides. Fl. & Fr.: July-Dec. *Vern.: Duradagondaku.* Naskal, *MSM & PRSR* 15121.

Note: Linnaeus f. when uniting *Croton hastatus* and *C. urens* (Sp. Pl. ed. 1: 1005. 1753) and transferring them to *Tragia*, failed to choose between one of these epithets. Hence his name *Tragia cannabina* becomes superflous and illegitimate. As the epithets *hastatus* and *urens* have since been preoccupied A. Radcliffe-Smith has named it after Leonard Plukenet (1642–1706).

Acalypha hispida Burm. f., *Codiaeum* spp., *Jatropha rosea* L. etc. are planted for ornamental purpose. *Ricinus communis* L. (*Vern.*: Amudam) is cultivated for castor oil.

URTICACEAE

1. Prostrate herb, leaves small, sub-succulent .... ***Pilea***
1. Erect herb, leaves large, non-succulent .... ***Pouzolzia***

PILEA Lindley *nom. cons.*

Pilea microphylla (L.) Liebm. Kongel. Danske. Vidensk. Selsk. Natur-Vidensk. Math. Afh. (Ser. 2) 5(2): 296. 1851; FBI 5: 551; C. Fischer 1379(965). *Parietaria microphylla* L. Syst. Nat. (ed. 10) 1308. 1759. *Urtica microphylla* Sw. Fl. Ind. Occid. 1: 305. 1797.

A monoecious delicate herb, up to 15 cm tall. Leaves elliptic-ovate, 4 × 2 mm, sub succulent, 1-nerved, base acute, margin entire, tip obtuse-subacute. Flowers in umbellate clusters. Achenes ellipsoid.

Rare in garden. Fl. & Fr.: All seasons. Central University, *MSM & TP* 15133.

POUZOLZIA Gaudichaud-Beaupre

Pouzolzia zeylanica (L.) Bennett in Bennett *et al.* Pl. Jav. Rar. 67. 1838. *Parietaria zeylanica* L. Sp. Pl. 1052. 1753. *P. indica* L. Mant. Pl. 128. 1767. *Pouzolzia indica* (L.) Gaudich. Voy. Bonite. Bot. 503. 1826; FBI 5: 581; C. Fischer 1382(967).

An annual erect herb, up to 20 cm tall, branchlets glabrescent. Leaves opposite, ovate, 2–4.5 × 0.8–2 cm, base cuneate, margin entire, tip acuminate, appressed pubescent on both sides, petiole to 1 cm. Flowers sessile in axillary clusters, bracts 2, concave, bracteoles lanceolate, strigose without. Achenes obovoid, winged.

Rare in forest under growth. Fl. & Fr.: June–Sept. Kusumasamudram beat, *MSM* 10334.

ULMACEAE

HOLOPTELEA J.E. Planchon

Holoptelea integrifolia (Roxb.) Planch. in Ann. Sci. Nat. Bot. (Ser. 3) 10: 266. 1848; FBI 5: 481; C. Fischer 1348(943); Khan in FFH 305. *Ulmus integrifolia* Roxb. Pl. Cor. t. 78. 1798.

A deciduous tree, up to 8 m tall, branchlets pubescent. Leaves elliptic-ovate, 8–12 × 3–6 cm, base rounded, margin entire, tip acuminate, glabrous on both sides. Flowers yellow in axillary racemes. Samaras 2.5 cm long, glabrous, winged, yellow.

Common in forests. Fl. & Fr.: Sept.–Jan. *Vern.:* Nemalinara. Kusumasamudram beat, *MSM & KH* 11036.

MORACEAE

1. Fruits achenes .... ***Ficus***
1. Fruits berries .... ***Streblus***

FICUS Linnaeus

1. Figs dense on leafless branchlets:
   2. Leaves alternate, glabrous, margin entire .... ***F. racemosa***
   2. Leaves opposite, pubescent, margin serrate or toothed .... ***F. hispida***
1. Figs solitary, paired, axillary on leafy branchlets:
   3. Leaf tip caudate-acuminate, blade shining .... ***F. religiosa***
   3. Leaf tip acute, blade not shiny:
      4. Branchlets fulvous tomentose. Midrib usually with an elongated gland at base .... ***F. mollis***
      4. Branchlets not fulvous tomentose. Midrib eglandular .... ***F. benghalensis***

Ficus benghalensis L. Sp. Pl. 1059. 1753; FBI 5: 499; C. Fischer 1361(952). *Urostigma benghalense* (L.) Gasp. Nov. Gen. Fic. 7. 1844; Khan in FFH 308.

A deciduous tree, up to 10 m tall, stem deeply fluted, buttressed, glabrous. Leaves elliptic-ovate, 10–17 × 7–10 cm, base rounded, margin entire, tip obtuse, glabrous on both sides. Figs red, monoecious, in axillary pairs. Achenes globose 1.5 cm long, tomentose.

Common in plains and forests. Fl. & Fr.: Dec.–May. *Vern.:* Marri. Mohammadabad beat, *MSM & KH* 10208.

Ficus hispida L.f. Suppl. Pl. 442. 1781; FBI 5: 522; Gamble 1367(956); Khan in FFH 311.

A deciduous tree up to 5 m tall, branchlets hispid. Leaves oblong-obovate, 10–20 × 7–12.5 cm, hispid-scabrid above and hispid pubescent below, shortly dentate. Figs fascicled on the stem or leafy branchlets, sessile, globose, hispid.

Common near canals and field hedges. Fl. & Fr.: Apr.–Aug. Rudraram RF, *MSM & TP* 12725.

Ficus mollis Vahl, Symb. Bot. 1: 82. 1790. *F. tomentosa* Roxb. ex Willd. Sp. Pl. 4: 1136. 1806; FBI 5: 501; C. Fischer 1361(952).

An evergreen tree, up to 10 m tall, branchlets fulvous-tomentose. Leaves elliptic-ovate, 6–17 × 3–9 cm, base cordate, margin entire, tip acute, scabrous above, rusty tomentose below. Figs brown, monoecious, in axillary pairs. Achenes 1 cm long, tomentose.

Occasional in forests. Fl. & Fr.: Jan.–July. *Vern.*: Banda jeevi. Anantasagar beat, *MSM & NV* 10409.

Ficus racemosa L.Sp. Pl. 1060. 1753. *F. glomerata* Roxb. Pl. Cor. 2: 13. 123. 1798; FBI 5: 535; Gamble 1364(954); Khan in FFH 312.

A deciduous tree up to 10 m tall, branchlets pubescent. Leaves elliptic-lanceolate, 5–10 × 2–4 cm, base rounded, margin entire, tip slightly notched. Figs on leafless branchlets, pyriform-subglobose, pubescent, red when ripe.

Rare along roadsides (Seshagiri Rao, 1986). Fl. Fr.: Dec.–March. *Vern.*: Medi.

Ficus religiosa L. Sp. Pl. 1059. 1753; FBI 5: 513; Gamble 1363(953); Khan in FFH 309.

An evergreen tree up to 15 m tall, branchlets glabrous. Leaves ovate, 5–10 × 6–8 cm, base cordate, margin entire, tip acuminate. Figs in axillary pairs, sessile, glabrous, reddish-yellow when ripe.

Occasional in villages. Fl. & Fr.: June–Dec. *Vern.:* Ravi. Central University, *MSM & TP* 15139.

STREBLUS Loureiro

Streblus asper Lour. Pl. Cochinch. 615. 1790; FBI 5: 489; C. Fischer 1353(947). *Epicarpurus orientalis* Bl. Bijdr. 488. 1825; Khan in FFH 305.

An evergreen treep, up to 10 m tall, stem with milky latex, glabrous. Leaves elliptic-obovate, 2–6 × 1–3 cm, base truncate, margin irregularly toothed, tip acuminate, glabrous above, scabrous below. Flowers white, male flowers in globose heads. Female flowers solitary. Berries 0.5 cm long, glabrous, yellow when ripe, enclosed in persistent perianth, 1-seeded.

Common in forests. Fl. & Fr.: Feb.–June. *Vern.*: Barinka. Gadirayal beat, *MSM & KH* 10333.

CERATOPHYLLACEAE

CERATOPHYLLUM Linnaeus

Ceratophyllum demersum L. Sp. Pl. 992. 1753; FBI 5: 369; C. Fischer 1391(973).

A submerged rootless aquatic herb, branchlets glabrescent. Leaves whorled, 1.5–2 cm, forked, segments filiform, margin denticulate, tip acute. Flowers axillary, solitary, unisexual. Nutlets ellipsoid, black when dry, 0.5 cm long, with 3 spurs, 1 apical and 2 lateral ones, straight. Seed globose.

Common in ponds and tanks. Fl. & Fr.: Sept.–Mar. Malkalcheruvu, *MSM & KH* 10567.

HYDROCHARITACEAE

1. Stems reduced, leaves radical .... ***Ottelia***
1. Stems not reduced, leaves cauline .... ***Hydrilla***

HYDRILLA L.C. Richard

Hydrilla verticillata (L.f.) Royle. Ill. Bot. Himal. t. 376. 1839; FBI 5: 659; C. Fischer 1396(977). *Serpicula verticillata* L.f. Suppl. Pl. 416. 1781.

A submerged herb, up to 60 cm tall, stem much branched often rooting at nodes. Leaves linear-oblong, 0.5–1.5 × 0.2–0.5 cm, base cuneate, margin entire, tip apiculate, glabrous on both sides. Flowers violet in axillary sessile spathes, unisexual. Fruit subulate, 0.5 cm long, 5-seeded.

Common in stagnant water ditches. Fl. & Fr.: Sept.–Jan. Dharmapur, *MSM* 10485.

OTTELIA Persoon

Ottelia alismoides (L.) Pers. Syn. Pl. 1: 400. 1805; FBI 5: 662; C. Fischer 1398(978). *Stratiotes alismoides* L. Sp. Pl. 535. 1753.

An aquatic submerged or partially floating succulent, acaulescent herb. Submerged leaves, radical, oblong, 5–7 × 1–2 cm, floating leaves ovate, 5–15 × 6–15 cm, membranous, undulate. Flowers cream, solitary in spathes, uni or bisexual. Fruit ellipsoid, 4 cm long, crowned, many-seeded.

Common in ponds and tanks. Fl. & Fr.: Dec.-Mar. Malkalcheruvu, *MSM & KH* 10540.

ORCHIDACEAE

1. Plants epiphytes .... ***Vanda***
1. Plants terrestrial .... ***Habenaria***

HABENARIA Willdenow

Habenaria roxburghii (Pers.) R. Br. Prodr. 1: 312. 1810. *Orchis roxburghii* Pers., Syn. Pl. 2: 503. 1807. *Habenaria platyphylla* (Willd.) Spreng. in L. Syst. Veg. ed. 16. 690. 1826; FBI 6: 140; C. Fischer 1470 (1027).

An annual erect herb up to 40 cm tall. Leaves in rosette, flat on the ground, elliptic oblong or orbicular, 5–7 × 4–6 cm, base amplexicaul, margin entire, tip acute, obtuse, fleshy, dark green. Flowers white in long terminal crowded spikes. Capsules many-seeded.

Occasional in open forest as undergrowth. Fl. & Fr.: Aug.–Dec. Anantagiri RF, *MSM & TP* 11080.

VANDA W. Jones ex Robert Brown

Vanda tessellata (Roxb.) Hook. ex Don in Lond. Hort. Brit. 372. 1830; C. Fischer 1445(1010). *Epidendrum tessellatum* Roxb. Pl. Cor. t. 42. 1795. *Vanda roxburghii* R. Br. Bot. Reg. 6: t. 506. 1820; FBI 6: 52.

An epiphytic shrub, stem up to 50 cm, roots simple or branching (velamen). Leaves conduplicate, obtusely keeled, linear-oblong, 7–22 × 0.5–1.5 cm, thick, coriaceous. Flowers yellow in axillary or terminal racemes. Capsules 8 mm long, 8–10-ribbed, many-seeded.

Occasional in forests. Fl. & Fr.: Dec.–May. Dharmapur gutta, *MSM & KH* 10290.

AMARYLLIDACEAE

CRINUM Linnaeus

Crinum defixum Ker-Gawl. in J. Sci. Arts(London) 3: 105. 1817; FBI 6: 281; C. Fischer 1504(1051).

A perennial acaulescent herb, bulb ovoid, 6 × 4 cm. Leaves linear, 30–70 × 1.5–3 cm, margin entire, tip obtuse, glabrous on both sides, canaliculate. Flowers white in umbels on long scapes. Capsules 4 cm long, 1-many-seeded.

Occasional along water courses and moist places. Fl. & Fr.: July–Dec. Mohammadabad beat, *MSM & NV* 10425.

HYPOXIDACEAE

CURUCULIGO Gaertner

Curculigo orchioides Gaertn. Fruct. 1: 63. t. 16. f. 11. 1788; FBI 6: 279; C. Fischer 1502(1050).

A perennial acaulescent herb, root stock stout. Leaves lanceolate, 20–35 × 0.5–1.5 cm, base sheathing, margin entire, tip acuminate. Flowers yellow in subcapitate clusters or long racemes, lower one bisexual, upper ones often male. Fruit oblong, succulent, 1–4-seeded.

Occasional in open forests. Fl. & Fr.: Aug.–Dec. Rudraram RF, *MSM & TP* 12712.

AGAVAECAE

AGAVE Linnaeus

Agave americana L. Sp. Pl. 323. 1753; FBI 6: 277; C. Fischer 1505(1052); Khan in FFH 319.

A perennial shrub up to 1.5 m tall. Leaves succulent in rosettes, lanceolate, 50–100 × 10–15 cm, margin with recurved prickles, tip spiny, brown, glaucous on both sides. Flowers cream in clusters of cymes. Capsules trigonous, oblong.

A common hedge plant. Fl. & Fr.: Sept.–Jan. Pargi, *MSM & PRSR* 15102.

Sansevieria roxburghiana Schult. & Schult. f. is grown as a pot herb for ornamental purpose.

DIOSCOREACEAE

DIOSCOREA Linnaeus

1. Leaves 1-foliate:
   2. Leaves ovate cordate .... ***D. bulbifera***
   2. Leaves elliptic-oblong .... ***D. oppositifolia***
1. Leaves 3-foliolate .... ***D. pentaphylla***

Dioscorea bulbifera L. Sp. Pl. 1033; C. Fischer 1511(1055). *D. sativa* Thunb. Fl. Jap. 15. 1784, non L. 1753; FBI 6: 295.

A climbing herb, up to 5 m, root tuber globose. Stem unarmed, glabrous, bulbiferous. Leaves ovate cordate, 5–15 × 4–12 cm, base cordate, margin entire, tip acuminate, 9-nerved. Flowers brown, male flowers in axillary, 10–15 cm long spikes, female flowers in axillary 10–20 cm long spikes. Capsules 3 cm long, glabrous, seeds winged.

Common in dry deciduous forests. Fl. & Fr.: Sept.–Jan. Anantasagar beat, *MSM* 10511.

Dioscorea oppositifolia L. Sp. Pl. 1033. 1753; FBI 6: 292; C. Fischer 1512 (1056).

An annual climbing herb, root stocks with cylindrical long roots, stem unarmed, glabrous. Leaves elliptic-oblong, 4–9 × 2–5.5 cm, base truncate, margin entire, tip acuminate. Male flowers green in dense spikes, female flowers green in long lax spikes. Capsules 2 cm long, glabrous. Seeds winged.

Occasional in forests. Fl. & Fr.: Jun.–Oct. Anantasagar beat, *MSM & NV* 10399, *MSM* 10509.

Dioscorea pentaphylla L. Sp. Pl. 1032. 1753; FBI 6: 289; C. Fischer 1511(1056).

A climbing herb up to 5 m, root tubers oblong, stem glabrous, prickly towards base. Leaves 3-foliolate, leaflets elliptic-lanceolate, 2.5–7 × 1.5–4 cm, base cuneate, margin entire, tip acuminate. Flowers pale green, male and female flowers in dense pedunculate racemes. Capsules angular, 3 cm long, beaked, glabrous, seeds winged.

Common in dry deciduous forests. Fl. & Fr.: June–Sept. Anantasagar beat, *MSM* 10515.

LILIACEAE

1. Climbers:
   2. Leaves spine-tipped, flowers white .... ***Asparagus***
   2. Leaves tendrillar, flowers orange-scarlet .... ***Gloriosa***
1. Acaulescent herb:
   3. Leaves succulent. .... ***Aloe***
   3. Leaves grassy-linear acicular:
      4. Leaves grass like. Flowers white .... ***Chlorophytum***
      4. Leaves not grass like. Flowers other than white:
         5. Root stock a bulb. Flowers on scapes:
            6. Flowers purple appearing with leaves .... ***Scilla***
            6. Flowers brown appearing before the leaves .... ***Drimia***
         5. Root stock a corm. Flowers solitary .... ***Iphiginea***

ALOE Linnaeus

Aloe vera (L.) Burm. f. Fl. Ind. 83. 1768; C. Fischer 1520 (1062). *A. perfoliata* L. var. *vera* L. Sp. Pl. 320. 1753.

A perennial acaulescent, succulent herb, up to 50 cm tall. Leaves radical in rosettes, ensiform, 25–50 × 2–5 cm, succulent, margin spiny, tip spiny. Flowers orange in long terminal racemes. Capsules ellipsoid, oblong.

Common in hedges and waste lands. Fl. & Fr.: Dec.–Mar. Central University, *MSM* 19246.

ASPARAGUS Linnaeus

Asparagus racemosus Willd. Sp. Pl. 2: 152. 1799; C. Fischer 1517 (1060); Khan in FFH 322.

A perennial armed climbing shrub, stem angular. Leaves scaly, to 0.8 cm, cladodes 2–6, linear-falcate, 0.6–2 × 0.1cm, in the axils of leaves, spine-tipped. Flowers white in axillary racemes. Berries globose, red when ripe, 3–6-seeded.

Occasional in open forest. Fl. & Fr.: June–Oct. Lingampalle, *MSM* 12738.

CHLOROPHYTUM Ker Gawler

Chlorophytum laxum R. Br. Prodr. 277. 1810; FBI 6: 336. C. Fischer 3: 1526(1066).

An annual erect herb, up to 30 cm tall. Leaves 6–12, grass like, spreading, linear-falcate, flat, acute. Scape slender, 2–6 cm long. Flowers greenish-white in slender simple raceme. Capsules obcordate, 3-winged, 1–4-seeded.

Rare in open forest. Fl. & Fr.: July–Nov. Rudraram RF, *MSM & TP* 12717.

DRIMIA Jacquin ex Willdenow

Drimia indica (Roxb.) Jessop in J.S. Afr. Bot. 43: 272. 1977. *Scilla indica* Roxb. Fl. Ind. 2: 147. 1832. *Urginea coromandeliana* (Roxb.) Hook. f. Fl. Brit. India. 6: 347. 1892; C. Fischer 1527(1066). *Scilla coromandeliana* Roxb. Fl. Ind. 2: 148. 1832. *Urginea wightiana* Hook. f. Fl. Brit. India. 6: 347: 1892.

A herb with an ovoid, thick bulb. Leaves appearing after flowers, 15–35 × 1–2 cm, linear, flat, acute. Scape erect, brittle, 30–40 × 0.5–0.6 cm. Flowers brown in lax-flowered racemes. Capsules ellipsoid, tapering at both ends, 6–9-seeded.

Common in moist and shady places (Seshagiri Rao, 1986). Fl. & Fr.: Sept.–Nov.

GLORIOSA Linnaeus

Gloriosa superba L. Sp. Pl. 305. 1753; FBI 6: 358; C. Fischer 1519(1061); Khan in FFH 322.

A perennial climbing herb, root stock with cylindrical tubers, branchlets terete, glabrous. Leaves ovate-lanceolate, 5–10 × 1–3 cm, tendrillar, base cordate, margin entire, tip acuminate. Flowers orange-scarlet in axillary solitary or sub-corymbs, perianth margin cuspid. Capsules 5 cm long, torulose, many-seeded.

Common in forests and hedges. Fl. & Fr.: June–Oct. *Vern.*: Nabhi. Kusumasamudram beat, *MSM & NV* 10373; Anantasagar beat, *MSM & NV* 10394, *MSM* 10513.

IPHIGINIA Kunth *nom. cons.*

1. Leaves linear, flowers brown-purple .... *I. indica*
1. Leaves acicular, flowers purple .... *I. mysorensis*

Iphiginea indica (L.) A. Gray ex Kunth Enum. Pl. 4: 213. 1843; FBI 6: 357; C. Fischer 1528(1067). *Melanthium indicum* L. Mant. Pl. 226. 1771.

An annual erect herb, up to 20 cm tall, corm globose with pale brown sheaths, stem flexuous. Leaves linear-lanceolate, 10–18 × 0.3–0.5 cm, base sheathing, margin entire, tip acuminate. Flowers brown-purple, axillary or terminal, solitary. Capsules 2 cm long, 3-grooved. Seeds sub-globose.0

Occasional in forests. Fl. & Fr.: Jan.–June. Kusumasamudram beat, *MSM & NV* 10375.

Iphiginea mysorensis Arekal & Swamy in Bot. Notis 125. 226. 1972.

A perennial slender herb, corm bulb like with brown sheaths, stem 15–20 cm long. Leaves 2 or 4, acicular, 10–12 × 0.5–0.15 cm. Flowers purple in terminal 2–4-flowered racemes. Capsules small, obovoid, loculicidal, many-seeded. Seeds trigonous, verrucose.

Rare in Central University campus (Seshagiri Rao, 1986). Fl. & Fr.: Sept.–Oct.

SCILLA Linnaeus

Scilla hyacinthina (Roxb.) Macbr. Contr. Gray. Herb. 56: 14. 1918. *Ledebouria hyacinthina* Roth, Nov. Pl. Spec. 195. 1821. *Barnardia indica* Wight, Icon. t. 2041. 1853. *Scilla indica* (Wight) Baker in Saunders Refug. Bot. 3(Appen.): 12. 1870. non Roxb. 1832; FBI 6: 348; C. Fischer 1527(1067).

A perennial aculescent herb, bulb globose, 4 x 2.5 cm, with narrow short neck. Leaves radical, oblong-oblanceolate, 5–15 × 0.5–5 cm, base tapering into sheath, margin entire, tip acute, glabrous on both sides. Flowers greenish-purple in terminal racemes on naked scapes. Capsules 0.6 cm long, glabrous, many-seeded.

Common in moist places. Fl. & Fr.: June-Oct. Kusumasamudram beat, *MSM & NV* 10374; Mohammadabad beat, *MSM & NV* 10226.

Allium cepa L. (*Vern.:* Ulligadda), *A. sativum* L. (*Vern.:* Vellipaya) are cultivated for their edible bulbs.

PONTEDERIACEAE

MONOCHORIA K.B. Presler

Monochoria vaginalis (Burm. f.) C. Presl. Rel. Haenk. . 128. 1827 in Obs; FBI 6: 363; C. Fischer 1529(1069). *Pontederia vaginalis* Burm. f. Fl. Indica 80. 1768.

An aquatic erect herb, up to 60 cm tall, rooting in mud. Leaves floating on surface, root-stock short. Leaves radical, ovate, 3.5–11 × 2.5–9 cm, base cordate, margin entire, tip acuminate, glabrous on both sides. Flowers blue, terminal, solitary. Capsules 1 cm long, glabrous, many-seeded.

Common in ponds. Fl. & Fr.: Dec.–March. Mohammadabad beat, *MSM & KH* 10434; Malkalcheruvu, *MSM & KH* [illegible]541.

COMMELINACEAE

1. Flowers enclosed in leaf sheaths .... ***Amischophacelus***
1. Flowers not enclosed in leaf sheaths:
   2. Fertile stamens 3, staminodes present:
      3. Flowers in cymes .... ***Murdannia***
      3. Flowers in spathes .... ***Commelina***
   2. Fertile stamens 6, staminodes absent .... ***Cyanotis***

AMISCHOPHACELUS Rao & Kammathy

Amischophacelus axillaris (L.) Rao & Kamm. in J. Linn. Soc. Bot. London 59: 306. 1966. *Commelina axillaris* L. Sp. Pl. 42. 1753. *Cyanotis axillaris* D. Don, Prodr. Fl. Nepal. 46. 1825; FBI 6: 388; C. Fischer 1550(1082).

An annual erect herb, up to 60 cm tall, stem glabrous. Leaves linear-lanceolate, 3–10 × 0.5–1 cm, base sheathing, margin entire, tip acute, glabrous on both sides. Flowers blue in axillary cymes. Capsules 0.5 cm long, glabrous, 6-seeded.

Common in moist places. Fl. & Fr.: All seasons. Mohammadabd beat, *MSM*, 10495.

COMMELINA Linnaeus

1. Prostrate herbs. Capsules 0.5 cm long .... ***C. haskarlii***
1. Erect herb. Capsule more than 1 cm long:
   2. Leaves ovate .... ***C. benghalensis***
   2. Leaves other than ovate:
      3. Spathes sessile:
         4. Capsule 3-valved .... ***C. paludosa***
         4. Capsule 2-valved .... ***C. erecta***
      3. Spathes peduncled:
         5. Succulent herb .... ***C. diffusa***
         5. Non-succulent herb .... ***C. attenuata***

Commelina attenuata J. Koenig ex Vahl Enum. Pl. 2. 168. 1806; FBI 6: 372. 1892; C. Fischer 1539(1074).

A diffuse herb, stem and branchlets angled. Stem 30–65 cm long. Leaves linear-lanceolate, 4–7 × 0.5–1 cm, base obtuse, margin entire, tip acuminate, pubescent above, glabrous below. Flowers blue in axillary slender peduncles. Spathes narrowly ovate. Capsules oblong. Seeds whitish, appendaged at the ends.

Common near streams. Fl. & Fr.: July–Nov. Lingampalle, *MSM* 12751.

Commelina benghalensis L. Sp. Pl. 1753; FBI 6: 370; C. Fischer 1539(1075).

An annual erect herb, up to 70 cm tall, stem glabrous. Leaves ovate-elliptic, 1.5–5 × 0.5–2.5 cm, base unequal, cuneate, margin undulate, tip obtuse. Flowers blue in spathes, spathes in clusters, funnel-shaped, pubescent, auricled. Capsules pyriform.

Common weed of wet lands. Fl. & Fr.: All seasons. Central University, *MSM* 19219.

Commelina diffusa Burm. f. Fl. Indica. 18. t. 7. f. 2. 1768. *C. nudiflora* auct. non. L., FBI 6: 369; C. Fischer 1538(1074).

A diffuse succulent herb. Leaves oblong-lanceolate, 3.5–6 × 0.5–1 cm. Flowers blue, spathes ovate-lanceolate, peduncled. Capsules 3-celled, oblong.

Occasional along streams (Seshagiri Rao, 1986). Fl. & Fr.: Dec.–May.

Commelina erecta L. Sp. Pl. 41. 1753; *C. undulata* R. Br. Prodr. 270. 1810; FBI 6: 373. *C. kurzii* Clarke in J. Linn. Soc. Bot. 11: 444. 1870; FBI 6: 373; Fischer 1540(1075).

A straggling herb. Leaves linear-lanceolate to oblong-lanceolate, 4–8 × 2–2.5 cm, pubescent above, glaucous below. Flowers violet, spathes in terminal clusters, funnel-shaped, subsessile. Capsule 3-celled.

Common along canals, streams (Seshagiri Rao, 1986). Fl. & Fr.: Jun.–Oct.

Commelina haskarlii Clarke, Comm. Crypt. Beng. 13. t. 3. 1874; FBI 6: 370; C. Fischer 1538(1074).

An annual prostrate herb, up to 40 cm, stem glabrous, striate. Leaves lanceolate, 1.5–4 × 0.5–1 cm, base sheathing, ciliate, margin entire, tip acute, glabrous on both sides. Flowers blue in solitary spathes. Capsules 0.5 cm long, membranous, 2–4-seeded.

Common in all moist places. Fl. & Fr.: All seasons. Gadirayal beat, *MSM* 10519.

Commelina paludosa Bl., Enum. 1: 2. 1827. *C. obliqua* Ham. ex D. Don, Prodr. Fl. Nepal. 45. 1825. non Vahl 1805; FBI 6: 372; C. Fischer 1539 (1075).

An annual erect herb, up to 70 cm tall, stem stout, glabrous. Leaves elliptic-lanceolate, 4–15 × 1.3–5 cm, base attenuate, margin entire, tip acuminate, sparsely scabrid. Flowers blue in solitary or clustered spathes. Capsules 1–2 cm long, 3-gonous, 3-valved, 3-seeded.

Occasional in forests and moist places. Fl. & Fr.; June–Oct. Kusumasamudram beat, *MSM & NV* 10363.; Anantagiri, *MSM* 11061.

CYANOTIS D. Don

1. Leaves cauline, silk cob-webby .... *C. fasciculata*
1. Leaves radical and cauline, not silk cob-webby:
   2. A prostrate herb. Capsule small, 3 mm .... *C. arachnoidea*
   2. An erect herb. Capsule large, 3 cm .... *C. tuberosa*

Cyanotis arachnoidea Clarke in DC. Mongr. Phan. 3; 250. 1881; FBI 6: 386; C. Fischer 1550(1081).

A prostrate herb with woolly stem. Leaves variable, basal oblong-elliptic, 6–12 × 2–3 cm, upper ones 6–8 × 0.5–0.8 cm, base obtuse, margin entire, tip rounded. Flowers blue in axillary and terminal racemes. Capsules small, 3 mm. Seeds terete, pitted.

Occasional in wet places (Seshagiri Rao, 1986). Fl. & Fr.: Oct.–Dec.

Cyanotis fasciculata (Roth) Schult. & Schult. f. in L. Syst. Veg. 7: 1152. 1830; FBI 6: 387; C. Fischer 1550(1082). *Tradescantia fasciculata* Heyne ex Roth, Nov. Pl. Spec. 189. 1821.

An annual erect herb, up to 15 cm tall, branchlets purple pubescent. Leaves cauline, linear-lanceolate, 1–3 × 0.5–0.7 cm, base cordate, margin entire, tip acute, silk cob-webby on both sides. Flowers blue in axillary and terminal cob-webby cymes, bracts-recurved. Capsules 0.4 cm long, glabrous, 6-seeded.

Common on moist rocky places. Fl. & Fr.: Sept.–Dec. Dharmapur, *MSM* 10489; Lingampalle, *MSM* 12751.

Cyanotis tuberosa (Roxb.) Schult. & Schult. f. in L. Syst. Veg. 7: 1153. 1830; FBI 6; 386; C. Fischer 1549(1081). *Tradescantia tuberosa* Roxb. Pl. Cor. t. 108. 1799.

A perennial erect herb, up to 50 cm tall, roots with fusiform tubers. Stem hirsute at base. Leaves radical as well as cauline, linear-ensiform, 15–30 × 0.5–2.5 cm, margin entire, tip acute, glabrous above, pubescent and purple below. Flowers blue in terminal cymes. Capsules 0.3 cm long, 4–6-seeded.

Common in forest undergrowth. Fl. & Fr.: July–Dec. Kusumasamudram beat, *MSM & NV* 10341; Central University, *MSM* 15180.

MURDANNIA Royle *nom. cons.*

Murdannia nudiflora (L.) Brenan in Kew Bull. 7: 189. 1952. *Commelina nudiflora* L. Sp. Pl. 41. 1753 *p.p. Aneilema nudiflorum* (L.) R. Br. Prodr. 271. 1810; FBI 6: 378; C. Fischer 1545 (1078).

An annual diffuse herb, up to 35 cm, stem glabrous. Leaves linear-lanceolate, 3–13 × 0.4–0.8 cm, base sheathing, margin entire, tip acute, glabrous on both sides. Flowers blue in axillary long peduncled corymbs. Capsules 0.5 cm long, shining, beaked, 6-seeded.

Common in moist places, ponds. Fl. & Fr.: All seasons. Mohammadabad beat, *MSM* 10443.

ARECACEAE

1. Leaves simple, fan like .... ***Borassus***
1. Leaves compound, 1-pinnate .... ***Phoenix***

BORASSUS Linnaeus

Borassus flabellifer L. Sp. Pl. 1187. 1753; FBI 6: 482; C. Fischer 1562(1090); Khan in FFH 328.

A tree, up to 20 m tall. Leaves compound, fan-shaped, petiole stout, spiny. Flowers yellowish green in interfoliar spadices. Drupes globose, black when ripe.

Occasional in waste lands on roadsides. Fl. & Fr.: Mar.–May. *Vern.:* Thati. Central University, *MSM* 19222.

PHOENIX Linnaeus

Phoenix sylvestris (L.) Roxb. Fl. Ind. 3: 787. 1832; FBI 6: 425; C. Fischer 1559(1088); Khan in FFH 326. *Elate sylvestris* L. Sp. Pl. 1189. 1753.

A tree, up to 10 m tall, trunk covered with persistent leaf bases. Leaves 2–3 m long, leaflets many, 10–20 × 1–2 cm, glabrous with spinous tip. Flowers yellow in axillary spadix. Drupes oblong, brown when ripe, 1-seeded.

Common near streams. Fl.& Fr.: Jan.–May. Central University, *MSM* 19221.

Caryota urens L. (*Vern.:* Jeeluga) is cultivated for ornamental purpose. *Cocos nucifera* L. (*Vern.*: Kobbari) for its fruits.

TYPHACEAE

TYPHA Linnaeus

Typha angustata Bory & Chaub., Exp. Sci. Moree Bot. 1: 338. 1833; FBI 6: 489; C. Fischer 1571(1096).

A perennial aquatic herb, up to 2 m tall. Leaves linear, 2–2.5 m long, 3.5 cm across, base sheathing, margin entire, tip acute. Flowers brown in terminal spikes. Spikes monoecious, male below, female above. Nutlets fusiform.

Common in streams and ponds. Fl. & Fr. All seasons. *Vern.:* Thunga. Central University, *MSM* 19223.

ARACEAE

1. Aquatic herb .... ***Pistia***
1. Terrestrial herb .... ***Arisaema***

ARISAEMA Martius

Arisaema leschenaultii Bl. Rhumpia 1: 93. 1836; FBI 6: 1893. C. Fischer 1585(1105); Malhotra & Balodi in J. Econ. Tax. Bot. 7: 585. 1985. *A pulchrum* N.E. Br. in J. Linn. Soc. 18. 252. t. 6. 1880.

A cormous herb. Corm subglobose. Leaf solitary, radiatisect, leaflets 5–12, lanceolate. Spathe 15 cm long. Spadix dioecious. Berries with 1-few globose seeds.

Rare in forest. Fl. & Fr.: Aug.–Dec. Rudraram RF, *MSM* 12716.

PISTIA Linnaeus

Pistia stratiotes L. Sp. Pl. 963. 1753; FBI 6: 497; C. Fischer 1573(1097).

An aquatic stoloniferous floating herb. Leaves clustered, obovate, 2–6 × 1.5–6 cm, base cuneate, margin entire, tip rounded-truncate, white pubescent below. Flowers creamish-white in spadix, male and female separated by disc. Berries ovoid, many-seeded.

Occasional in tanks and ditches(Seshagiri Rao, 1986). Fl. & Fr. All seasons. *Vern.*: Antharatamara.

ALISMATACEAE

LIMNOPHYTON Miquel

Limnophyton obtusifolium (L.) Miq. Fl. Ned. Ind. 3: 242. 1855; FBI 6: 560; C. Fischer 1595 (1113). *Sagittaria obtusifolia* L. Sp. Pl. 993. 1753.

An aquatic scapigerous acaulescent herb. Leaves all radical, deltoid, sagittate, 5–10 × 7–20 cm, base with broad sinus, margin entire, tip obtuse, basal lobes tapering into acuminate tip, glabrous on both sides. Flowers white in whorls on stout panicles. Male in upper whorls, bisexual in lower whorls. Achenes 2 cm long, 12–16-seeded.

Common in ponds and tanks. Fl. & Fr.: Dec.–Mar. Mohammadabad beat, *MSM & KH* 10542.

APONOGETONACEAE

APONOGETON Linnaeus

Aponogeton natans (L.) Engl. & Krause in Engl. Pflanzenr. 4. 1324: 11. 1906; C. Fischer 1598 (1115). *Saururus natans* L. Mant. Pl. 227. 1771. *Aponogeton monostachyon* L.f. Suppl. Pl. 214. 1781; FBI 6: 564.

A perennial aquatic herb, up to 1 m, rhizome tuberous. Leaves oblong, 6–12 × 0.8–1.5 cm, base rounded, margin entire, tip acute-rounded, glabrous on both sides. Flowers violet in terminal spikes. Fruit oblong, 0.3 cm long, beaked, glabrous, 2–8-seeded.

Common in ponds and tanks. Fl. & Fr.: All seasons. Malkalcheruvu, *MSM & KH* 10539.

ERIOCAULACEAE

ERIOCAULON Linnaeus

Eriocaulon quinquangulare L. Sp. Pl. 87. 1753; FBI 6: 582; C. Fischer 1620 (1128).

An annual erect herb, up to 15 cm tall, stem glabrous, striate. Leaves lanceolate, purplish-red, 3–5 × 0.2–0.4 cm, base broad, rounded, margin entire, tip acute, glabrous on both sides. Flowers white, in terminal heads. Involucral bracts obovate, floral bract hyaline, obovate-cuneate.

Occasional in forests and marshy places. Fl. & Fr.: Sept.–Dec. Gadirayal beat, *MSM* 10522; Kusumasamudram, *MSM* 11032; 11033.

CYPERACEAE

1. Glumes show distichous arrangement:
   2. Herbs bulbous-thick, spikelets a few in number (1–3) .... ***Abildgaardia***
   2. Herbs not bulbous, spikelets many:
      3. Style 2-fid. Nuts flattened bilaterally:
         4. Rachilla jointed .... ***Kyllinga***
         4. Rachilla not jointed:
            5. Nuts compressed radial to rachilla .... ***Pycreus***
            5. Nuts compressed at 90° to rachilla .... ***Juncellus***

3. Style 3-fid. Nuts flattened dorsiventrally:
   6. Rachilla not jointed, persistent .... ***Cyperus***
   6. Rachilla jointed, deciduous .... ***Mariscus***
1. Glumes show spiral arrangement:
   7. Spikelets dark brown .... ***Lipocarpha***
   7. Spikelets not dark brown:
      8. Style base swollen, not continuous with nut:
         9. Nut beaked:
            10. Leaves absent .... ***Eleocharis***
            10. Leaves present .... ***Rhynchospora***
         9. Nut beakless .... ***Fimbristylis***
      8. Style base not swollen, continuous with nut:
         11. Leaves reduced to sheaths .... ***Schoenoplectus***
         11. Leaves well developed:
            12. Leaves with a collar around the stem .... ***Fuirena***
            12. Leaves without a collar around the stem .... ***Rikliella***

ABILDGAARDIA Vahl

Abildgaardia ovata (Burm. f.) Kral in Sida 4: 71. 1971. *Carex ovata* Burm. f. Fl. Ind. 194. 1768. *Cyperus monostachyos* L. Mant. 2: 180. 1771. *Abildgaardia monostachyos* (L.) Vahl, Enum. 2: 296. 1805. *Fimbristylis monostachyos* (L.) Hassk. Fl. Jav. Rav. 61. 1848. (*monostachya*); FBI 6: 649; C. Fischer 1660(1152). *Fimbristylis ovata* Kern in Blumea 15: 126. 1987 & in Steenis Fl. Males 1, 7: 595. 1974.

An annual herb, up to 40 cm tall. Stem filiform, trigonous. Leaves filiform. Spikelets solitary, compressed, ovate, shining. Involucral bracts 2. Nut obovoid, trigonous, stipitate.

Occasional in wet land. Fl. & Fr.: Dec.–Mar. Mohammadabad RF, *MSM & KH* 10505.

CYPERUS Linnaeus

1. Spikelets in heads:
   2. Nuts obovate:
      3. Spikelets digitate in clusters of 3–15 .... ***C. tenuispica***
      3. Spikelets many (20–30) .... ***C. difformis***
   2. Nuts oblong:
      4. Spikelets black .... ***C. castaneus***
      4. Spikelets rusty brown .... ***C. cuspidatus***
1. Spikelets in spikes:
   5. Rachilla of spikelets winged:
      6. Spikes cylindrical, spikelets upto 20 .... ***C. exaltatus***
      6. Spikes elliptic, spikelets up to 30:
         7. Rhizomatous or stoloniferous herbs:
            8. Stems thick, leaf sheaths bladeless:
               9. Stems obtuse-trigonous, bracts shorter than corymb .... ***C. corymbosus***
               9. Stems acute-trigonous, bracts longer than corymb .... ***C. pangorei***
            8. Stems slender, leaf sheaths with elongated blades:

10. Stems up to 30 cm tall, longest bract 15 cm long .... ***C. rotundus***
10. Stems up to 60 cm tall, longest bract 10 cm long .... ***C. bifax***
7. Non-rhizomatous, non-stoloniferous herbs:
11. Spikelets linear-lanceolate, glumes closer .... ***C. tenuiculmis***
11. Spikelets filiform, glumes spaced:
12. Spikelets spreading, glumes rounded .... ***C. distans***
12. Spikelets erect, glumes emarginate .... ***C. nutans***
5. Rachilla of spikelets not winged:
13. Annual herbs, non-stoloniferous:
14. Spikelets 1–2.5 cm long, glumes acute at tip .... ***C. compressus***
14. Spikelets 1 cm long, glumes obtuse at tip:
15. Rachilla of spike glabrous, keel not winged, smooth .... ***C. iria***
15. Rachilla of spike scabrid, keel winged. serrate .... ***C. alulatus***
13. Perennial herbs, stoloniferous .... ***C. pilosus***

Cyperus alulatus Kern in Reinwardt. 1: 463–466 f. 1. 1952. *C. iria var. rectangularis* Kuek. in Engl. Pflanzenr 101. 152. 1935. *C. rectangularis* (Kuek.) Bennet in Indian For. 95. 692. 1969.

An annual tufted herb, up to 60 cm tall. Stems triquetrous. Leaves linear, margin serrate near tip. Inflorescence terminal, compound umbel. Spikelets greenish-yellow. Involucral bracts 8. Nuts obovoid, triquetrous.

Occasional near ditches. Fl. & Fr.: Aug.–Nov. Central University, *MSM* 15174, 15175.

Cyperus bifax Clarke in Kew Bull. 8: 13. 1908. *C. retzii* Nees in Wight Contrib. Bot. Ind. 82. 1834. *C. tuberosus* sensu Clarke in Hook. f. FBI 6: 616.

A perennial rhizomatous stoloniferous herb, up to 70 cm tall. Stems trigonous. Leaves few, flat. Spikes bearing 15 spikelets. Spikelets 20-flowered, linear. Involucral bracts 4. Nuts obovoid, trigonous.

Occasional in wet lands near tanks. Fl. & Fr.: July–Dec. Anantasagar beat, *KH & MSM* 10928.

Cyperus castaneus Willd. Sp. Pl. 1: 278. 1797; T. Koyama in Gard. Bull. Singapore 30: 143. 1977.

An annual erect herb, up to 10 cm tall, stems tufted or solitary, glabrous. Leaves linear. Spikelets 5–30-flowered, stellate, clustered. Spikes 3–5, terminal. Involucral bracts 2–5, surpassing the inflorescence. Nuts obovoid, brown, puncticulate.

Occasional in marshy places in forests. Fl. & Fr.: Sept.–Dec. Gadirayal beat, *MSM* 10525.

Cyperus compressus L. Sp. Pl. 46. 1753; FBI 6: 605; C. Fischer 1640 (1140).

An annual erect herb, up to 60 cm tall. Stems tufted, glabrous. Leaves linear, spikelets 20–40-flowered, linear-oblong, much compressed, 4–8 in clusters on 4–7 cm long peduncles. Involucral bracts 3–5, leaf like. Nuts obovoid, black.

Common in moist places. Fl. & Fr.: June–Dec. Kusumasamudram beat, *MSM & NV* 10382; Mohammadabad beat, *MSM* 10478.

Cyperus corymbosus Rottb. Descr. Icon. Rar. 42. t. 7. f. 4. 1773; FBI 6: 612; C. Fischer 1641 (1140).

A perennial rush like herb, up to 90 cm long, rhizome creeping. Stems trigonous, glabrous. Leaves reduced to sheaths. Spikes in terminal corymbs. Spikelets 15–20-flowered. Involucral bracts 3. Nuts obovoid, trigonous.

Common in marshy areas. Fl. & Fr.: Aug.–Dec. Mohammadabad, *KH & MSM* 10922.

Cyperus cuspidatus Kunth, Nov. Gen. Pl. 1: 204. 1815; FBI 6: 598. *C. uncinatus* auct. non Poir.: C. Fischer 1639 (1139).

An annual erect herb, up to 10 cm long. Stems tufted, filiform, glabrous. Leaves filiform. Spikelets 12–24-flowered, compressed, reddish brown, 5–15 digitately arranged in heads. Heads on 1–3 cm long peduncles, 5–15 in umbels. Central head sessile. Involucral bracts 3–5, filiform. Nuts obovoid, brown.

Rare in moist places. Fl. & Fr.: June–Dec. Kusumasamudram beat, *MSM & NV* 10379; Gadirayal beat, *MSM* 10530.

Cyperus difformis L. Cent. Pl. 2: 6. 1756; FBI 6: 599; C. Fischer 1640 (1139).

An annual erect herb. Stems tufted, weak, up to 60 cm tall. Leaves linear, spikelets 10–30-flowered, 25–50 in dense globose spicate heads. Heads 3–10 in umbels. Involucral bracts 2–4, leaf like. Nuts ellipsoid, yellowish brown.

Common in forests and wet places. Fl. & Fr.: May–Sept. Mohammadabad beat, *MSM* 10479; *MSM & KH* 5584; Lingampalle, *MSM* 12733.

Cyperus distans L.f. Suppl. 103. 1781; FBI 6: 607; C. Fischer 1640 (1140).

A perennial erect herb, up to 1 m tall. Stems not tufted, trigonous. Leaves flat. Inflorescence brown, decompound. Spikes ovate, 15–20-spikelets, spikelets distant, linear, 13–15-flowered. Nuts ellipsoid, triquetrous.

Occasional in wet places. Fl. & Fr.: Aug.–Dec. Tandur, *MSM & TP* 12705.

Cyperus distans L.f. *var.* **pseudonutans** Kuekenth. in Engl. Pflanzen. 101. 140. 1935. *Cyperus nutans* auct. non. Vahl; FBI 6: 607.

A perennial rhizomatous herb, up to 90 cm tall. Spikes densely spikeletted. Spikelets crowded, erect or suberect. Nut ellipsoid, triquetrous.

Occasional in wet places. Fl. & Fr.: Sept.–Dec. Rudraram RF, *MSM & TP* 12730.

Cyperus exaltatus Retz. Obs. Bot. 5: 11. 1788; FBI 6; 617; C. Fischer 1642 (1141).

An annual erect herb, up to 1.5 m tall. Stems tufted, glabrous. Leaves linear, flat, spikelets 50-flowered, oblong, lanceolate, reddish brown. Spikes in terminal umbel, compound spikelets spicate, at right angles to rachis. Involucral bracts 3–5, variable in length. Nuts ellipsoid, trigonous, smooth, apiculate, tapering at both ends.

Rare in forests. Fl. & Fr.: Sept.–Dec. Kanmankalva beat, *MSM* 10470; *KH & MSM* 10914.

Cyperus iria L. Sp. Pl. 45. 1753; FBI 6: 606; C. Fischer 1640 (1140).

An annual erect herb, up to 60 cm tall. Stems tufted or solitary, glabrous. Leaves linear. Spikelets 4–15-flowered, linear-oblong, 15–20 in 3–5 cm long spikes. Spikes 2–10, sessile or stalked on 8–15 cm long peduncle, 5–10 in terminal umbellate clusters. Involucral bracts 3–5, 25 cm long. Nuts obovoid, brown.

Common weed of all places. Fl. & Fr.: June–Dec. Kusumasamudram beat, *MSM & NV* 10377; Kanmankalva beat, *MSM* 10466; *KH & MSM* 5572, 5579.

Cyperus nutans Vahl, Enum. Pl. 2: 363. 1806; FBI 6: 697; C. Fischer 1640 (1140). ssp. **elusinoides** (Kunth) T. Koyama in Gard Bull. Singapore 30: 136. 1977. *Cyperus elusinoides* Kunth, Enum. Pl. 2: 39. 1937; FBI 6: 608; C. Fischer 1640 (1140).

A perennial herb, up to 90 cm tall. Stems solitary, trigonous. Leaves as long as stems. Inflorescence decompound. Spikes oblong, densely spikeletted. Spikelets linear. Nuts obovoid, trigonous.

Occasional in wet places. Fl. & Fr.: Aug.–Dec. Mohammadabad RF, *KH & MSM* 10921.

Cyperus pangorei Rottb. Descr. Icon. Rar. 31. 7. f. 3. 1773; C. Fischer 1641 (1140). *C. tegetum* Roxb. Fl. Ind. 1: 211. 1820; FBI 6: 613. *C. corymbosus* Rottb. *var. pangorei* (Rottb.) Clarke in J. Linn. Soc. 21: 292. 1884 & in FBI 6: 613. 1893.

A perennial glaucous green robust herb, up to 1 m tall. Stems trigonous rigid, ribbed. Leaves flat. Umbel compound. Spikes corymbose, 3–10-spikeletted. Spikelets narrow oblong, purplish-brown. Involucral bracts 3. Nuts ellipsoid, trigonous, yellowish brown.

Occasional along streams. Fl. & Fr.: Sept.–Dec. Mohammadabad RF, *KH & MSM* 5577; Rudraram RF, *MSM* 12730.

Cyperus pilosus Vahl, Enum. Pl. 2: 354. 1805; FBI 6: 609; C. Fischer 1641 (1140).

A perennial rhizomatous herb, up to 90 cm tall. Stems triquetrous, rigid. Leaves flat canaliculate. Umbel compound. Spikes 15–20-spikeletted, spikelets oblong-narrow, grey-purple, 5–10-flowered. Involucral bracts 3–7. Nuts ellipsoid, triquetrous, dark grey.

A common weed in cultivated fields and wet places (Seshagiri Rao, 1986). Fl. & Fr.: June–Dec.

Cyperus rotundus L. Sp. Pl. 45. 1753; FBI 6: 614; C. Fischer 1641(1140).

A perennial stoloniferous herb, up to 30 cm, stems solitary, glabrous. Leaves linear. Spikelets 10–45-flowered, linear, reddish brown, 2–5 in short spikes on 1–6 cm long peduncles, 3–6 in terminal umbels. Involucral bracts 3, variable in length. Nuts obovoid, greyish black.

Common in moist places. Fl. & Fr.: June–Dec. Kusumasamudram beat, *MSM & NV* 10381; Mohammadabad, *KH & MSM* 3571.

Cyperus tenuiculmis Boeck. in Linneae 36. 286. 1870; T. Koyama in Gard. Bull. Straits Settlem. 30. 135. 1977. *C. zollingeri* auct. non Steud.: FBI 6: 613. 1893; C. Fischer 1641(1141).

An annual erect herb, stems solitary, up to 50 cm tall, stoloniferous. Leaves canaliculate. Spikelets 10–15-flowered, narrow, oblong. 2–10-spikeletted, spikes at right angles to rachis, involucral bracts 3, overtopping. Nuts obovoid, trigonous, apiculate, stipitate.

Occasional in dried ponds. Fl. & Fr.: July–Oct. Malkalcheruvu, *KH & MSM* 5582.

Cyperus tenuispica Steud. Syn. Pl. Glum. 2. 11. 1855; C. Fischer 1640 (1139). *C. flavidus* auct non Retz.; FBI 6: 600.

An annual erect herb, up to 4 cm tall, stems tufted, weak. Leaves linear, 14 × 0.5 cm, glabrous. Spikelets 10–20-flowered, linear, lanceolate, 0.5 × 0.1 cm, sessile, stellate, arranged in heads. Heads pedunculate in compound umbels. Involucral bracts 2–3. Nuts obovoid, light yellow.

Common in moist places in forests. Fl. & Fr.: May–Sept. Kanmankalva beat, *MSM* 10452; Gadirayal beat, *MSM* 10532.

ELEOCHARIS Robert Brown

1. Style 2-fid:
    2. Stems up to 10 cm long .... ***E. atropurpurea***
    2. Stem up to 40 cm long .... ***E. geniculata***
1. Style 3-fid .... ***E. congesta***

Eleocharis atropurpurea (Retz.) C. Presl, Rel. Haenk. 1: 196. 1828; FBI 6: 627; C. Fischer 1648 (1144). *Scirpus atropurpureus* Retz. Obs. Bot. 5; 14. 1788.

An annual erect herb, stems dwarf, up to 10 cm, smooth. Leaf sheaths purplish. Spikelets ovoid, obtuse. Nuts obovoid, biconvex, apically truncate, base stipitate.

Common in ponds. Fl. & Fr.: Dec.–Mar. Anantasagar beat, *MSM & KH* 10927.

Eleocharis congesta D. Don, Prodr. Fl. Nepal 41. 1825; FBI 6; 630; C. Fischer 1648(1145).

An annual erect herb, stems solitary or tufted, up to 40 cm tall. Leaves purplish, spikelets ovate. Nuts obovoid, trigonus, smooth.

Common in ponds and ditches. Fl. & Fr.: Dec.–Mar. Mohammadabad beat, *MSM & KH* 10920.

Eleocharis geniculata (L.) Roem. & Schult. in L. Syst. Veg. 2: 150. 1817. *Scirpus geniculatus* L. Sp. Pl. 48. 1753. *Eleocharis capitata* R. Br. Prodr. 225. 1810; FBI 6: 627; C. Fischer 1648(1145).

An annual erect herb, stems tufted, up to 40 cm tall, angular, smooth. Leaves linear, sheaths obliquely acute, purple. Spikelets globose. Nuts obovoid, trigonous, smooth.

Common in ponds. Fl. & Fr.: Dec.–Mar. Ànantasagar beat, *MSM & KH* 10929.

FIMBRISTYLIS Vahl

1. Nutlets oblong-cylindrical. Stigmas exclusively 2-fid .... *F. tetragona*
1. Nutlets obovate-orbicular. Stigmas 2–3-fid:
    2. Stigmas 3. Nutlets trigonous:
        3. Leaves with blades:
            4. Leaves with ligule .... *F. complanata*
            4. Leaves without ligule .... *F. tenera*
        3. Leaves blade less .... *F. miliacea*
    2. Stigmas 2 (3 occasionally in *F. cymosa*). Nutlets biconvex:
        5. Inflorescence a corymb:
            6. Styles glabrous .... *F. cymosa*
            6. Styles fimbriate:
                7. Leaves ligulate:
                    8. Glumes pubescent .... *F. sieberiana*
                    8. Glumes glabrous:
                        9. Spikelets oblong .... *F. bisumbellata*
                        9. Spikelets ovate:
                            10. Leaves half as long as stems, spikelets green, bracts 2 .... *F. alboviridis*
                            10. Leaves as long as stem, spikelets brown, bracts 6 .... *F. dichotoma*
                7. Leaves without ligules .... *F. aestivalis*
        5. Inflorescence a head:
            11. Inflorescence with many spikelets, style glabrous .... *F. argentea*
            11. Inflorescence with solitary spikelet, style fimbriate:
                12. Nuts yellow .... *F. schoenoides*
                12. Nuts purple .... *F. polytrichoides*

Fimbristylis aestivalis (Retz.) Vahl, Enum. Pl. 2: 288. 1806; FBI 6: 637. *Scirpus aestivalis* Retz. Obs. Bot. 4. 12. 1786.

An annual erect herb, up to 10 cm tall. Stems tufted, trigonous. Leaves flat. Corymbs compound. Spikelets solitary, lanceolate-ovate, 40-flowered. Involucral bracts 3. Nuts obovate, biconvexed, smooth.

Occasional in wet and moist sandy lands. Fl. & Fr.: Jan.–Apr. Rudraram RF, *MSM & TP* 12727.

Fimbristylis albo-viridis Clarke in Hook. f. Fl. Brit. India 6. 638. 1893. *F. annua* Roem. & Schult. *var. pluristriata* Back. & Bakh. Fl. Java (em. ed) 101.: 121. 1949. fam. 246. p. 21 (non *F. diphylla* Vahl *var. pluristriata* C.B. Clarke 1893).

An annual erect herb, up to 20 cm tall. Stems tufted, slender. Leaves half as long as stems. Inflorescence with several spikelets. Spikelets solitary, ovoid, many-flowered. Involucral bracts 2. Nuts obovate, biconvex, shortly stipitate.

Rare in grass lands. Fl. & Fr.: Oct.–Dec. Mohammadabad, *MSM & KH* 10529.

Fimbristylis argentea (Rottb.) Vahl, Enum. Pl. 2. 294. 1806; FBI 6: 640; C. Fischer 1659 (1151). *Scirpus argenteus* Rottb. Descr. Icon. Rar 51. t. 17. 1773 & Descr. Pl. Rar. 27. 1777.

An annual spreading herb, up to 9 cm. Stems filiform, glaucous. Leaves filiform, flexuous. Spikelets 10–20-flowered, oblong, sessile, 5–20 in each terminal globose head. Involucral bracts 3–4. Nuts obovoid.

Common in forests and wet places. Fl. & Fr.: Jan.–July. Kusumasamudram beat, *MSM & NV* 10378; Mohammadabad, *KH & MSM* 5574.

Fimbristylis bisumbellata (Forssk.) Bubani Dodecanthea 30. 1850; C. Fischer 1658 (1151). *Scirpus bisumbellatus* Forssk. Fl. Aegypt.-Arab 1; 15. 1775. *Fimbristylis dichotoma* auct. non Vahl, FBI 6: 635.

An annual erect herb, stems tufted, up to 30 cm tall. Leaves flat, spikelets many-flowered, solitary, oblong. Involucral bracts 3–5. Nuts obovoid, trabeculate. Shortly stipitate, dull stramoneous.

Occasional in ponds. Fl. & Fr.: July–Oct. Malkalcheruvu, *KH & MSM* 5578, 5580.

Fimbristylis complanata (Retz.) Link, Hort. Berol. 1: 292. 1827; FBI 6: 646; C. Fischer 1659 (1151). *Scirpus complanatus* Retz. Obs. Bot. 5: 14. 1791.

A perennial rhizomatous herb, up to 40 cm tall. Stems tufted. Leaves crowded at base. Umbels decompound. Spikelet solitary, ovoid-oblong, 12-flowered. Involucral bracts 2. Nuts obovoid, trigonous.

Common on river banks and in forest. Fl. & Fr.: Aug.–Apr. Tandur, *MSM & TP* 12704.

Fimbristylis cymosa R. Br. Prodr. 228. 1810. *F. spathacea* Roth, Nov. Pl. Spec. 24. 1821; FBI 6: 640; C. Fischer 1659(1151).

An annual erect herb, stems tufted, up to 20 cm tall, rigid, trigonus. Leaves flat, scabrid, tip acute. Spikelets solitary, globose. Involucral bracts 2–3. Nuts obovoid.

Common in moist places. Fl. & Fr.: July–Oct. Rangammagudem beat, *MSM & KH* 5585.

Fimbristylis dichotoma (L.) Vahl, Enum. Pl. 2. 287. 1806; FBI 6: 635; C. Fischer 1658 (1151); ssp. **podocarpa** (Nees & Meyen) T. Koyama, Micronesica 1: 87. 1964. *Scirpus dichotomus* L. Sp. Pl. 50. 1753. *Fimbristylis podocarpa* Nees & Meyen in Wight, Contrib. Bot. India 98. 1834.

An annual erect herb, up to 60 cm tall. Stems solitary, glabrous. Leaves narrowly linear, spikelets solitary or 2–10, ovoid, 5–15-flowered, ovate, brown, involucral bracts 2–7, the lowest 1 or 2 leaflike, remaining setaceous. Nuts obovate, tuberculate on shoulder, cream coloured.

Common in marshy places in forests. Fl. & Fr.: Sept.–Dec. Kanmankalva beat, *MSM* 10455.

Fimbristylis miliacea (L.) Vahl, Enum. Pl. 2: 287. 1806; FBI 6: 644; C. Fischer 1660 (1152). *Scirpus miliaceus* L. Syst. Nat (ed. 10) 2: 868. 1759. *Fimbristylis littoralis* Gaudich, Voy. Bonite. Bot. 413. 1826.

An annual erect herb, stem solitary, up to 40 cm, caespitose. Leaves flat, scabrid, tip acuminate. Spikelet solitary, globose, purplish. Involucral bracts 3–5, not overtopping. Nuts obovoid, trigonous, verruculose, umbonulate.

Common in ponds and ditches. Fl. & Fr.: July–Oct. Malkalcheruvu, *KH & MSM* 5583.

Fimbristylis polytrichoides (Retz.) R. Br. Prodr. 226. 1810; FBI 6: 632. 1893; C. Fischer 1658 (1150). *Scirpus polytrichoides* Retz. Obs. Bot. 4: 11. 1786 (8).

A perennial herb, up to 25 cm tall. Stems tufted. Leaves canaliculate. Inflorescence a single terminal spikelet. Spikelets ellipsoid, many-flowered. Involucral bracts 2. Nuts obovoid, biconvex, purple at maturity.

Occasional in open forest. Fl. & Fr.: Sept.–Mar. Rudraram RF, *MSM & TP* 12726.

Fimbristylis schoenoides (Retz.) Vahl, Enum. Pl. 2: 286. 1805; FBI 6: 634; C. Fischer 1658 (1150). *Scirpus schoenoides* Retz. Obs. Bot. 5: 14. 1788.

An annual erect herb, up to 60 cm tall. Stems tufted, slender, smooth, striate. Leaves filiform. Spikelets 1–4, ovoid, many-flowered, orange brown when living, pale when dried. Involucral bracts 1 or 2. Nuts obovate, biconvex, yellow.

Occasional in marshy places at foot hills. Fl. & Fr.: Sept.–Dec. Mohammadabad beat, *MSM* 10483.

Fimbristylis sieberiana Kunth, Enum. Pl. 2. 237. 1837. *F. ferruginea* Decne in Nouv. Ann. Mus. Hist. Nat. Paris 3: 357. 1834 non (L.) Vahl 1806.

A perennial tufted rhizomatous herb. Stems up to 45 cm tall. Leaves flat, margins inrolled. Inflorescence simple. Spikelets solitary. Involucral bracts 3–5, lowest overtopping. Nuts obovoid, biconvex, stipitate.

Occasional in marshy places. Fl. & Fr.: June–Nov. Gacchibowli, *MSM* 15191.

Fimbristylis tenera Schult. Mant. 2. 57. 1825; FBI 6: 642. 1893; C. Fischer 3: 1660 (1152). *F. monticola* Hochst ex Steud. Syn. Pl. Cyp. 3: 111. 1855.

An annual slender caespitose herb, up to 30 cm tall. Stems filiform, trigonous, striate. Leaves filiform, glaucous green. Inflorescence compound umbel. Spikelets oblong, dark brown. Involucral bracts setiform. Nuts stipitate, globose-obovoid, trigonous.

Occasional in moist places. Fl. & Fr.: Aug.–Nov. Central University, *MSM* 15177.

Fimbristylis tetragona R. Br. Prodr. 226. 1810; FBI 6: 631; C. Fischer 1658 (1150).

A perennial erect herb, up to 70 cm tal. Stems tufted, glabrous. Leaves bladeless, sheaths 1–5 cm. Spikelet solitary, terminal, ovoid, brown, densely many-flowered. Involucral bracts absent. Nuts oblong, terete, short stipitate at base, cream-coloured.

Occasional in marshy places of forests. Fl. & Fr.: Sept.–Dec. Anantasagar beat, *MSM* 10502; *KH & MSM* 10931.

FUIRENA Rottboll

1. Spikelets squarrulose. Nut obovoid .... ***F. ciliaris***
1. Spikelets ellipsoidal. Nut rhombic oval .... ***F. capitata***

Fuirena capitata (Burm. f.) T. Koyama in Dassanayake & Fosberg. Rev. Handb. Fl. Ceylon 5: 151. 1985. *Scirpus capitatus* Burm. f. Fl. Ind. 21. 1768. non L. 1753. *Fuirena uncinata* (Willd.) Kunth. Enum. Pl. 2. 184. 1837; FBI 6: 666. C. Fischer 1669 (1158). *Scirpus uncinatus* Willd. Sp. Pl. 1: 300. 1798.

An annual herb, up to 20 cm tall. Stems tufted, ribbed. Leaves basal and cauline, linear-lanceolate, flat. Inflorescence capitate, bearing 5 glomerules of spikelets. Spikelets sessile, ellipsoidal. Nuts rhombic oval, triquetrous.

Common in moist places. Fl. & Fr.: Sept.–Mar. Shaikpalli, *KH & MSM* 10934.

Fuirena ciliaris (L.) Roxb. Fl. Ind. 1: 184. 1820. *Scirpus ciliaris* L. Mant. Pl. 182. 1771. *Fuirena glomerata* Lam. Tabl. Encycl. 1: 150. 1791; FBI 6: 666: C. Fischer 1669 (1158). *F. rottboelli* Nees in Wight Contrib. Bot. India 94. 1834.

An annual erect herb, stems tufted, up to 40 cm tall. Leaves flaccid. Spikelets many-flowered, terete, squarrose. Nuts obovoid, fuscous, smooth.

Common in dried ponds. Fl. & Fr.: Dec.–Mar. Malakalacheruvu, *MSM & KH* 10924.

JUNCELLUS (Griseb) Clarke

1. Stems robust. Inflorescence umbelled .... ***J. alopecuroides***
1. Stems tufted. Inflorescence capitate:
    2. Non-rhizomatous herb. Involucral bracts 6 .... ***J. pygmaeus***
    2. Rhizomatous herb. Involucral bracts 2 .... ***J. laevigatus***

Juncellus alopecuroides (Rottb.) Clarke in Hook. f., Fl. Brit. India 6: 595. 1893; C. Fischer 1629(1133). *Cyperus alopecuroides* Rottb. Descr. Pl. Rar. Progr. 20. 1772.

A perennial, erect herb, up to 1 m tall. Stems rounded below and trigonous above. Leaves flat. Inflorescence compound umbel bearing oblong spikes densely covered with spikelets. Spikelets spicate, 15–30-flowered, elliptic-lanceolate. Involucral bracts 5. Nuts elliptic, light brown.

Common in marshy low land and along banks of lakes, canals (Seshagiri Rao, 1986). Fl. & Fr.: Sept.–Dec.

Juncellus laevigatus (L.) Clarke in Hook. f. Fl. Birt. India 6: 596. 1893; C. Fischer 1629 (1133). *Cyperus laevigatus* L. Mant. Alt. 179. 1771.

A perennial rhizomatous herb, up to 30 cm tall. Stems tufted, fleshy. Spikelets 30, ovate, straw coloured, dotted with brown spots. Involucral bracts 2. Nuts obovoid, brown, stipitate.

Occasional in moist localities (Seshagiri Rao, 1986). Fl. & Fr.: Sept.–Dec.

Juncellus pygmaeus (Rottb.) Clarke in Hook. f. Fl. Brit. India 6: 596. 1893; C. Fischer 1629 (1133). *Cyperus pygmaeus* Rottb. Descr. Icon. 20. t. 14. f. 5. 1753.

An annual erect herb. Stems tufted up to 4 cm tall. Leaves canaliculate 3 × 0.1 cm. Spikelets 10–20-flowered, broadly ovate, incurved, pale in heads. Involucral bracts 5–7, glumes oblong, membraneous. Nuts ellipsoid, trigonous, 0.1 cm, yellow, apiculate, stipitate.

Occasional in moist places. Fl. & Fr.: July–Oct. Gadirayal beat, *MSM & KH* 5575.

KYLLINGA Rottboell

1. Inflorescence a solitary head .... ***K. melanosperma***
1. Inflorescence a 3-lobed spike .... ***K. bulbosa***

Kyllinga bulbosa Beauv. Fl. d'Oware & Benin 1: 11. t. 8. f. 1. 1804. *Kyllinga triceps* Rottb. Descr. Icon. Rar. 14. t. 4. f. 6. 1773; FBI 6: 587; C. Fischer 1623 (1130).

An annual erect herb, up to 20 cm tall. Stems tufted, glabrous. Leaves linear. Spikelets 1-flowered, oblong, many, aggregated into dense globose spikes. Spikes terminal, usually 3 together, middle one being the largest. Involucral bracts 3–4, leaflike, 6 cm. Nuts oblong, 0.2 cm long, yellowish brown.

Common weed of all places. Fl. & Fr.: June–Oct. Kusumasamudram beat, *MSM & NV* 10383.

Kyllinga melanosperma Nees in Wight. Contrib. Bot. India 91. 1834; FBI 6: 588; C. Fischer 1624 (1130).

A rhizomatous herb, up to 10 cm long. Leaves reduced to bladeless sheaths. Sheaths 4, membraneous, 18 cm long, greenish. Inflorescence a solitary head. Spikelets elliptic-oblong Involucral bracts 3. Nut elliptic-oblong, apiculate.

Occasional in forest. Fl. & Fr.: July–Jan. Central University, *MSM* 15188.

LIPOCARPHA Robert Brown

Lipocarpha sphacelata (Vahl) Kunth, Enum. Pl. 2: 267. 837; FBI 6: 667. *Hypaelytrum sphacelatum* Vahl, Enum. Pl. 2: 283. 1866. *Tunga triceps* Roxb. Fl. Ind. 1: 183. 1820. *Lipocarpha triceps* (Roxb.) Nees in Wight Contrib. Bot. India 92. 1834; C. Fischer 1670(1158).

An annual erect herb, up to 40 cm tall. Stems solitary, glabrous. Leaves linear, canaliculate, spikelets 2–5 in a head, dark purple brown, ellipsoidal, many-flowered, involucral bracts 2, leaflike. Nuts oblong-elliptic, trigonous, light brown, puncticulate, conical, beaked at apex.

Occasional in sandy places at foot hills. Fl. & Fr.: Sept.–Dec. Gadirayal beat, *MSM* 10521; Shaikpalli, *KH & MSM* 10930.

MARISCUS Vahl

1. Stems stoloniferous, leaves linear .... ***M. clarkei***
1. Stems not stoloniferous, leaves flaccid .... ***M. sumatrensis***

Mariscus clarkei (T. Cooke) T. Koyama in J. Jap. Bot. 51: 313. 1976. *Cyperus clarkei* T. Cooke. Fl. Bombay 2: 873 (387). 1908. *Mariscus bulbosus* Clarke in FBI 6: 620. non Steuedel 1885: C. Fischer 1644(1142).

A perennial erect herb, up to 15 cm tall, rhizome short, stolons slender. Culm solitary, smooth. Leaves linear, leaf sheaths brown. Spikelets cylindrical, sessile, involucral bracts 5–9, spreading to reflexed, surpassing the inflorescence. Nuts oblong-elliptic, tetragonous, dark brown.

Occasional in sandy places in forests. Fl. & Fr.: June–Nov. Kusumasamudram beat, *MSM & NV* 10380; Mohammadabad beat, *MSM* 10430.

Mariscus sumatrensis (Retz.) T. Koyama in Gard. Bull. Straits. Settlem. 30: 154. 1977. *Kyllinga sumatrensis* Retz. Obs. Bot. 4: 13. 1786. *Mariscus siberianus* Nees ex Clarke in Hook. f. Fl. Brit. India 6: 622. 1893; C. Fischer 1645 (1143).

An annual erect herb, stems tufted, up to 40 cm tall. Leaves flaccid. Spikelets compressed in umbellate spikes. Involucral bracts 7, overtopping. Nuts trigonus, apex beaked, base tapering.

Occasional in moist places. Fl. & Fr.: May–Dec. Dharmapuram, *MSM & KH* 10933.

PYCREUS Palisot de Beauvois

1. Leaves distant. Glumes furrowed on both sides ..... ***P. sanguinolentus***
1. Leaves basal. Glumes not furrowed:
   2. Rachilla not winged ..... ***P. flavidus***
   2. Rachilla winged:
      3. Glumes less than 1.5 mm long ..... ***P. pumilus***
      3. Glumes less than 2 mm long ..... ***P. polystachyos***

Pycreus flavidus (Retz.) T. Koyama in J. Jap. Bot. 51(10): 313. 1976. var. **flavidus.** *Cyperus capillaris* Koenig ex Roxb. Icon. 1: 198. 1820. *Pycreus globosus* (All.) Reich. Fl. Germ. Excurs. 140. 1830; *C. Fischer* 1627 (1132). *P. capillaris* Nees ex Clarke in Hook. f. Fl. Brit. India 6: 591. 1893.

An annual erect herb, stems tufted, up to 40 cm tall. Leaves flat, purple. Spikelets 30-flowered, oblong in 10–15-spikeleted pale spikes. Involucral bracts 3, overtopping. Nuts obovoid, fucous, apiculate, stipitate.

Occasional in ponds and ditches. Fl. & Fr.: July–Oct. Rangammagudem beat, *MSM* 10477; Mohammadabad beat, *KH & MSM* 5573 & 10923.

Pycreus polystachyos Beauv. Fl. de Oware & Benin 2: 48. t. 86. f. 2. 1807; FBI 6: 592. *var.* **laxiflorus** (Benth) Clarke in Hook. f. Brit. India 6: 592. 1893. *Cyperus polystachyos var. laxiflorus* Benth. Fl. Austr. 7: 261. 1878. *Cyperus polystachyos* Rottb. Descr. Ic. Rar. Nov. Pl. 39. t. 11. f. 1. 1773. *Pycreus odoratus* Urban Symb. Antill 2: 164. 1900; C. Fischer 3: 1627 (1132).

A perennial tufted herb, up to 60 cm tall. Stems trigonous. Leaves flat, linear. Inflorescence simple head. Spikelets compressed, yellow to straw coloured linear-lanceolate. Involucral bracts 8. Nuts oblong, biconvex, dark brown.

Occasional in marshy places. Fl. & Fr.: July–Dec. Central University, *MSM* 15178.

Pycreus pumilus (L.) Nees ex Clarke in Hook. f. Fl. Brit. India 6: 591. 1893; C. Fischer 1627 (1132). *Cyperus pumilus* L. Cent. Pl. 2. 6. 1756. *C. nitens* Retz. Obs. Bot. 5: 13. 1789. *Pycreus nitens* (Retz.) Nees, Nov. Actorum Acad. Caes. Leop. Carol. Nat. Cur. 19 (Suppl) 1: 53. 1843; FBI 6: 591.

An annual erect herb, up to 20 cm tall. Stems densely tufted, triquetrous, glabrous. Leaves linear, spikelets contracted to a head like cluster, oblong, flattened, 8–35-flowered, light greenish. Involucral bracts 3 to 5. Nuts elliptic, laterally flattened, puncticulate, brownish.

Common in sandy places in forest. Fl. & Fr.: Sept.–Dec. Rangammagudem beat, *MSM* 10475.

Pycreus sanguinolentus (Vahl) Nees ex Clarke in Hook. f. Fl. Brit. Ind. 6: 590. 1893; C. Fischer 1627 (1132). *Cyperus sanguinolentus* Vahl, Enum. Pl. 2: 351. 1806.

An annual erect herb. Stems tufted, up to 70 cm tall. Leaves flat. Spikelets 3-flowered, 2–7 in a cluster. Involucral bracts overtopping. Nuts obovoid, compressed, brown, puncticulate rugose.

Occasional in marshy places. Fl. & Fr.: May–Dec. Mohammadabad beat, *MSM & KH* 10925 & 10926.

RHYNCHOSPORA Vahl *nom. cons.*

Rhynchospora wightiana Steud. Syn. Pl. Glum. 2: 148. 1855; FBI 6: 669; C. Fischer 1672 (1160).

An erect herb, up to 30 cm tall. Stems tufted, glabrous. Leaves clustered at the base, linear. Spikelets 1–2-flowered, lanceolate, yellow, densely clustered in terminal solitary heads. Involucral bracts 6. Nuts oblong, brown.

Occasional in forests. Fl. & Fr.: May–Sept. Mohammadabad, *MSM* 10480; Central University, *MSM* 15168.

RIKLIELLA J. Raynal

Rikliella squarrosa (L.) Raynal in Adansonia (Ser. 2) 13L 154, 1973. *Scirpus squarrosus* L. Mant. Pl. 2: 181. 1771; FBI 6: 663; C. Fischer 1666 (1156).

An annual slender erect herb, up to 10 cm tall. Stems tufted, striate, glabrous. Leaves filiform, spikes 1–3 in a pseudolateral head, brown, sessile, ovate, squarrose, densely many-flowered. Bracts 1 or 2, lower leaflike, surpassing the inflorescence. Nuts obovate, minutely puncticulate.

Occasional in sandy places in forests. Fl. & Fr.: Sept.–Dec. Gadirayal beat, *MSM* 10527.

SCHOENOPLECTUS (Reichb.) Palla

1. Leaves cauline, linear .... ***S. maritimus***
1. Leaves radical or reduced to sheaths:
    2. Spikelets solitary or in clusers bristels absent .... ***S. juncoides***
    2. Spikelets corymbose, bristles present:
        3. Inflorescence arises from the middle of stem:

4. Stems stout, spongy, jointed .... *S. articulatus*
4. Stems slender, not spongy, not jointed .... *S. senegalensis*
3. Inflorescence arises above the middle of stem .... *S. supinus*

Schoenoplectus articulatus (L.) Palla in Bot. Jahrb. Syst. 10: 299. 1888. *Scirpus articulatus* L. Sp. Pl. 47. 1753; FBI 6: 656; C. Fischer 1666(1156).

An annual erect herb. Stems tufted, up to 75 cm tall, septate, spongy. Sheaths brown. Spikelets 60 in a cluster, ovoid-oblong, pink, fulvous. Involucral bracts stems like. Nuts obovoid, trigonous, dark, smooth, apiculate.

Common in moist places. Fl. & Fr.: Sept.–Dec. Kothapalli beat, *MSM & KH* 10915, Malkalcheruvu, *MSM & KH* 10916.

Schoenoplectus juncoides (Roxb.) Palla in Bot. Jahrb. Syst. 10: 299. 1889. *Scirpus juncoides* Roxb. Fl. Ind. 1: 218. 1820. *S. erectus* auct. non Poir. 1829; FBI 6: 656; C. Fischer 1666(1156).

An annual erect herb, stems caespitose, sheaths grey-fuscous, spikelets 1–3 in a cluster, ovoid-terete, fulvous. Involucral bract to 10 cm. Nuts broadly obovoid, fulvous, dark, undulate, apiculate.

Common in marshy places. Fl. & Fr.: Sept.–Dec. Mohammadabad beat, *MSM & KH* 10917.

Schoenoplectus maritimus (L.) Lye in Blythia 29. 145. 1971. *Scirpus maritimus* L. Sp. Pl. 75. 1753; FBI 6: 658. C. Fischer 1666(1156).

A perennial rhizomatous creeping herb, up to 75 cm. Stems triquetrous. Leaves flat, linear. Inflorescence terminal umbel. Spikelets ovoid. Involucral bracts, 3–5, leaflike. Nuts obovate, trigonous, smooth.

Occasional on moist land in forest. Fl. & Fr.: Oct.–Mar. Gadirayal beat, *KH & MSM* 5576.

Schoenoplectus senegalensis (Hochst. ex Steud.) Palla in Bot. Jahrb. 10: 299. 1888. *Isolepis senegalensis* Hochst. ex Steud. Syn. Pl. Glum. 2: 96. 1855. *Scirpus jacobii* Fischer in Kew Bull. 1931: 103. 1931; C. Fischer 1666(1156).

A perennial caespitose herb, up to 30 cm tall. Stems slender, transversely septate. Inflorescence pseudolateral, capitate, spikelets 20, ovoid-subglobose. Involucral bracts stem like, greater than 20 cm long. Nuts obovoid, triquetrous angled sharply.

Occasional in marshy places in forest. Fl. & Fr.: Aug.–Dec. Mohammadabad RF, *KH & MSM* 10918.

Schoenoplectus supinus (L.) Palla, Sitzber. Zool. Bot. Ges. Wien. 38: 49. 1888. *Scirpus supinus* L. Sp. Pl. 49. 1753. FBI 6: 655; C. Fischer 1666 (1156).

An annual herb, up to 30 cm tall. Stems tufted, trigonous. Inflorescence pseudo lateral, head like. Spikelets 15 to an inflorescence, ovate-oblong, becoming straw coloured. Involucral bract 1. Nuts elliptic, triangular.

Common in marshy and wet places. Fl. & Fr.: Aug.–Dec. Mohammadabad RF, *KH & MSM* 10919.

POACEAE (*nom. alter.* GRAMINEAE)

KEY TO SUB-FAMILIES

1. Mature spikelets falling entire; 2-heteromorphous florets .... ***Panicoideae***
1. Mature spikelets breaking up partly; not heteromorphous florets .... ***Pooideae***

KEY TO THE TRIBES OF PANICOIDEAE

1. Spikelets in pairs or threes, one sessile other pedicelled, those of each pair homo or heterogamous, glumes rigid .... ***Andropogoneae***
1. Spikelets solitary more or less similar, glumes membranous .... ***Paniceae***

KEY TO THE GENERA OF THE TRIBE ANDROPOGONEAE

1. Spikelets in each pair dissimlar:
    2. Inflorescence a solitary raceme:
        3. Spikes awned .... ***Heteropogon***
        3. Spikes unawned .... ***Mnesithea***
    2. Inflorescence of racemes more than one pair:
        4. Racemes supported by spathes:
            5. Involucral spikelets present .... ***Iseilema***
            5. Involucral spikelets absent:
                6. Basal joints of racemes bulbous .... ***Apluda***
                6. Basal joints of racemes not bulbous:
                    7. Joints thickened upwards, toothed .... ***Andropogon***
                    7. Joints not thickened upwards, not toothed .... ***Cymbopogon***
        4. Racemes not supported by spathes:
            8. Upper lemma awned from back .... ***Arthraxon***
            8. Upper lemma awned from apex:
                9. Upper lemma well developed .... ***Ischaemum***
                9. Upper lemma reduced .... ***Dicanthium***
1. Spikelets in each pair similar:
    10. Spikelets arranged in panicles and enclosed in silky hair .... ***Imperata***
    10. Spikelets are in several pairs of racemes not enclosed in silky hair ... ***Lophopogon***

KEY TO GENERA OF THE TRIBE PANICEAE

1. Spikelets unawned:
    2. Inflorescence of spiciform racemes:
        3. Ligule absent .... ***Echinochloa***
        3. Ligule present:
            4. Lower glume well developed
                5. Lower glume turned away from axis .... ***Paspalidium***
                5. Lower glume turned towards axis .... ***Brachiaria***
            4. Lower glume absent:

6. Spikelets elliptic-oblong .... *Digitaria*
6. Spikelets orbicular-obovate .... *Paspalum*
2. Inflorescence of panicle:
7. Spikelets covered by bristles:
8. Bristles rigid barbed .... *Setaria*
8. Bristles plumose .... *Pennisetum*
7. Spikelets not covered by bristles:
9. Spikelets without annular callus:
10. Spikelets globose .... *Sacciolepis*
10. Spikelets not globose .... *Panicum*
9. Spikelets with annular callus .... *Eriochloa*
1. Spikelets awned:
11. Upper lemma muticous, lower lemma awned, upper glume awned... *Echinochloa*
11. Upper lemma aristulate, lower lemma muticous, upper glume unawned.. *Alloteropsis*

KEY TO THE TRIBES OF SUB-FAMILY POOIDEAE

1. Woody shrubs or trees, leaf blades articulate with sheaths .... *Bambuseae*
1. Herbs, leaf blades not articulate on the sheaths:
2. Inflorescence solitary or digitate, spikes sessile .... *Chorideae*
2. Inflorescence panicles or spike like, sessile or pedicelled:
3. Spikelets with 1-fertile floret:
4. Glumes minute, lemma and palea similar in texture .... *Oryzeae*
4. Glumes well developed, lemma and palea not similar.
5. Spikelets with 2 florets:
6. Glumes equal, upper lemma unawned .... *Isachneae*
6. Glumes unequal, upper lemma awned .... *Arundinelleae*
5. Spikelets with 3 florets:
7. Spikelets awned:
8. Spikelets 1-flowered, awn single .... *Perotideae*
8. Spikelets many-flowered, awn trifid .... *Aristideae*
7. Spikelets unawned:
9. Inflorescence panicle, spikelets ellipsoid .... *Sporoboleae*
9. Inflorescence spike, spikelets laterally compressed ...... *Zoysieae*
3. Spikelets with 2 or more fertile florets .... *Eragrosteae*

KEY TO THE GENERA OF TRIBE CHLORIDEAE

1. Spikelets awned:
2. Inflorescence a solitary spike .... *Melanocenchris*
2. Inflorescence with many (8–10) racemes .... *Chloris*
1. Spikelets unawned:
3. Spikes solitary, terminal .... *Oropetium*
3. Spikes of terminal umbellate raceme .... *Cynodon*

KEY TO THE GENERA OF TRIBE ERAGROSTEAE

1. Spikes awned:
2. Inflorescence of digitate racemes or spikes .... *Dactyloctenium*
2. Inflorescence panicle or spiciform racemes .... *Dinebra*
1. Spikes unawned:

3. Inflorescence of panicle, lax or contracted:
    4. Upper glumes 3-nerved .... ***Eragrostis***
    4. Upper glume 1-nerved .... ***Leptochloa***
3. Inflorescence of digitate racemes or spikes:
    5. Spikes solitary .... ***Eragrostiella***
    5. Spikes 2–10 .... ***Eleusine***

The following tribes contain only one genus.

1. Aristideae - *Aristida*
2. Arundinelleae - *Arundinella*
3. Bambuseae - *Dendrocalamus*
4. Isachneae - *Isachne*
5. Oryzeae - *Oryza*
6. Perotideae - *Perotis*
7. Sporoboleae - *Sporobolus*
8. Zoysieae - *Tragus*

ALLOTEROPSIS J.S. Presl. ex C.B. Presl.

Alloteropsis cimicina (L.) Stapf in Prain Fl. Trop. Africa. 9. 487. 1919; C. Fischer 1766 (1233); Khan in FFH 339; Bor 276. *Milium cimcianum* L. Mant. Alt. 184. 1771

An annual erect herb, up to 30 cm tall. Stems tufted, glabrous. Leaf sheaths 2–4 cm, pilose-glabrous, blade ovate-lanceolate. Spicate racemes 3–9, terminal 5–10 cm long, pedunculate. Spikelets compressed. Caryopsis elliptic-oblong, dorsally compressed.

Common in waste lands and forests. Fl. & Fr.: Aug.–Dec. Anantagiri RF, *MSM & TP* 11079, Rudraram RF, *MSM & TP* 12715.

ANDROPOGON Linnaeus

Andropogon pumilus Roxb. Fl. Ind. 1: 277. 1820; FBI 7: 170; C. Fischer 1753 (1215); Bor 93.

An annual stoloniferous herb, up to 20 cm. Stems tufted, glabrous. Leaf sheaths 1–3 cm, glabrous, blade linear. Racemes 2, digitate, pedunculate. Spikelets 2, sessile hermaphrodite, pedicelled staminate or neuter. Caryopsis ellipsoid.

Occasional in forest (Seshagiri Rao, 1986). Fl. & Fr.: June–Dec.

APLUDA Linnaeus

Apluda mutica L. Sp. Pl. 82. 1753; C. Fischer 1750 (1212); Khan in FFH 337; Bor 93. *A. varia* Hackel, ssp. *aristata* (L.) Hackel in Monogr. Androp. 6. 196. 1889; FBI 7: 150. *A. aristata* L., Amoen. Acad. 4. 303. 1759; C. Fischer 1750(1212).

A perennial erect herb, up to 2.5 m tall, stems tufted, glabrous. Leaf sheaths 3 cm, glabrous, blade linear-lanceolate, panicles 3–80 cm, terminal, spikelets in threes, one sessile, hermaphrodite, the rest two pedicelled.

Fl. & Fr.: July–Dec. Central University, *MSM* 12737 & 12750.

ARISTIDA Linnaeus

1. Glumes muticous .... *A. adscensionis*
1. Glumes aristate:
    2. Panicles effuse, rachis angular .... *A. hystrix*
    2. Panicles contracted, rachis terete .... *A. setacea*

Aristida adscensionis L. Sp. Pl. 82. 1753; FBI 7: 224; Khan in FFH 342; *var.* **adsensionis**; Bor 407. *A. depressa* Retz. Obs. Bot. 4. 22.1786; C. Fischer 1809 (1252).

An annual erect herb, up to 70 cm tall. Stems tufted, glabrous. Leaf sheaths 5–6 cm, glabrous, blade linear-convolute. Panicles 9–16 cms, contracted. Spikelets 1–6-flowered. Caryopsis linear.

Common in open forest. *Fl. & Fr.:* June–Dec. Anantasagar, *MSM & KH* 11009.

Aristida hystrix L.f. Suppl. Pl. 113. 1781; FBI 7: 225; C. Fischer 1809 (1252); Bor 410.

An annual erect herb, up to 30 cm tall, stems tufted, glabrous. Leaf sheaths 3–5 cm, pilose at collar, blade linear, scaberulous on margin. Panicles effuse. Caryopsis linear.

Common in open forest. Fl. & Fr.: All seasons. Anantasagar beat, *MSM & KH* 11017.

Aristida setacea Retz. Obs. Bot. 4: 22. 1786; FBI 7: 225; C. Fischer 1809 (1252); Bor 412.

A perennial herb, up to 1 m tall, stems tufted, glabrous. Leaf sheaths 2–8 cm, glabrous, blade linear, convolute. Panicles contracted, 25 cm. Spikelets linear, 1-flowered, pedicelled. Glumes linear, convolute, awned. Caryopsis linear, prickly.

Common in open forest. Fl. & Fr.: July–Jan. Chevella, *TP & MSM* 19254.

ARTHRAXON Beuvois

Arthraxon lanceolatus (Roxb.) Hochst. var. **echinatus** (Nees) Hack. Mon. Androp. 348. 1889. *Bathratherum echinatum* Nees in Edinb. New Phil. J. 18. 181. 1835. *Arthraxon echinatus* (Nees) Hochst in Fl. 39: 188. 1856; C. Fischer 1728 (1198); Bor 99. *A. spathaceus* Hook. f. Fl. Brit. India 7: 145. 1896.

An annual erect herb up to 40 cm tall. Stems tufted. Leaf sheaths 2–3 cm, tuberculate, ciliate-glabrous, blade ovate-lanceolate. Racemes 3–5 cm, terminal. Spikelets alternate. Caryopsis narrowly linear.

Occasional near moist places. Fl. & Fr.: June–Dec. Lingampalle, *MSM* 12764.

ARUNDINELLA Raddi

Arundinella pumila (Hochst) Steud. Syn. Pl. Glum. 1. 114. 1854; Bor 423; C. Fischer 1801 (1247). *Acratherum pumilum* Hochst ex A. Rich. Tent. Pl. Abyss. 2. 414. t. 100. 1851.

An annual erect herb up to 30 cm tall. Stems slender, tuberculate-pilose. Leaf sheaths 2–8 cm, glabrous, blade ovate-lanceolate. Panicles effuse, capillary, branched. Spikelets pedicelled. Caryopsis narrow ellipsoid.

Common in forest (Seshagiri Rao, 1986). Fl. & Fr.: July–Jan.

BRACHIARIA Grisebach

1. Spikelets imbricate:
    2. Rachis flat, margins scabrid ..... ***B. distachya***
    2. Rachis trigonous or terete:
        3. Panicles linear ..... ***B. eruciformis***
        3. Panicles oblong-lanceolate ..... ***B. reptans***
1. Spikelets distant ..... ***B. remota***

Brachiaria distachya (L.) Stapf in Oliver Fl. Trop. Africa 9: 565. 1919; C. Fischer 1769 (1226); Bor 281. *Panicum distachyum* L. Mant. Pl. 183. 1771; FBI 7: 37.

An annual erect herb, up to 15 cm tall. Stems solitary, glabrous. Leaf sheaths 1–3 cm, glabrous, blade linear-lanceolate, glabrous, cartilaginous on margins. Racemes with 2 or 3 branches. Spikelets ellipsoid. Caryopsis ovoid, compressed.

Common in ponds and marshy places. Fl. & Fr.: July–Dec. Malkalcheruvu, *MSM & KH* 10568.

Brachiaria eruciformis (Smith) Griseb. in Ledeb. Fl. Ross. 4: 469. 1853; C. Fischer 1769 (1226); Bor 283; *Panicum eruciforme* Smith in Sibth. & Smith Fl. Graeca 1: 44. t. 59. 1806. *P. isachne* Roth ex Roem. & Schult. in L. Syst. Veg. 2: 458. 1817; FBI 7: 28.

A perennial herb, up to 40 cm tall. Stems tufted. Leaf sheaths 2–4 cm, blade lanceolate. Panicles 6 cm. Spikelets ellipsoid. Caryopsis ovoid, dark.

Common in fields. Fl. & Fr.: July–Dec. Mekavarpalli, *MSM & TP* 11046.

Brachiaria remota (Retz.) Haines Bot. Bih. Orissa 1005. 1924; C. Fischer 1770 (1226); Bor 785. *Panicum remotum* Retz. Obs. Bot. 4: 17. 1786; FBI 7: 38.

An annual herb, up to 50 cm tall. Stem tufted, glabrous. Leaf sheaths 9 cm, glabrous, blades linear. Panicles 10 cm, open, branches spreading. Spikelets ellipsoid, distant. Caryopsis ellipsoid.

Occasional in shady places of forest. Fl. & Fr.: July–Nov. Rudraram RF, *MSM & TP* 12720.

Brachiaria reptans (L.) Gardn. & Hubb. in Hook. Icon. Fl. Sub. t. 3363. 1938; Bor 285. *Panicum reptans* L. Syst. Nat. (ed. 10) 870. 1759. *P. prostratum* Lam. Tabl. Encycl. 1: 171. 1791; FBI 7: 33. *Urochloa reptans* (L.) Stapf in Oliv. Fl. Trop. Africa 9: 701. 1920; C. Fischer 1775(1230).

An annual prostrate herb up to 40 cm. Stems tufted. Leaf sheaths 2–3 cm, glabrous, blades ovate-lanceolate. Panicles of 5–9 racemes, 7 cm, spikelets ovoid, alternate, shortly pedicelled. Caryopsis oblong, flattened.

Common in cultivated and waste lands (Seshagiri Rao, 1986). Fl. & Fr.: July–Dec.

CHLORIS Swartz

1. Stems stoloniferous. Spikes 6 cm .... ***C. virgata***
1. Stem not stoloniferous. Spikes 7 cm .... ***C. barbata***

Chloris barbata (L.) Swartz, Fl. Ind. Occid. 1: 200. 1797; FBI 7: 292; C. Fischer 1838(1272); Khan in FFH 344; Bor 465. *Andropogon barbatus* L. Mant. Pl. Alt. 302. 1771. non Linn. 1759.

An annual erect herb, up to 50 cm tall. Stems tufted, glabrous, leaf sheaths 4–7.5 cm, ciliate on collar, blade oblong, scabrid within. Spikes purplish. Spikelets 3-flowered. Caryopsis ellipsoid.

Common in moist places. Fl. & Fr.: All seasons. Mohammadabad beat, *MSM & KH* 10536.

Chloris virgata Swartz. Fl. Ind. Occ. 1: 203. 1797; FBI 7: 291; C. Fischer 1838 (1272); Bor 468.

An annual ascending herb, up to 50 cm tall. Stems stoloniferous, glabrous. Leaf sheaths 2–8 cm, glabrous, blades narrowly linear. Spikes 4–8, 6 cm long, digitate, crowded at the apex of peduncle, spiklets ovate-lanceolate, perfect only one. Caryopsis ellipsoid, compressed.

Common in open forest (Seshagiri Rao, 1986). Fl. & Fr.: July–Dec.

CYMBOPOGON Sprengel

1. Blades cordate and amplexicaul at base .... ***C. martinii***
1. Blades narrow and rounded at base:
   2. Pedicel of the spikelet swollen .... ***C. flexuosus***
   2. Pedicel of the spikelet not swollen .... ***C. coloratus***

Cymbopogon coloratus (Nees) Stapf, Bull. Misc. Inform. 1906. 321. 1906; C. Fischer 1756 (1217); Bor 126. *Andropogon coloratus* Nees in Wight. Cat. Ind. Pl. n. 1703. 1883. *A. nardus* L. var. *coloratus* Hook. f. in Fl. Brit. India 7: 206. 1896.

A perennial herb, up to 2 cm tall. Stems tufted. Leaf sheaths 5–15 cm, glabrous, tomentose, blade linear-lanceolate. Panicles 8 cm terminal with many spiciform racemes, 2-nate. Spikelets yellowish brown, one sessile, other pedicelled.

Common in open forests. Fl. & Fr.: Oct.–Jan. Lingampalle, *MSM* 12735.

Cymbopogon flexuosus (Nees ex Steud.) Wats. in Atkins. Gaz. N.W. Prov. India. 392. 1882; *var* **flexuosus**; C. Fischer 1756 (1216); Bor 127. *Andropogon flexuosus* Nees ex Steud. Syn. Pl. Glum. 1: 388. 1854. *A. nardus* L. *var. flexuosus* Hack. in DC. Monogr. Phan. 6: 603. 1889; FBI 7: 207.

A perennial herb, up to 2 m tall. Stems tufted, glabrous. Leaf sheaths 6–12 cm, glabrous, blades lanceolate, glabrous. Panicles decompound, lax, 30 cm long, spikelets 2-nate. One sessile, other pedicelled.

Common in open forests. Fl. & Fr.: Aug.–Dec. Mekavarpalli, *MSM & TP* 11052.

Cymbopogon martinii (Roxb.) Wats. in Atkins. E.T. Gaz. N. .W. Prov. India. 392. 1882; C. Fischer 1756 (1217); Khan in FFH 338; Bor 129. *Andropogon martinii* Roxb. Fl. Ind. 1: 280. 1820. *A. schoenanthus* L. *var. martinii* (Roxb.) Hook. f. FBI : 7: 204. 1896.

A perennial erect scented herb, up to 2 m tall. Stems tufted, glabrous. Straw coloured. Leaf sheaths 7–15 cm glabrous, blades linear-lanceolate. Panicles 25 cm, one sessile other pedicelled.

Occasional in forest. Fl. & Fr.: Aug.–Dec. Rangammagudem, *MSM* 10580.

CYNODON Richard *nom. cons.*

1. Stoloniferous .... ***C. dactylon***
1. Non-stoloniferous .... ***C. barberi***

Cynodon barberi Rang. & Tad. in J. Bombay Nat. Hist. Soc. 24: 846. 1916; C. Fischer 3: 1836 (1270); Bor 469.

A perennial herb, up to 30 cm tall, stems tufted, glabrous. Leaf sheaths glabrous. Blades short-flat, narrow. Spikes 3–5, 6 cm digitate. Spikelets closely appressed to rachis. Caryopsis oblong.

Occasional in forest. Fl. & Fr.: June–Dec. Mekavarpalli, *MSM & TP* 11054; Vikarabad, *MRS* 116 (DD).

Cynodon dactylon Pers. Syn. Pl. 1: 85. 1805; FBI 7: 288; C. Fischer 1835(1270); Khan in FFH 343; Bor 469. *Panicum dactylon* L. Sp. Pl. 58. 1753.

A perennial prostrate herb, stems rhizomatous or stoloniferous, tufted, glabrous. Leaf sheaths, 0.3–3 cm, glabrous, blades linear, spikes 4 cm, spikelets subsessile, laterally compressed. Glumes subequal. Caryopsis oblong.

Common in plains and hills. Fl. & Fr.: June–Dec. Malkalacheruvu, *MSM* 10570.

DACTYLOCTENIUM Willdenow

1. Plants stoloniferous, spikes 2–4 cm long .... ***D. aegyptium***
1. Plants non-stoloniferous, spikes up to 2 cm long .... ***D. aristatum***

Dactyloctenium aegyptium (L.) P. Beauv. Ess. Agrost. Expl. t. 15. 1812; C. Fischer 1840 (1273); Khan in FFH 344; Bor 489. *Cynosurus aegyptius* L. Sp. Pl. 72. 1753. *Eleusine aegyptica* (L.) Desf. Fl. Atlant. 1: 85. 1798; FBI 7: 295.

An annual erect herb, up to 20 cm tall. Stems solitary, nodes glabrescent. Leaf sheaths 1.5–5 cm, hispid, blade tuberculate, hispid without, margin ciliate. Spikes digitate, oblong. Caryopsis obovate.

Common in marshy places. Fl. & Fr.: Oct.–Jan. Mohammadabad beat, *MSM & KH* 10551; Rangammagudem, *MSM &KH* 10583.

Dactyloctenium arisatum Link Hort. Berol 1: 59. 1827; Bor 489. *Eleusine aristata* Hook f. Fl. Brit. India 7: 296. 1896.

An annual herb, up to 15 cm tall. Stems tufted, glabrous. Leaf sheaths 0.5–1.5 cm. Blades linear-lanceolate. Spikes solitary, terminal. Spikelets laterally compressed. Caryopsis globose with rugose margins.

Vary rare in plains. Fl. & Fr.: July–Dec. Central University, *MSM* 15176.

DENDROCALAMUS Nees

Dendrocalamus strictus (Roxb.) Nees in Linnaea 9: 476. 1835; FBI 7: 404; C. Fischer 1858 (1286). *Bambusa stricta* Roxb. Pl. Cor. t. 80. 1798.

A perennial unarmed grass up to 10 m tall. Stems tufted. Leaf sheaths 3–15 cm, glabrous, blades linear-lanceolate. Inflorescence a compound panicle. Spikelets in globose heads of spike. Spikelets spiny, 1–3-flowered. Caryopsis ovoid, subglobose.

Common in forests. Fl. & Fr.: July–Dec. Central University, *MSM* 15172.

DICANTHIUM Willemet

Dicanthium caricosum (L.) A. Camus in Bull. Mus. Hist. Nat. (Paris) 27: 549. 1921; C. Fischer 1741 (1206); Bor 134; Deshpande in Fasc. Fl. Ind. 15: 7. 1984. *Andropogon caricosum* L. Sp. Pl. 2: 1480. 1763; FBI 7. 196.

An annual erect herb, up to 60 cm tall, stems solitary, nodes hairy. Leaf sheaths 4–10 cm, glabrous, blades pubescent. Racemes 2 or 3, sessile spikelets 0.4 cm. Pedicelled spikelet 0.4 cm. Caryopsis obtuse.

Occasional in forest undergrowth. Fl. & Fr.: Oct.–Dec. Mohammadabad beat, *MSM & KH* 10554.

DIGITARIA Heister ex Fabricius

1. Spikelets 2-nate .... *D. ciliaris*
1. Spikelets 3-nate:
   2. Spikelets tomentose with clavate hairs .... *D. stricta*
   2. Spikelets not as above .... *D. abludens*

Digitaria abludens (Roem. & Schult.) Veldekamp in Blumea 21: 53. 1773. *Panicum abludens* Roem. & Schult. in Syst. Veg. 2: 457. 1817. *Paspalum granulare* Trin. ex Spr. Neue. Entd. 2: 47. 1821. *Digitaria granularis* (Spr.) Henr. Mongr. Digitaria 302. 1950; Bor 301. *Paspalum pedicellare* Trin. ex Hook. f. Fl. Brit. India 7: 19. 1896. *Digitaria pedicellaris* (Hook. f.) Prain Bengal Pl. 1182. 1903; C. Fischer 1765(1223).

An annual herb, up to 50 cm tall. Stems tufted, glabrous. Leaf sheaths 1–3 cm, glabrous, blades glabrous. Spikes 3–10, 9 cm long, digitate. Spikelets ternate. Caryopsis oblong.

Common near wet areas. Fl. & Fr.: Sept.–Dec. Vikarabad, *MRS* 84 (DD); Anantagiri RF, *MSM & TP* 11065.

Digitaria ciliaris (Retz.) Koel. Descr. Gram. 27. 1802. *Panicum ciliare* Retz. Obs. Bot. 4: 16. 1786. *P. adscendens* H.B. & K. Nov. Gen. Pl. 1: 97. 1816. *Digitaria adscendens* (H.B.K.) Henrard in Blumea 1: 92. 1934; Bor 238. *D. marginata* Link, Enum. Hort. Berol. 1: 102. 1821; C. Fischer 1764 (1222).

An annual erect herb, up to 50 cm tall, stems tufted, glabrous, nodes glabrescent. Leaf sheaths lax, blades oblong-linear. Racemes digitate, spikelets oblong, smooth. Caryopsis oblong.

Common in forest undergrowth. Fl. & Fr.: Oct.–Dec. Kondapur beat, *MSM & KH* 10591; Mohammadabad beat, *MSM & KH* 10593.

Digitaria stricta Roth ex Roem. & Schult. in L. Syst. Veg. 2: 474. 1817. *var. denudata* (Link) Henr. Monogr. Digitaria 175. 1950; Bor 306. *D. denudata* Link Hort. Berol. 1: 222. 1827. *Paspalum royleanum* Nees ex Thw. Enum. Pl. Zeyl. 358. 1864; FBI 7: 18.

A perennial herb, up to 50 cm tall. Stems tufted, glabrous. Leaf sheaths 2–5 cm, hairy, blades linear, flat, hairy. Racemes 2–10, erect, 10 cm, spikelets ellipsoid, secund tomentose. Caryopsis oblong.

Rare in forest. Fl. & Fr.: Sept.–Dec. Vikarabad, *MRS* 324 (DD).

DINEBRA N.J Jacquin

Dinebra retroflexa (Vahl) Panzer, Denkschr. Kongl. Akad. Wiss. Munchen. 270. t. 12. 1814; C. Fischer 1841 (1274); Bor 491. *Cynosurus retroflexus* Vahl, Symb. Bot. 2: 20. 1791. *Dinebra arabica* Jacq. Fragm. Bot. 77. t. 121. f. 1. 1807; FBI 7: 297; Bor 279.

An annual erect herb, up to 75 cm tall, stems tufted, glabrous. Leaf sheaths 2.5–8 cm, glandular, blades flat, base cordate. Spikes 20 × 40 cm, oblong. Spikelets 2-seriate, sessile, 2- or 3-flowered. Caryopsis oblong.

Common in marshy places. Fl. & Fr.: July–Oct. Anantasagar beat, *MSM & KH* 11015.

ECHINOCHLOA P. Beauvois

1. Spikelets unawned .... ***E. colona***
1. Spikelets awned .... ***E. crusgalli***

Echinochloa colona (L.) Link, Hort. Berol 2: 209. 1883; C. Fischer 1776 (1230); Khan in FFH 340; Bor 308. *Panicum colonum* L. Syst. Nat. (ed. 10) 870. 1759; FBI 7: 32.

An annual ascending herb, up to 70 cm tall, stems tufted. Leaves linear, 10–20 × 0.2–0.8 cm, glabrous, acuminate, sheaths 8 cm, glabrous. Inflorescence 8–15 cm long panicle, composed of 1–3 cm long racemes, racemes appressed to the rachis. Spikelets crowded, 2-flowered, ovoid. Caryposis ellipsoid.

Common weed of rice fields. Fl. & Fr.: May–Sept. Mohammadabad beat, *MSM* 10500.

Echinochloa crusgalli (L.) Beauv. Ess. Agrost. 53: 161. 1812; Bor 310; C. Fischer 1777 (1231). *Panicum crusgalli* L. Sp. Pl. 56. 1753; FBI 7: 30.

An annual herb, up to 70 cm tall, stems tufted, glabrous. Leaf sheaths 5–20 cm, glabrous, blades linear-lanceolate. Panicles 10 cm, racemes spreading, spikelets ovoid, crowded. Caryopsis ellipsoid.

Common in cultivated and wet fields. Fl. & Fr.: Sept.–Dec. Mohammadabad, *MSM* 10537.

ELEUSINE Gaertner

Eleusine indica (L.) Gaertn. Fruct. 1: 18. 1788; Bor 493; FBI 7: 293; C. Fischer 1839 (1273); *Cynosurus indicus* L. Sp. Pl. 72. 1753.

An annual herb, up to 50 cm tall. Stems tufted, glabrous. Leaf sheaths 4–10 cm, glabrous, blades linear, convolute. Spikes 2–7 or more, 10 cm, digitate. Spikelets 3–6-flowered. Caryopsis oblong.

Common in plains usually in rice fields. Fl. & Fr.: Sept.–Dec. Lingampalle, *MSM* 12732; Central University, *MSM* 15186.

ERAGROSTIELLA Bor

Eragrostiella brachyphylla (Stapf) Bor in Indian Forester 66: 270. 1940 & Grass Ind. 494. *Eragrostis brachyphylla* Stapf in FBI 7; 327. 1896; C. Fischer 1828 (1265).

A perennial herb, up to 40 cm tall. Stems tufted. Leaf sheaths 4–10 cm, glabrous, blades linear, flat. Spike 10–20 cm, spikelets ovoid-oblong, slightly compressed, 5–20-flowered. Caryopsis slightly ellipsoid.

Common in open forest. Fl. & Fr.: Sept.–Dec. Anantagiri RF, *MSM & TP* 11076.

ERAGROSTIS P. Beauvois

1. Rachilla of spikelets tough, persistent, breaking up from base upwards:
    2. Paleas falling or deciduous:
        3. Spikelets ovate-oblong .... ***E. unioloides***
        3. Spikelets oblong-linear:
            4. Caryopsis ellipsoid .... ***E. atrovirens***
            4. Caryopsis oblong .... ***E. gangetica***
    2. Paleas persistent:
        5. Spikelets fascicled .... ***E. nutans***
        5. Spikelets scattered .... ***E. cilianensis***
1. Rachilla of spikelets jointed, breaking from above downwards:
    6. Lemmas ciliate on margins .... ***E. coarctata***
    6. Lemmas not ciliate on margin:
        7. Panicles spiciform, dense:
            8. Stems wiry .... ***E. riparia***
            8. Stems slender .... ***E. ciliaris*** var. ***ciliaris***
        7. Panicles effuse .... ***E. viscosa***

Eragrostis atrovirens (Desf.) Trin. ex Steud. Nom. Bot. ed. 2. 1: 562. 1840; Bor 503. *Poa atrovirens* Desf. Fl. Atlant. 1: 73. t. 14. 1798.

A perennial herb, up to 80 cm tall. Stems tufted, glabrous. Leaf sheaths 4–10 cm glabrous, blades linear, involute. Panicles 30 cm, effuse, spikelets oblong, secund, 4–20-flowered. Caryopsis ellipsoid.

Common in waste lands. Fl. & Fr.: Sept.–Dec. Mohammadabad, *MSM* 10538; Kondapur, *MSM & KH* 10588.

Eragrostis cilianensis (All.) Vign. in Malpighia 18: 286. 1903; C. Fischer 1327 (1264); Bor 503. *Poa cilianensis* All. Fl. Pedem. 2: 246. t. 91. f. 2. 1785. *Eragrostis major* Hort. Icon. Descr. Gram. Austr. 4: 14. t. 24. 1809; FBI 7; 320.

An annual herb, up to 80 cm tall. Stems tufted, glabrous. Leaf sheaths 2–4 cm, blades lanceolate, flat. Panicle 20 cm, spreading or contracted. Spikelets oblong, 10–30-flowered. Caryopsis subglobose.

Common along hill slopes. Fl. & Fr.: Aug.–Dec. Vikarabad, *MRS* 105 (DD).

Eragrostis ciliaris (L.) R. Br. in Tuckey. Narr. Exp. Congo. App. 478. 1818. *var. ciliaris*; FBI 7: 314. C. Fischer 3: 1825 (1263); Bor 506. *Poa ciliaris* L. Syst. Nat. ed. 10. 2. 875. 1759.

An annual herb, up to 40 cm tall. Stems tufted, glabrous. Leaf sheaths 2.5 cm, glabrous, blades linear, covolute. Panicles 8 cm. Spikelets oblong, subsessile. Caryopsis ovoid.

Occasional in forest (Seshagiri Rao, 1986). Fl. & Fr.: Sept.–Dec.

Eragrostis coarctata Stapf apud. Hook. f. FBI 7: 313. 1896; C. Fischer 1824 (1263); Bor 507.

A perennial herb, up to 50 cm tall. Stems tufted, glabrous. Leaf sheaths 4–10 cm, hairy, blade linear-lanceolate. Panicles compact, 10 cm, spikelets oblong, compressed. Caryopsis ovoid.

Occasional near streams. Fl. & Fr. Sept.--Dec. Mohammadabad, *MSM* 10550.

Eragrostis gangetica (Roxb.) Steud. Syn. Pl. Glum. 1: 266. 1854; C. Fischer 1826 (1264); Bor 508. *Poa gangetica* Roxb. Fl. Ind. 1: 341. 1820. *Eragrostis stenophylla* Hochst. ex Miq. in Verh. Konink. Nederl. Inst. 3, 4, 39, 1851; FBI 7: 318, *p.p.*

A perennial herb, up to 50 cm tall. Stems tufted, glabrous. Leaf sheaths 2–10 cm, glabrous, mouth bearded, blades very narrow, convolute. Panicles 15 cm, spreading, spikelets oblong, 8–10-flowered. Caryopsis oblong.

Common in moist places. Fl. & Fr.: Sept.–May. Kusumasamudram, *MSM & KH* 11029.

Eragrostis nutans (Retz.) Nees ex Steud. Nom. Bot. (ed. 2) 1: 563. 1840; C. Fischer 1826 (1264); Bor 511. *Poa nutans* Retz. Obs. 4. 19. 1786. *Eragrostis stenophylla* auct. non Hochst. 1851; FBI 7: 318.

A perennial herb, up to 50 cm tall. Stems tufted, glabrous. Leaf sheaths 5–12 cm, glabrous, blades linear, flat. Panicles 10 cm, contracted, branches subwhorled. Spikelets linear-oblong, 8–14-flowered. Caryopsis globose.

Common in marshy places and slopes of hills. Fl. & Fr.: June–Dec. Kondapur, *MSM* 10589.

Eragrostis riparia (Willd.) Nees in C. Martius Fl. Brass. Enum. Pl. 2(1): 512. 1829; C. Fischer 1825 (1263); Bor 513. *Poa riparia* Willd. Ges. Natur. Freunde Berlin, Neue Schr. 4: 185. 1803. *E. tenella* P. Beauv. var. *riparia* Stapf in Hook. f. Fl. Brit. India 7: 315.

An annual erect herb, up to 20 cm tall. Stems tufted or solitary. Leaf sheaths 2–4 cm, purplish, blade glabrous. Panicles oblong, spikelets oblong, 0.1–0.2 cm, yellow-purplish, 8–10-flowered. Caryopsis linear.

Common in forest undergrowth. Fl. & Fr.: Oct.–Dec. Mohammadabad beat, *MSM & KH* 10550.

Eragrostis unioloides (Retz.) Nees ex Steud. Syn. Pl. Glum. 1: 264. 1854; C. Fischer 1826(1264); Khan in FFH 342; Bor 515. *Poa unioloides* Retz. Obs. Bot. 5. 19. 1788. *Eragrostis amabilis* auct. non L. FBI 7: 317.

An annual erect herb, up to 50 cm tall, stems solitary or tufted. Leaf sheaths 1–4 cm, blade oblong, flat, sub-amplexicaul at base. Panicles oblong-ovate, spikelets ovoid to oblong, 0.5–1 cm long, straw-coloured, 10–35-flowered. Caryopsis ovoid.

Common in forest undergrowth. Fl. & Fl.: July–Oct. Mohammadabad beat, *MSM & KH* 10438, 10533.

Eragrostis viscosa (Retz.) Trin. Mem. Acad. Imp. Sci. St. Petersbourg. Ser. 6, Sci. Math. 1: 397. 1830; C. Fischer 1826(1624); Bor 515. *Poa viscosa* Retz. Obs. Bot. 4: 20. 1786–87. *Eragrostis tenella* P. Beauv. var. *viscosa* (Retz.) Stapf in Hook. f. Fl. Brit. India 7: 315. 1896.

An annual prostrate viscous grass, up to 30 cm, stems tufted. Leaf sheaths 1–2.5 cm, blade glabrous, sticky. Panicle oblong, glandular. Spikelets oblong, 8–10-flowered, pale yellow. Caryopsis ovoid.

Common in forest undergrowth. Fl. & Fr.: July–Dec. Rangammagudem beat, *MSM & KH* 10584.

ERIOCHLOA Humboldt. Bonpland & Kunth

Eriochloa procera (Retz.) C.E. Hubb. Bull. Misc. Inform. 1930. 256. 1930; C. Fischer 1767(1224); Bor 312. *Agrostis procera* Retz. Obs. Bot. 4. 19. 1786. *Eriochloa polystachya* Hook. f. in FBI 7: 20. 1896.

A perennial herb, up to 50 cm tall, stems tufted, glabrous. Leaf sheaths 4–15 cm, glabrous, blades linear, flat. Panicles 10 cm, lax. Spikelets ellipsoid. Caryopsis oblong-dorsally compressed.

Common near streams. Fl. & Fr.: Aug.–Oct. Mohammadabad, *MSM* 10594.

HETEROPOGON Persoon

Heteropogon contortus (L.) P. Beauv. ex Roem. & Schult. in L. Syst. Veg. 2: 836. 1817; C. Fischer 1743(1208); Khan in FFH 336; Bor 163. *Andropogon contortus* L. Sp. Pl. 1045. 1753; FBI 7: 199.

An annual erect herb, up to 70 cm tall, stems tufted, nodes glabrous. Leaf sheaths 4 cm, glabrescent, blade lanceolate, scaberulous. Racemes solitary. Spikelets lower homogamous, upper heterogamous, female sessile, male pedicelled. Sessile spikelets 0.2 cm. Pedicelled spikelets 0.6 cm. Caryopsis linear, terete.

Common in forest undergrowth. Fl. & Fr.: All seasons. Peddakunta, *MSM & KH* 10571.

IMPERATA Cyrillo

Imperata cylindrica (L.) Raeusch. Nom. Bot. (ed. 3) 10. 1797. *var. major* (Nees) C.E. Hubb. Grass. Maur. 96. 1940; Bor 170. *Lagurus cylindricus* L. Syst. Nat. ed. 10. 878. 1759. *Imperata koenigii* (Retz.) Beauv. var. *major* Nees, Fl. Afr. Austral. 89. 1891. *I. cylindrica* (L.) Raeucsh. var. *koenigii* (Retz.) T.A. Durand & Schinz. Consp. Fl. Afric. 5: 694; C. Fischer 1707 (1184). *I. arundinacea* Cyrillo Pl. Rar. Neapol. 2: 27. 1792; FBI 7: 106.

A perennial herb, up to 60 cm tall. Stems tufted, bearded. Leaf sheaths overlapping, glabrous. Ciliate near mouth, blades lanceolate. Spike like panicle 15 cm. Spikelets 2-nate. Caryopsis oblong.

Common in open forest and waste lands (Seshagiri Rao, 1986). Fl. & Fr.: July–Oct.

ISACHNE Robert Brown

Isachne globosa (Thunb.) O. Kuntze, Rev. Gen. Pl. 2: 478. 1891; Bor 580. *Milium globosum* Thunb. Fl. Jap. 49. 1784. *Isachne australis* R. Br. Prodr. 196. 1810; FBI 7; 24. *I. dispar* auct. non. Trin. 1828; C. Fishcer 3: 1797 (1244). 1934; Bor 580. *p.p.* 1960; FBI 7: 26. 1896, *p.p.*

An annual herb, up to 25 cm tall. Stems tufted, villous. Leaf sheaths 2–5 cm, margin ciliate, blades lanceolate. Panicles 8 cm. Spikelets globose, sessile or pedicelled. Caryopsis ellipsoid.

Common in marshy places (Seshagiri Rao, 1986). Fl. & Fr.: July–Dec.

ISCHAEMUM Linnaeus

Ischaemum pilosum (Klein ex Willd.) Wight in Madras Lit. Sci. J. 138. 1835; FBI 7: 130; C. Fischer 1723 (1194); Bor 183. *Andropogon pilosus* Klein ex Willd. Sp. Pl. 4: 920. 1806.

A perennial stoloniferous grass. Stems up to 60 cm tall. Leaf sheaths 5–20 cm long, glabrous, blades 7–10 x 0.5–0.7 cm, linear, base rounded, tip acuminate. Spikelets 2-nate. Caryopsis lanceolate.

Common in cultivated soils. Fl. & Fr.; Oct.–Jan. Mekavarpalli, *TP & MSM* 11051.

ISEILEMA Anderson

1. Total prostrate herb .... ***I. prostratum***
1. Decumbent or ascending herb:
    2. Pedicels of the spikelets longer than broad .... ***I. laxum***
    2. Pedicels of the spikelets almost as broad as long .... ***I. anthephoroides***

Iseilema anthephoroides Hack. in A. & C. DC. Monogr. Phan. 6: 683. 1889; FBI 7: 219; C. Fischer 1748(1211); Bor 187.

An annual decumbent herb, up to 60 cm, stems tufted, nodes glabrous. Leaf seaths 1–2.5 cm, blades lanceolate, flat. Panicles 4 cm, homogamous spikelet lanceolate, 0.4 cm. Sessile spikelet lanceolate, 0.3 cm. Caryopsis ellipsoid. Pedicelled spikelet similar to homogamous.

Common in forest undergrowth. Fl. & Fr.: Oct.–Dec. Malkalcheruvu, *MSM & KH* 10569; Rangammagudem, *MSM* 10582.

Iseilema laxum Hack. in DC. Monogr. Phan. 6: 682. 1889; FBI 7: 219. 1896; C. Fischer 1748(1211); Bor 188.

A perennial herb, up to 30 cm tall. Stems tufted, glabrous. Leaf sheaths 2–4 cm, glabrous, blades cauline, lanceolate, flat. Panicles 15 cm, flexuous, interrupted. Spikelets pinkish and green, spikelets 3, one sessile, other two pedicelled. Caryopsis ellipsoid.

Common in waste lands. Fl. & Fr.: Aug.–Jan. Mekavarpalli, *MSM & TP* 11053.

Iseilema prostratum (L.) Anders. in Nov. Act. Soc. Sci. Upsaal. Ser. 3. 2: 251. 1856; C. Fischer 1748(1211); Bor 188. *Andropogon prostratum* L. Mant. Alt. 304. 1771. *Iseilema wightii* Anders. I.c. 1856; FBI 7: 218.

A perennial herb, up to 60 cm tall. Stems tufted, bearded. Leaf sheaths 3–5 cm, glabrous, blades linear, flat. Panicle 30 cm, lax, spreading, sessile and pedicelled. Spikelets brownish-violet in whorls. Caryopsis oblong.

Occasional in forest. Fl. & Fr.: Nov.–June. Mohammadabad, *MSM* 10595.

LEPTOCHLOA P. Beauvois

Leptochloa chinensis (L.) Nees in Syst. Ratisb. 1: 4. 1824; FBI 7: 299; C. Fischer 1820 (1260); Bor 516. *Poa chinensis* L. Sp. Pl. 69. 1753.

An annual marshy herb, up to 65 cm tall. Stems tufted, glabrous. Leaf sheaths 5–10 cm, glabrous, blades linear-lanceolate, convolute. Panicle 20 cm, spikelets 2–7-flowered. Caryopsis oblong.

Occasional in rice fields (Seshagiri Rao, 1986). Fl. & Fr.: July–Dec. Central University, *KSR* 83.

LOPHOPOGON Hackel

Lophopogon tridentatus (Roxb.) Hack. in Engl. Nat. Pflanzenf. 2(2): 26 1887; FBI 7: 149; C. Fischer 1724(1195); Bor 190. *Andropogon tridentatus* Roxb. Fl. Ind. 1: 261. 1820.

A perennial herb, up to 30 cm tall. Stems tufted, glabrous. Leaf sheaths 1–4 mm, glabrous, blades linear, flat or convolute. Racemes 3 cm, solitary, spikelets 2-nate, one sessile other pedicelled. Caryopsis ellipsoid.

Occasional in abandoned plains. Fl. & Fr.: July–Oct. Lingampalle, *MSM* 12754.

MELANOCENCHRIS C.G.D. Nees

Melanocenchris jacquemontii Jaub. ex Spach. Illus. Pl. Or. 4: 36. 325. 1851; Bor 473. *M. royleana* Nees ex Steud. Syn. Pl. Glum. 1: 218. 1854; C. Fischer 1832 (1268). *Gracilea royleana* Hook. f. in FBI 7: 284.

An annual herb up to 20 cm tall. Stems tufted, glabrous. Leaf sheaths 1–2 cm, blades linear, convolute. Spikes 6 cm. Spikelets 8 mm, clustered, 2-flowered. Caryopsis ellipsoid.

Common in dry sandy soils. Fl. & Fr.: July–Jan. Anantagiri RF, *MSM & TP* 11072.

MNESITHEA Kunth

Mnesithea granularis (L.) Koning & Sosef in Blumea 31: 295. 1986. *Cenchurs granularis* L. Mant. 2. App. 575. 1771. *Manisuris granularis* (L.) L.f. Nov. Gram. Gen. 40. 1979; FBI 7; 159. *Hackelochloa granularis* (L.) O. Kuntze, Rev. Gen. Pl. 2. 776. 1891; C. Fischer 1758 (1218); Bor 159.

An annual herb, up to 45 cm tall. Stems tufted, bearded. Leaf sheaths 2–6 cm, blades lanceolate, flat. Racemes 4 cm, spiciform. Spikelets globose 2-nate. One sessile and other pedicellate. Caryopsis ovoid.

Common in hills. fl. & Fr,: July–Sept. Vikarabad, *MRS* 323 (DD); Tandur, *MSM & TP* 12706.

OROPETIUM Trinius

Oropetium thomaeum (L.f.) Trin. Fund. Agrost. 98. t. 3. 1822; FBI 7: 366. C. Fischer 1830(1267); Bor 474. *Nardus thomaea* L.f. Suppl. Pl. 105. 1781.

An annual herb, up to 10 cm tall. Stems densely tufted. Leaf sheaths small, compressed, blades filiform, sparsely ciliate. Spike 4 cm, solitary. Spikelets solitary. Caryopsis oblong.

Rare in forest. Fl. & Fr.: Aug.–Dec. Vikarabad, *MRS* 87 (DD).

ORYZA Linnaeus

Oryza rufipogon Griff. Notul. 3. 5. 1851; Bor 605.

An annual erect aquatic herb, up to 1.5 cm tall. Stems stoloniferous, rooting from nodes, spongy below. Leaf sheaths glabrescent, blade scaberulous. Panicles drooping. Spikelet oblong, awned, awn 8–9 cm long, barbellate. Caryopsis oblong.

Occasional in ponds and wet places. Fl. & Fr.: Oct.–Dec. Anantasagar beat, *MSM & KH* 11006.

PANICUM Linnaeus

1. Leaf blade linear-lanceolate. Panicles 1.5 cm long .... ***P. trypheron***
1. Leaf blade flat not lanceolate. Panicles contracted .... ***P. sumatrense***

Panicum sumatrense Roth ex Roem. & Schult. Syst. Veg. 2: 434. 1819; Bor 701. *P. miliare* Lam. Tabl. Encycl. 1: 173. 1791; FBI 7: 46; C. Fischer 1782 (1234); Bor 329.

An annual herb, up to 80 cm tall. Stems tufted, glabrous. Leaf sheaths 4–10 cm, glabrous, blades flat, panicle 25 cm, thyrsiform or contracted. Spikelets ellipsoid. Caryopsis ovoid.

Occasional in cultivated lands. Fl. & Fr.: Oct.–Dec. Mohammadabad, *MSM* 10552; 10555.

Panicum trypheron Schult. Syst. Veg. 2. Mant. 744. 1824; FBI 7: 47. 1896; C. Fischer 1783(1235); Bor 331.

An annual erect herb, up to 50 cm tall. Stems tutted. Leaf sheaths 2–8 cm, ciliate, blades linear-lanceolate. Panicles 15 cm long. Spikelets ellipsoid. Caryopsis ovoid.

Common grass in ponds and moist places. Fl. & Fr.: Oct.–Mar. Central University, *MSM & TP* 15142.

PASPALIDIUM Stapf

1. Annual herb, ligule absent. Spikelets gibbose .... ***P. flavidum***
1. Perennial herb, ligule present. Spikelets not gibbose .... ***P. geminatum***

Paspalidium flavidum (Retz.) A. Camus in Lecomte Fl. Gen. Indochine. 7: 419. 1922; C. Fischer 1774(1229); Khan in FFH 340; Bor 333. *Panicum flavidum* Retz. Obs. Bot. 4. 15. 1786; FBI 7: 28.

An annual herb, up to 70 cm tall. Stems tufted, glabrous. Leaf sheaths 6–15 cm, glabrous, blades linear-lanceolate. Paniculate racemes 20 cms. Spikelets broadly ovate, secund in 2 rows, lower neutral or male, upper hermaphrodite. Caryopsis ovate-oblong.

Common in open forest and waste land. Fl. & Fr.: July–Oct. Rudraram RF, *MSM & TP* 12719.

Paspalidium geminatum (Forssk.) Stapf in Oliver. Fl. Trop. Africa 9: 583. 1920; C. Fischer 1774 91229); Bor 333. *Panicum geminatum* Forssk. Fl. Aegypt.-Arab. 18. 1775. *P. paspaloides* Pers. Syn. Pl. 1: 81. 1805; FBI 7: 30.

A perennial erect herb, up to 15 cm tall. Stems tufted, leaf sheaths 4–10 cm, blade linear. Racemes 10–30-flowered. Spikelets 10–30 per raceme, ovoid. Caryopsis ellipsoid.

Common in marshy places. Fl. & Fr.: : July–Sept. Anantasagar beat, *MSM & KH* 11002.

PASPALUM Linnaeus

Paspalum scrobiculatum L. Mant. Pl. 1: 29. 1767; FBI 7: 10; C. Fischer 1772(1227); Khan in FFH 339; Bor 340.

An annual erect herb, up to 1 m tall, stems tufted, hollow, glabrous. Leaf sheaths 2–8 cm, glabrous, blade lanceolate-linear, flat. Racemes 2 or 3, spikelets 2-ranked. Caryopsis plano-convex.

Occasional in dried ponds. Fl. & Fr.; Oct.–Dec. Kondapur beat, *MSM & KH* 10590. Rudraram RF, *MSM* 12729.

PENNISETUM Richard ex Persoon

1. Panicles purplish, densely woolly .... *P. pedicellatum*
1. Panicles other than purple, not woolly .... *P. hohenackeri*

Pennisetum hohenackeri Hochst. ex Steud. Syn. Pl. Glum. 1: 103. 1854; C. Fischer 1792(1241); Bor 344. *P. alopecuros* Steud. loc. cit. 102. non Jacq.; FBI 7: 84.

An annual erect herb, up to 50 cm tall, stems tufted, nodes glabrous. Leaf sheath 8–10 cm, blade linear, rigid, margin scabrid. Panicles compact, solitary. Spikelets lanceolate. Involucral bristles 0.2 cm, scabrid. Caryopsis oblong.

Occasional in forest undergrowth. Fl. & Fr.: Sept.–Dec. Rangammagudem beat, *MSM & KH* 10579.

Pennisetum pedicellatum Trin. in Mem. Acad. Imp. Sci. Petersb. ser. 6. Sci. Math. Nat. 3; 184. 1834; FBI 7: 86; C. Fischer 1792(1241); Bor 346.

An annual herb, up to 1 m tall. Stems tufted. Leaf sheaths glabrous, blades linear, flat. Racemes 13 cm, cylindric, densely flowered. Spikelets solitary. Caryopsis ellipsoid.

Occasional in plains. Fl. & Fr.; June–Dec. Lingampalle, *MSM* 12734.

PEROTIS W. Aiton

Perotis indica (L.) O. Kuntze Rev. Gen. Pl. 2: 787. 1891; C. Fischer 1814 (1256); Bor 611. *Anthoxanthum indicum* L. Sp. Pl. 28. 1753. *Perotis latifolia* Aiton in Hort. Kew. 1: 85. 1789; FBI 7: 98.

A perennial herb, up to 50 cm tall. Stems tufted, glabrous. Leaf sheaths 1–7 cm, glabrous, blades ovate-lanceolate. Spike like racemes 15 cm, terminal. Spikelets linear, subsessile. Caryopsis linear.

Common in sandy land in open forest. Fl. & Fr.: July–Jan. Rudraram RF, *MSM & TP* 12722.

SACCIOLEPIS Nash

1. Stems spongy, creeping, spikelets in interrupted crowded panicles ...... ***S. interrupta***
1. Stems slender, spikelets in continuous crowded panicles .... ***S. indica***

Sacciolepis indica (L.) Chase. Proc. Biol. Soc. Wash. 21: 8. 1908; C. Fischer 1787(1238); Bor 357. *Aira indica* L. Sp. Pl. 63. 1753 & in errata 63. n. 1. *Panicum indicum* L. Mant. Pl. 2: 184. 1771; FBI 4: 41.

An annual herb, up to 60 cm tall. Stems tufted, glabrous. Leaf sheaths 2–5 cm, glabrous, blades linear. Spike like panicle 70 cm, contracted, spikelets linear. Caryopsis linear.

Occasional in villages and forest. Fl. & Fr.: Aug.–Oct. Mohammadabad, *MSM* 10553.

Sacciolepis interrupta (Willd.) Stapf in Prain. Fl. Trop. Africa. 9: 757. 1920; C. Fischer 1787(1238); Bor 358. *Panicum interruptum* Willd. Sp. Pl. 1: 341. 1798; FBI 7: 40.

An aquatic perennial herb, up to 80 cm tall. Stems tufted, spongy, creeping, glabrous. Leaf sheaths 5–15 cm, glabrous, blades linear-lanceolates, flat. Spike like panicle 20 cm, contracted, interrupted. Spikelets elliptic-lanceolate. Caryopsis linear.

Occasional in marshy places. Fl. & Fr.: Jun.–Dec. Anantasagar, *MSM & KH* 11002.

SETARIA P. Beauvois *nom. cons.*

1. Involucral bristles retrorsely barbellate .... ***S. verticillata***
1. Involucral bristles antrorsely barbellate:
    2. Panicles not continuous, interrupted .... ***S. intermedia***
    2. Panicles continuous .... ***S. pumila***

Setaria intermedia Roem. & Schult. Veg. 2. 489. 1817; FBI 7: 79. C. Fischer 1789(1239). *Panicum tomentosum* Roxb. Fl. Ind. 1: 303. 1820. *Setaria tomentosa* (Roxb.) Kunth Rev. Gram 1: 47. 1829; Bor 365.

An annual erect herb, up to 50 cm tall. Stems tufted, glabrous. Leaf sheaths 4–10 cm, ciliate, blade linear-lanceolate. Panicle lax, interrupted. Caryopsis ellipsoid.

Common grass in moist places. Fl. & Fr.: July–Dec. Central University, *MSM* 15171.

Setaria pumila (Poir.) Roem. & Schult. in L. Syst. Veg. 2: 891. 1817. *Panicum pumilum* Poir. Encycl. (Suppl.) 4: 273. 1816. *Setaria pallide-fusca* (Schum.) Stapf & C.E. Hubb. in Bull. Misc. Inform. 1930. 259. 1930; C. Fischer 1789(1239); Bor 363. *Panicum pallide-fuscum* Schum. Beskr. Guinpl. 58. 1827. *Setaria glauca* Hack. Bol. Soc. Brot. 3: 135. 1884; FBI 7: 78.

An annual erect herb, up to 50 cm tall. Stems tufted, glabrous, nodes glabrescent. Leaf sheaths 2–4 cm, glabrous, blade lanceolate, flaccid, flat. Spike pilose, involucral bristles 9–11 per whorl, barbed. Spikelets 3 or solitary, ovoid. Caryopsis ellipsoid.

Common in wet places. Fl. & Fr.: Sept.–Dec. Mohammadabad beat, *MSM & KH* 10534, 10592; Rangammagudem, *MSM & KH* 10581.

Setaria verticillata (L.) Beauv. Ess. Agrost. 51. 178. 1812; FBI 7: 80. C. Fischer 1789(1239); Bor 365. *Panicum verticillatum* L. Sp. Pl. ed. 2. 82. 1762.

An annual herb up to 50 cm tall. Stems tufted. Leaf sheaths 3–8 cm long, glabrous, blade lanceolate, 3–10 × 0.5–1 cm. Panicles spiciform. Spikelets ovate-oblong. Caryopsis ellipsoid.

Occasional in moist places. Fl. & Fr.; Nov.–Jan. Central University, *MSM* 15183.

SPOROBOLUS Robert Brown

Sporobolus indicus (L.) R. Br. Prodr. 170. 1810. *var. diander* (Retz.) Jov. & Gued. in Bull. Centre, Et. Rech. Sc. Biarritz 7: 65. 1968. *Agrostis diandra* Retz. Obs. Bot. 5. 19. 1789. *Sporobolus diander* (Retz.) Beauv. Ess. Agrost. 26: 147, 178. 1812; FBI 7: 247; C. Fischer 1817(1258); Bor 629.

A perennial herb, up to 60 cm tall. Stems tufted, glabrous. Leaf sheaths 3–10 cm, glabrous, blades linear, flat or convolute. Panicle 25 cm, contracted. Spikelets linear. Caryopsis obovoid.

Common in moist places. Fl. & Fr.: June–Dec. Mohammadabad, *MSM & KH* 10535.

TRAGUS Heller *nom. cons.*

Tragus roxburghii Panigr. in Kew. Bull. 29. 495. 1974. *Lappago biflora* sensu Roxb. Fl. Ind. 1: 284. 1820 quoad. descrip. excl. Type. *Tragus biflorus* Sensu Schult. Man. 2: 205. 1824; C. Fischer 1813(1255); Bor 682. *T. racemosus* auct. non. Scop. (1777); FBI 7: 97.

A perennial herb, up to 20 cm tall. Stems tufted, glabrous. Leaf sheaths 2–4 cm, glabrous, blades lanceolate, flat. Racemes 4 cm, solitary. Spikelets lanceolate, 2 in subsessile clusters facing each other, 1-flowered. Caryopsis ellipsoid.

Common in waste lands. Fl. & Fr.: June–Oct. Rudramram RF, *MSM & TP* 12711.

Oryza sativa L. var. *sativa* (*Vern.:* Vari), *Saccharum officinarum* L. (*Vern.:* Cheraku) are most widely cultivated crops in the district. *Pennisetum americanum* (L.) K. Schum (*Vern.:* Sajja), *Eleusine coracana* (L.) Gaertn.

(*Vern.:* Ragulu), *Zea mays* L. (*Vern.:* Mokkajonna), *Setaria italica* Beauv. (*Vern.:* Korra) etc. are occasionally cultivated crops in the district.

FERN AND FERN ALLIES

KEY TO FAMILIES

1. Aquatic plants:
   2. Rhizomatous, leaves lobed .... ***Marsileaceae***
   2. Non-rhizomatous, leaves not lobed .... ***Azollaceae***
1. Terrestial plants:
   3. Sporangia borne on ordinary or special fronds .... ***Actinopteridaceae***
   3. Sporangia borne on strobili .... ***Selaginellaceae***

ACTINOPTERIDACEAE

ACTINOPTERIS Link

Actinopteris radiata (Sw.) Link Pl. Sp. 80. 1841; Bedd. Ferns S. India t. 124. 1865; Pichisermolli in Webbia 17: 318. 1963. *Asplenium radiatum* Sw. in Schrad. J. Bot. 1800. 50. 1801.

A terrestrial fern. Rhizome very short. Lamina fan-shaped, 3 cm, petioles 8 cm, all arising from rhizome. Pinnae 3–5, clustered, deeply dissected, lobes linear, glabrous. Sori arranged in two rows on the lower side of the pinnae lobes.

Common under rocks on hills, often in open forest. Sori: Sept.–Dec. Kondapur beat, *MSM & KH* 10319.

AZOLLACEAE

AZOLLA Lamarck

Azolla pinnata R. Br. Prodr. Fl. Nov. Holl. 167. 1810; Holtum. Fl. Mal. 2: 621. 1954. Dixit, Cens. Ind. Pterid. 174. 1984.

A small aquatic herb. Stems horizontal, branched zig-zag, bearing roots densely covered by hairs. Leaves alternate, rectangular, sessile, margin entire. Micro sporocarps globose enclosing micro sporangia, mega sporocarps enclosing a single mega sporangium.

Rare in ponds. Central University, *MSM & TP* 19226.

SELAGINELLACEAE

SELAGINELLA Palisot de Beauvois

Selaginella bryopteris (L.) Baker in J. Bot. Brit. & For. 22. 376. 1884.
Lycopodium bryopteris L. Sp. Pl. 1103. 1753.

An erect or prostrate herb, up to 10 cm. Rhizome globose, stems dichotomously branched. Leaves heterophyllous, arranged in four rows, central two rows smaller than lateral two rows, blade ovate, margin serrate, tip mucronate, green above, white below. Strobili 0.4 cm long.

Common in forest on rocks in the form of mats. Strobili: Aug.–Sept. Kondapur beat, *MSM & KH* 10321.

MARSILEACEAE

MARSILEA Linnaeus

Marsilea quadrifolia L. Sp. Pl. 1099. 1753; Gupta Marsilea Monogr. 22. 1962.

A rhizomatous aquatic herb. Stems cylindrical, rooting at nodes. Leaves 4-lobed, each leaflet or lobe obovate, 1–2 × 0.5–1 cm, base narrowed, margin entire, tip rounded. Sporocarps brown on long stalks.

Occasional in marshy places. Sporocarps: July–Sept. Rudraram RF, *MSM & TP* 20710.

BIBLIOGRAPHY

Aiton, W. 1789. *Hortus Kewensis*. (ed. 1) London.

Aiton, W.T. 1810–1813. *Hortus Kewensis*, (ed. 2) London.

Ali, S.I. 1967. The genus *Goniogyna* DC. *Taxon* **16**: 463–464.

Ali, S.I. 1968. Revision of the genus *Indigofera* L. from West Pakistan and North-West Himalayas. *Bot. Notiser* **3:** 543–577.

Anderson, T. 1867. An enumeration to Indian species of Acanthaceae. *Journ. Linn. Soc. Bot.* **9**: 425–526.

Austin, D.F. 1977. *Ipomoea carnea* Jacq. vs. *Ipomoea fistulosa* Mart. ex Choisy. *Taxon* **26**: 235–238.

Backer, C.A. and R.C. Bakhuizen van den Brink. 1973–68. *Flora of Java*. 3 Vols. Netherlands.

Beauvois, A.M.F.J.P. de. 1812. *Essaiu d'une nouvelle Agrostographie*, Paris.

Beddome, R.H. 1868–74. *Icones Plantarum Indiae Orientalis*. Madras.

Beddome, R.H. 1869–74. *The Flora Sylvatica for Southern India*. 2. Vols. Madras

Beddome, R.H. 1883. *Handbook of the Ferns of British India, Ceylon and the Malay Peninsula*. Calcutta.

Bennet, S.S.R. 1986. *Name changes in Flowering Plants of India and adjoining regions*. Dehra Dun.

Bentham, G. 1835. *Scrophularineae Indicae, A synopsis of the Eastern Indian Scrophulariaceae*. London.

Bentham, G. and J.D. Hooker. 1862–1883. *Genera plantarum*. 3 vols. London.

Biswas, K. and C. Calder. 1937. *Handbook of common water and marshy Plants of India and Burma*. Delhi. (ed. 2, 1955, Calcutta).

Blatter, E. & C. McCann. 1935. *The Bombay grasses*. Delhi.

Blume, C.L. Von. 1836–1849. *Rumphia, sive Commentationes botanicae imprimis de plantis Indiae Orientalis*. Leiden.

Bradley, H. 1849. Statistics of the Circar of Dowlutabad, Hyderabad, Deccan, *Madras Journ. Lit. & Sci.* **15**: 481–551.

Brandis, D. 1906. *Indian Trees*. New Delhi.

Bremekamp, C.E.B. 1944. Materials for a Monograph of the Strobilanthinae (Acanthaceae). *Vern. Nederl. Akad. Wet.* **41**: 1–306.

Bruggen, H.W.E. Van. 1970. Revision of the genus *Aponogeton* (Aponogetonaceae), *Blumea* **16**: 457–487.

Buchanan-Hamilton, F. 1807. *A Journey from Madras through the countries of Mysore, Canara and Malabar*. 3 vols. London.

Burman, N.L. 1768. *Flora indica*, Leiden.

Campbell, A.C. 1898. *Glimpses of the Nizam's Dominions*. London.

Candolle, A.P. de 1824–1874. *Prodromus Systematis Naturalis Regni vegetabilis*. 17 vols. & Index. Paris. (vol. 7–17 by A.L.P.P. de Candolle).

Chakravarthy, H.C. 1959. Monograph of the Indian Cucurbitaceae. *Rec. Bot. Surv. India* **17**: 1–234.

Chakravarthy, H.C. 1982. Cucurbitaceae: *Fascicles of Flora of India*. No. 11, Howrah.

Champion, H.G. and S.K. Seth. 1968. *A revised survey of the forest types of India*. New Delhi.

Clarke, C.B. 1876. *Compositae Indica descriptae et Secus genera Benthamii ordinatae*. Calcutta.

Cooke, T. 1901–1908. *The Flora of the Presidency of Bombay*. London (repr. ed. 1958. Calcultta).

Cramer, L. 1989. The *Hygrophila* complex (Acanthaceae) in India and Ceylon. *Nord. Journ. Bot.* **9**: 261–263.

Delile, A.R. 1813. *Description de l'Egypto Historoire naturelle*, Vol. 2. Paris.

Desfontaines, R.L. 1798–1799. *Flora Atlantica* etc. 2 vols. Atlas, Paris.

Don, D. 1825. *Prodromus Florae Nepalensis*. London.

Don, G. 1831–1838. *A General History of Dichlamydeous plants*. London.

Engler, H.G.A. 1900–1937. *Das Pflanzenreich Regni vegetabilis Conspectus*. Leipzig.

Engler, H.G.A and K.A.E. Prantl. 1887–1915. *Die Naturalichen Pflanzenfamilien* etc. 23 vols. ed. 1. Leipzig.

Forster, J.R. and J.G.A. Forster. 1775. *Characteres Generum Plantarum* etc. London.

Forsskal, P. 1775. *Flora Aegyptiaco-arabica*. Copenhagen.

Gaertner, J. 1788–1807. *De fructibus et Siminibus Plantarum*. 3 vols. Stuttgart.

Gamble, J.S. and C.E.C. Fisher. 1915–1935. *Flora of the Presidency of Madras*. London. (repr. ed. 1957. Calcutta).

Gmelin, J.F. 1791–1792. *Linnaeus systema Naturae* ed. 13. Leipzig.

Greuter, W. (ed.) 1988. *International Code of Botanical Nomenclature*. Konigstein (Germany).

Hackel, E. 1989. *Monographia Andropogonearum*. Paris.

Haines, H.M. 1921–25. *The Botany of Bihar and Orissa,* 3 vols. Calcutta. (repr. ed. 1961).

Henry, A.N., K. Vivekanathan and N.C. Nair. 1978. Rare and threatened flowering plants of South India. *Journ. Bombay Nat. Hist. Soc.* **75**: 684–697.

Henry, A.N., V. Chitra and N.P. Balakrishnan. 1989. *Flora of Tamilnadu, India Ser.* **1**: Analysis, vol. 3. Coimbatore.

Hooker, J.D. 1872–1897. *Flora of British India.* 7 vols. London.

Hooker, J.D. and T. Thomson. 1856. *Flora Indica.* London.

Hooker, W.J. and G.A.W. Arnott. 1830–1841. *The Botany of Captain Beechey's voyage.* London.

Humboldt. F.H.A. and A.J.W. Bonpland and C.B. Kunth. 1816–1825. *Nova Genera et Species Plantarum* etc. Paris.

Jacquin, N.J. 1787–1797. *Collectanae and Botanicam, Chemiam et Historium naturalem Spectantia.* 5 vols. Wien.

Jain, S.K. 1978. Welcome address. 125th anniversary Souvenir of Madras Herbarium. Botanical Survey of India. Southern circle, Coimbatore, p. 1–2.

Jain, S.K. 1982. *Flora of India—Tasks before 2000 A.D.* Panchanan Maheshwari Memorial Medal lecture, Indian Botanical Society.

Jain, S.K. 1989. Botanical regions and Flora of India. *Everyman Science* **24:** 213–223.

Jain, S.K. and U.R. Deshpande. 1978. Transfer of some Indian species of *Bothriochloa* and *Capillipedium* to *Dichanthium. Bull. Bot. Surv. India.* **20:** 133–135.

Jain, S.K. and R.R. Rao. 1977. *Field and Herbarium methods.* New Delhi.

Jeffrey, C. 1980. Further notes on Cucurbitaceae of the Indian subcontinent. *Kew Bull.* **34**: 789–810.

Jussieu, A.L. de. 1789. *Genera Plantarum Secundum Ordines Naturales Disposita* (ed. 2, 1791). Paris.

Kapoor, S.L. and L.D. Kapoor. 1973. Further contribution to the flora of the Karimnagar district of Andhra Pradesh. *Bull. Bot. Surv. India.* **15**: 76–84.

Karthikeyan, S., M.P. Nayar and R.S. Raghavan. 1981. *An annotated Bibliography of Taxonomy Botany of Peninsular India 1959–1978.* BSI, Howrah.

Karthikeyan, S., S.K. Jain, M.P. Nayar and M. Sanjappa. 1989. *Florae Indicae Enumeratio*: *Monocotyledonae.* Flora of India. Ser. Publication. 4. BSI.

Khan, M.S. 1953. *Forest Flora of Hyderabad State.* Hyderabad.

Koyama, T. 1985. Cyperaceae: In: "*A Revised Handbook of Flora of Ceylon*" (ed. Dassanayake and Fosberg), vol. 5, 125–405.

Krishnamurthy, A.V.R.G. 1960. *Forest types of Andhra Pradesh*, Mss.

Kunth, C.S. 1833–1850. *Enumeratio Plantarum omnium hucusque cognitarum,* Stuttgart.

Kuntze, O. 1891–1898. *Revisio Generum Plantarum Vascularium Omnium,* Leipzig.

Lamarck, J.B.A.and P.M. de. 1783–1817. *Encyclopedique et methodique; Botanique.* 8 vols. & 5 suppl., Paris (vol. 5–13 by J.L.M. Poiret).

Lamarck, J.B.A. and P.M. de. 1791–1823. *Tableau Encyclopedique et methodique; Botanique.* 3 vols. & Suppl. Paris (vol. 2, part 5, 2 & vol. 3 by J.L.M. Poiret).

Lewis, W.H. 1961. Merger of North American *Houstonia* and *Oldenlandia* under *Hedyotis*. *Rhodora* **63**: 316–223.

Linnaeus, C. 1753. *Species plantarum.* 2 vols. Stockholm.

Linnaeus, C. 1755–56. *Centuria Plantarum,* 2 vols. Stockholm.

Linnaeus, C. 1758–59. *Systema Naturae.* ed. 10. vol. 2. Botany. Stockholm.

Linnaeus, C. 1767. *Mantissa Plantarum,* pages 1–142. Stockholm.

Linnaeus, C. 1771. *Mantissa Plantarum Altera,* pages 143–587. Stockholm.

Linnaeus, C. 1781. *Supplementum Plantarum Systematis vegetabilium* etc. Braunschweig.

Merrill, E.D. 1917. *An Interpretation of Rumphius Herbarium Amboinense.* Manila.

Michaux, A. 1803. *Flora boreali-americana,* 2 vols. Paris.

Miller, P. 1768. *The gardener's Dictionary* (ed. 8). London.

Nair, N.C. and A.N. Henry. 1983. *Flora of Tamil Nadu, India,* vol. 1 and 2. Coimbatore.

Narasimha Rao, G. 1986. *Studies on the floristics of Angiosperms and the dermotypes of Acanthaceae of Narsapur taluk (Medak district, Andhra Pradesh),* Ph.D. Thesis, Osmania University, Hyderabad.

Nees, C.G.D. 1841. *Florae Africae Australioris (Illustrationes monographieae).* Glogan.

Nees, C.G.D. 1847. Acanthaceae in DC., *Prodr.* vol. 11: 46–519.

Patridge, E.A. 1911. *Forest Flora of H.H. the Nizam's Dominions, Hyderabad, Deccan,* Hyderabad.

Pennell, F.W. 1935. Scrophulariaceae of Eastern temperate North America. *Acad. Nat. Sci. Philad. Monogr.* **1**: 137–159.

Prain, D. 1963. *Bengal plants* (repr. ed.) 2 vols. Calcutta.

Prain, L.C. 1909. *Botany in "Imperial Gazetteer of India" Provincial series, Hyderabad State.* Calcutta.

Pullaiah, T. 1997. *Flora of Andhra Pradesh,* vol. III. Jodhpur.

Pullaiah, T. and D. Ali Moulali. 1997. *Flora of Andhra Pradesh,* vol. II. Jodhpur.

Pullaiah, T. and E. Chennaiah. 1997. *Flora of Andhra Pradesh*, vol. I. Jodhpur.

Pullaiah, T., C. Prabhakar and B. Ravi Prasad Rao. 1998. *Flora of Medak District*. Delhi.

Pullaiah, T. and B. Ravi Prasad Rao. 1995. *Flora of Nizamabad*. Dehra Dun.

Pullaiah, T. and P. Suryaprakash Babu. 1998. *Flora of Andhra Pradesh*, vol. IV - Illustrations. Jodhpur.

Pullaiah, T., P.V. Prasanna, and G. Obulesu. 1992. *Flora of Adilabad District*, C.B.S. Publishers, New Delhi.

Raghava Rao, S.R. 1989a. *Flora of Mahaboobnagar district, Andhra Pradesh*. Ph.D. Thesis, Osmania University, Hyderabad.

Raghava Rao, S.R., T. Rajagopal a S.T. Ramachandrachary. 1989b. *Alysicarpus mahabubnagarensis* Sp. nov. (Fabaceae) from India. *Asian Journ. Plant Science* **1**: 97–99.

Rajagopal, T. 1973. *Flora of Hyderabad including study of the foliar epidermal characters of species as an aid to taxonomy*. Ph.D. Thesis, Osmania University, Hyderabad.

Raju, V.S. 1984. New Angiospermous taxa described from the State of Andhra Pradesh after its formation. *J. Econ. Tax. Bot.* **5**: 469–472.

Raju, V.S. and A. Ragan. 1988. Additions to the Cyperaceae of Andhra Pradesh. *J. Swamy Bot. Cl.* **5**: 115.

Raju, V.S. and A. Ragan. 1990. Additions to the genus *Cyperus* L. in Andhra Pradesh. *J. Econ. Tax. Bot.* **14**: 467–468.

Ramachandrachary, S.T. 1980. *Floristic study of Achampet taluk (Mahaboobnagar Dist., Andhra Pradesh)*. Ph.D. Thesis, Osmania University, Hyderabad.

Ramachandrachary, S.T. and N. Ramayya. 1986. Additions to the forest flora of Telangana region in Andhra Pradesh. *Ind. Journ. Bot.* **9:** 156–158.

Ramachandrachary, S.T. and J.J. Wood. 1981. A new species of *Habenaria* (Orchidaceae) from India. *Kew Bull.* **36**: 235–237.

Ramachandran, A. 1978. Inaugural address. 125th Anniversary Souvenir of Madras Herbarium. Botanical Survey of India. Southern circle. Coimbatore, pp. 9–12.

Ramayya, N. and T. Rajagopal. 1973. Taxonomy, forest ecology and anatomy of the flowering plants of the erstwhile state of Hyderabad, a review. *Ind. Nat. Sci. Acad.* **45**: 302–313.

Rao, P.N., V.S. Raju and R.S. Rao. 1983. Floristic Research work in Andhra Pradesh. *Bull. Bot. Surv. India.* **23**: 218–220.

Rao, R.S. 1956. *Parthenium hysterophorus* L., A new record for India. *Journ. Bombay Nat. Hist. Soc.* **54:** 218–220.

Rao, R.S. and R.V. Kammathy. 1966. Notes on Indian Commelinaceae - V. *Journ. Linn. Soc. (London) Bot.* **59**: 305–308.

Ravishankar, T. and N. Rama Rao. 1989. Additions to the Flora of Karimnagar district, Andhra Pradesh, India. *Indian J. Bot.* **12**: 72–77.

Reddy, R.D., M.K. Prasad and K. Venkaiah. 1991. *Forest Flora of Andhra Pradesh (Vernacular names).* Hyderabad.

Retzius, A.J. 1779–1791. *Observationes botanicae.* Leipzig.

Ridsdale, C.E. 1978. A revision of the tribe Naucleae s.s. (Rubiaceae). *Blumea* **24**: 307–366.

Roemer, J.A. and J.A. Schultes. 1817–1830. *Linnaeus systema vegetabilium* ed. 15. 7 vols. Stuttgart.

Roth, A.W. 1827. *Enumeratio plantarum Phanerogamarum* etc. Leipzig.

Roxburgh, W. 1795–1819. *Plants of the coast of Coromandel.* 3 vols. London.

Roxburgh, W. 1832. *Flora Indica.* 3 vols. Calcutta.

Saldanha, C.J. 1963. The genus *Striga* Lour. in Western India. *Bull. Bot. Surv. India.* **5**: 67–70.

Saldanha, C.J. 1986. *Flora of Karnataka.* Vol. 1. New Delhi.

Saldanha, C.J. and D.H. Nicolson, 1976. *Flora of Hassan district.* New Delhi.

Santapau, H. 1954. Annual excursion of the Indian Botanical Society, Hyderabad, Deccan, *Journ. Ind. Bot. Soc.* **33**: 78–92.

Santapau, H. 1955. *Botanical collector's manual.* Calcutta.

Santapau, H. 1958. *Flora of Purandhar.* New Delhi.

Santapau, H. 1967. The Flora of Khandala on the Western Ghats of India. (ed. 3). *Rec. Bot. Surv. India.* **16:** 1–372.

Santapau, H. and S.K. Wagh. 1964. Nomenclatural changes in Indian Plants. *Bull. Bot. Surv. India.* **5**: 105–109.

Satyanarayana Reddy. 1985. *Floristic studies of Warangal and epidermology of Cyperaceae, as an aid to taxonomy.* Ph.D. Thesis, Osmania University, Hyderabad.

Sayeeduddin, M. 1935. Some of the common flowering plants of Hyderabad State, their distribution, economic and medicinal importance. *Journ. Asiat. Soc. Bengal Sci.* **1**: 9–22.

Sayeeduddin, M. 1938. A further contribution to some of the common flowering plants of Hyderabad State, their distribution and economic importance. *Journ. Bombay Nat. Hist. Soc.* **40**: 191–212.

Sayeeduddin, M. 1941a. Preliminary notes of a recent tour to Amrabad forest reserve. *Journ. Bombay Nat. Hist. Soc.* **41**: 907–910.

Sayeeduddin, M. 1941b. Additions to our knowledge of the flowering plants of H.E.H. the Nizam dominions, Hyderabad, Deccan. *Journ. Bombay Nat. Hist. Soc.* **42:** 903–924.

Sayeeduddin, M. 1954. A brief sketch of the Flora of Hyderabad, Deccan, *Souvenir, 41st session, Indian Science Congress*, Hyderabad.

Sebastine, K.M. and A.N. Henry. 1966. A contribution to the flora of Pakhal and surrounding areas of Narsampet taluk, Warangal district, Andhra Pradesh. *Bull. Bot. Surv. India* **8**: 304–311.

Sebastine, K.M., K. Thothathri and N.P. Balakrishnan. 1960. Observations on the flora of Narsapur, Medak district, Andhra Pradesh. *Bull. Bot. Surv. India* **2**: 275–285.

Seshagiri Rao, R. 1986. *Systematic enumeration of University of Hyderabad Plant Resources*. M.Phil. dissertation. Univ. Hyderabad.

Sprengel, C.P.J. 1824–1828. *Linnaeus Systema vegetabilium* ed. 16. Gottingen.

Stapf. O. 1917–1930. Gramineae; in Prain, *Fl. Trop. Africa* **9**: 1–768.

Subba Rao, G.V. and G.R. Kumari. 1967. Contribution to the flora of Karimnagar district, A.P. *Bull. Bot. Surv. India* **9**: 95–113.

Subramanyam, K. 1962. *Aquatic angiosperms*. New Delhi.

Suxena, M.R. 1947. A study of the grasses of the H.E.H., the Nizam's dominions, Hyderabad, Deccan, *Osmania Univ. Journ. Res.* **12**: 28–65.

Thothathri, K. 1961. Studies in Leguminosae: 2. The Genus *Pongamia* Vent. in India. *Bull. Bot. Surv. India* **3**: 418.

Thothathri, K. 1964. The Flora of Nagarjunakonda Valley and the surrounding hills. *Indian For.* **90**: 539–546.

Thunberg, C.P. 1794–1800. *Prodromus plantarum Capensium*. Uppsala.

Thwaites, G.H.W. 1858–64. *Enumeratio plantarum Zeylaniae*. London.

Tirvengadum, D.D. 1978a. Re-establishment of genus *Catunaregam* Wolf (Rubiaceae). A contribution towards the plants described in Rheed's *Hortus Malabaricus*. *Taxon* **27**: 512–517.

Trimen, H. 1894. *A handbook to the flora of Ceylon*. London.

Trimen, H. and J.D. Hooker. 1893–1900. *A handbook to the Flora of Ceylon*, vol. 6 (Suppl.) London (1931 edition by Alston).

Vahl, M. 1790–1794. *Symbolace botanicae,* 9 vols. Copenhagen.

Van Steenis, C.G.G.J. (ed.) 1950–1984. *Flora Malesiana*, 10 vols.

Walker, A.L. 1849. Statistical report on the Circar of Warangal (Hyderabad, Deccan), *Madras Journ. Lt. & Sci.* **15**: 219–301.

Wallich, N. 1829–32. *Plantae Asiaticae Rariores*. 3 vols. London.

Wight, R. 1834. *Contribution to the Botany of India*. London.

Wight, R. 1838–1853. *Icones Plantarum Orientalis*. 6 vols. Madras.

Wight, R. 1840–50. *Illustrations to Indian Botany*. London.

Wight, R. and U.G.A. Walker-Arnott. 1934. *Prodromus Florae Peninsulae Indiae Orientalis*. London.

Wight, R. and C.L. Willdenow. 1830. *Species Plantarum* (ed. 4 of L. Sp. Pl), 4 vols. Berlin..

Willis, J.C. 1973. *A dictionary of flowering plants and ferns*. (rev. ed. 8th by H.K. Airy Shaw). Cambridge.

INDEX TO FAMILIES, GENERA AND SPECIES

(Basionyms and Synonyms are in italics)

C